Electronics
A First Course

Electronics
A First Course

Third Edition

Owen Bishop

AMSTERDAM • BOSTON • HEIDELBERG • LONDON • NEW YORK • OXFORD • PARIS
SAN DIEGO • SAN FRANCISCO • SINGAPORE • SYDNEY • TOKYO

Newnes is an imprint of Elsevier

Newnes is an imprint of Elsevier
The Boulevard, Langford Lane, Oxford OX5 1GB, UK
30 Corporate Drive, Suite 400, Burlington, MA 01803, USA

First edition 2002
Second edition 2006
Reprinted 2006
Third edition 2011

British Library Cataloguing in Publication Data
A catalogue record for this book is available from the British Library

Library of Congress Cataloging-in-Publication Data
A catalog record of this book is available from the Library of Congress

ISBN: 978-1-85617-695-8

Printed and bound in Italy

11 12 13 10 9 8 7 6 5

Contents

Introduction

This is a complete introductory textbook intended for courses leading to GCSE Electronics. It also caters for the Electronics sections of GCSE Design and Technology (Electronic Products). It assumes no previous knowledge of electronics or of the electronic aspects of physics.

Most of the text is divided into numerous short Topics. The topics are dealt with thoroughly, with simple explanations and plenty of examples and illustrations. This presentation has two advantages. It allows students to confine their attention to the particular topics found in a given specification. It also presents the student with self-contained, easily assimilable and readily testable segments of knowledge.

The book is student-centred and it features:

- Frequent 'Self Test' questions to allow students to assess their progress.
- Sets of questions at numerous key stages in the book, linking together the material of consecutive Topics. Answers to these and the Self Test questions are given at the end of the book, if they are numerical or of few words.
- An abundance of practical examples, with numerous circuit diagrams. Detailed instructions for simple constructional techniques, and for test procedures appear in the 'In the Lab' sections.
- An underlying emphasis on electronic systems.
- 'Design time' pages provide a wealth of practicable suggestions for circuits and projects that students can design for themselves. These are intended to help students to prepare coursework projects for the examination, as well as to promote understanding of electronics theory.
- The book is copiously illustrated with halftone and line drawings, the circuit symbols following the guide set out in the AQA GCSE Electronics specification. There are numerous photographs, including close-ups of electronics components, and illustrations of constructional techniques.

While updating the material for this edition we have taken into account the increased emphasis on electronic systems in the latest GCSE specifications. We have also tried to match the content of the book to the basic requirements for a number of units of the EDEXCEL level 2 BTEC First examinations in Electronics.

WHAT IS ELECTRONICS?

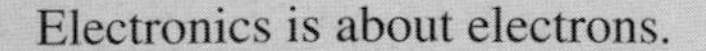

Electronics is about electrons.

It is about the ways we use electrons to do useful and interesting things.

The photos on this page show some of the things that we can do with electrons. Can you work out what these things are?

Electronics makes a big difference to all our lives. Without electronics, our lives would be less comfortable, less safe, less interesting and less fun. Or do you disagree? Talk about it with other students.

HOW TO USE THIS BOOK

Most of the book is set out as brief Topics. When you open the book at one of these Topics, the pages tell you all you need to know about one electronics topic. As well as the text and pictures, there may be:

- **Things to do** that help you to learn more.
- **Design tips** to help you design and build your electronics project.
- **Self tests** to find out how you are getting on. Answers are at the back of the book.

Most students will need to work on most of the spreads. These cover all the essential topics for Electronics exams at Level 2. Your teacher will tell you which spreads, if any, you can leave out.

In the Lab and **Design Time** pages provide you with advice and ideas for your project.

After every few spreads there is a batch of questions. The answers to these are on our website:

http://www.elsevierdirect.com/companions/9781856176958.

In addition, there are several features on the website to help you study electronics more effectively. These include animated diagrams of electronic circuits in action. The interactive multiple-choice questions on the website are a novel and effective way for you to test your understanding of the subject.

Also after each group of spreads are some pages that give you more details or cover extra topics. These are mostly for students taking the Higher Tier papers in the exam. They also cover topics that are required by only one or two of the exam boards. Ask your teacher which topics to study.

Part 1

Electricity

Topic 1

Electrons

Put some small pieces of kitchen foil on the workbench. You can use small pieces of cork, instead. Rub a plastic pen with a dry woollen cloth. Rub hard for ten or twenty seconds. Hold the pen a few millimetres above the pieces of foil. They jump up and stick to the pen. Some of them may jump up and down again several times.

FIGURE 1.1

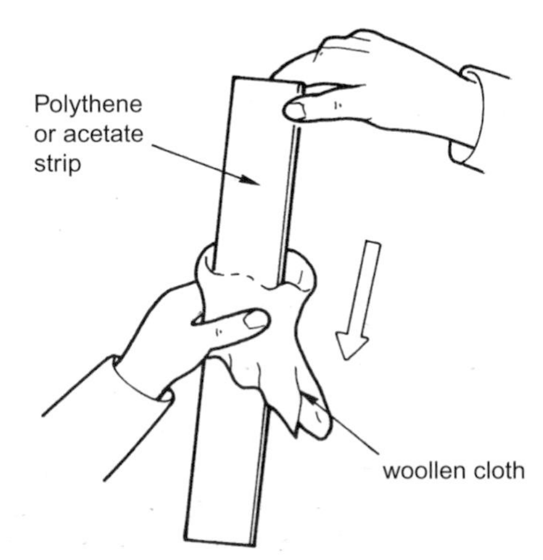

FIGURE 1.2

The reason that the pieces jump is that they are attracted by electrons on the pen. Rubbing the cloth on the pen has made electrons from the cloth transfer to the pen. We say that the pen is **charged** with electrons. It has an **electric charge**.

Some other things can be charged by rubbing. Rub a balloon with a cloth (or against your clothes). Then place it in contact with the wall of the room. It does not fall down to the floor but stays where you put it, on the wall. The electric charge has produced an **electric force** that holds the balloon against the wall.

FIGURE 1.3

Things to do

You need two strips of polythene, about 30 cm by 2 cm, and a soft dry cloth. Put the strips on the workbench and rub them briskly with the cloth. Pick up the strips by one end, one in each hand. Hold them about 50 cm apart. Then slowly move them together.

Repeat this, using one strip of polythene and one strip of acetate sheet. What do the strips do now?

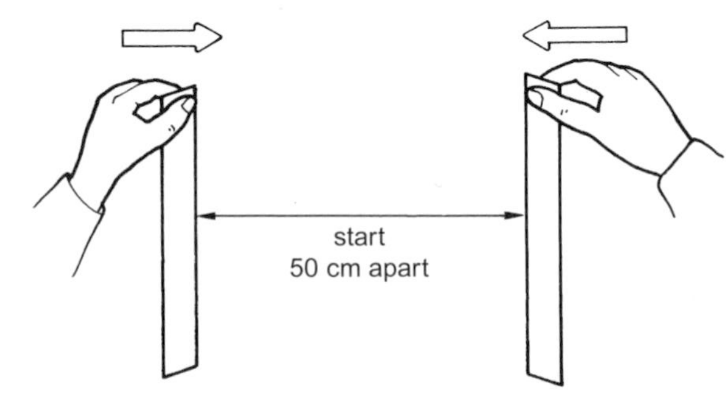

FIGURE 1.4

> **Self Test**
> What do you expect will happen if you try to bring two charged acetate strips together?

KINDS OF CHARGE

You have found that:

- Two charged polythene strips **repel** each other. They try to stay apart.
- A polythene strip and an acetate strip **attract** each other. They try to come together.

It seems that the charge on acetate is different from that on polythene, so:

There are two kinds of charge

Two charged polythene strips repel each other, so:

Like charges repel

Two differently charged strips attract each other, so:

Unlike charges attract

> **Self Test**
> Pieces of foil jump up to a charged plastic pen. Then some of them jump down again. Why does this happen?

POSITIVE AND NEGATIVE CHARGE

The two kinds of charge are called **positive charge** and **negative charge**. These names do not mean that positive charge has something that negative charge does not have. They just mean that the charges are of opposite kinds.

Rubbing a polythene strip with a cloth transfers some of the electrons from the atoms in the cloth on to the strip. Electrons have negative charge, so the strip becomes negatively charged. Also, the atoms of the cloth have now lost some electrons. This makes the cloth positively charged.

Rubbing an acetate strip with a cloth does the opposite. It *removes* electrons from the strip, leaving it positively charged. The cloth gains electrons and becomes negatively charged.

USING ENERGY

Positive and negative charges always attract each other. They try to come together. When you rub the cloth on the plastic, you separate the negative charge from the positive. It takes energy to pull them apart when they are trying to come together. This energy comes from the muscles of your arm.

ELECTRONS

Electrons are too small to see, even with a powerful microscope.

Electrons are too light to weigh. You need 1 000 000 000 000 000 000 000 000 000 000 electrons to weigh 1 kg (an amazing fact that you do not need to remember).

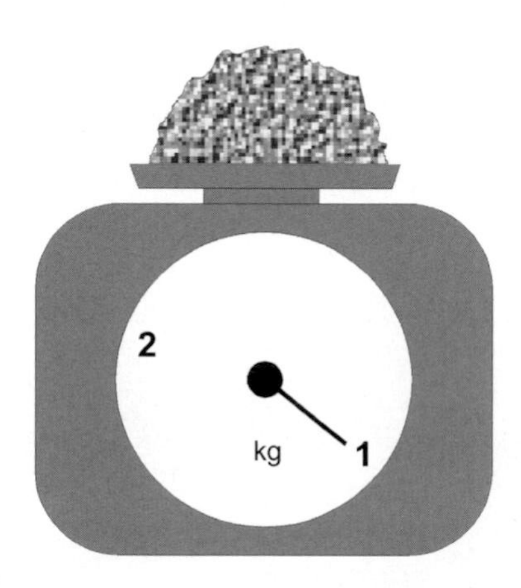

FIGURE 1.5

The most important fact about electrons is that they carry **negative electric charge**. The charge on a single electron is extremely small. But, if you have enough of them (as on the pen or the charged polythene), you can show the force that their charge causes. There are lots more things that we can do with electrons, as you will find out as you work through the book.

> **Self Test**
> Why is it impossible to have a *pile* of electrons, like that shown in the drawing?

ELECTRONS AND ATOMS

Matter is made up of molecules of many different kinds. Molecules are made up of one or more atoms. Atoms are made up of electrons (negatively charged), protons (positively charged) and neutrons (uncharged).

The simplest possible atom consists of one electron and one proton. The proton is at the centre of the atom and the electron is circling around it, in orbit.

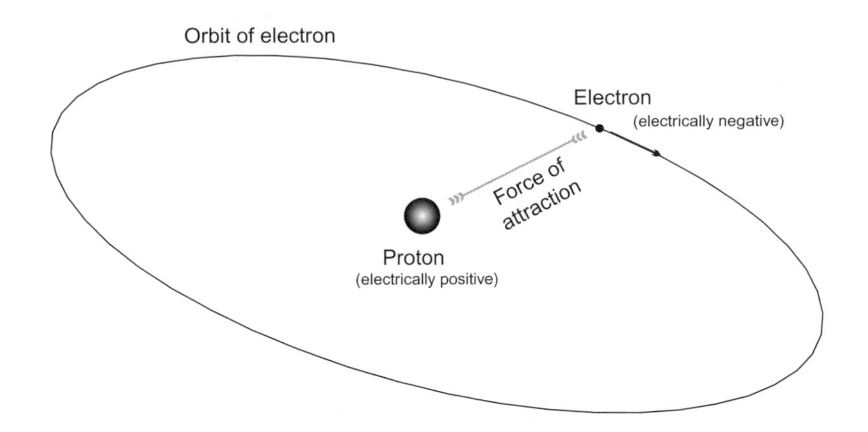

FIGURE 1.6

With one unit of negative charge on the electron and one unit of opposite but equal charge on the proton, the atom as a whole is uncharged.

The electron is circling at high speed around the proton, like a planet orbiting the Sun. There must be a force to keep it in orbit. In the case of a planet the force is gravity, the attraction between the masses of the Sun and the planet. In the case of the electron the force is the electrical attraction between oppositely charged bodies. The experiments on pages 3–4 demonstrated this.

OTHER KINDS OF ATOM

There are more than a hundred different elements in nature, including hydrogen, helium, copper, zinc, iron, mercury and oxygen, to name only a few.

Each element has its own distinctive structure, the atoms being made up of fixed numbers of electrons and protons.

In spite of these differences, all elements have the same basic plan. There is a central part, called the nucleus, where most of the mass of the atom is concentrated. The nucleus is surrounded by a cloud of circling electrons.

Atoms other than hydrogen have more than one proton and also some neutrons in the nucleus.

The protons in the nucleus give the nucleus a positive charge — one unit of charge for each proton. The number of electrons in the cloud equals the number of protons, so the cloud as a whole has an equal but negative charge which balances the charge on the nucleus.

The electrons of an atom are on the outside. They can be removed by friction, heating and electric fields. This is how we obtain the supply of electrons to use in the electronic circuits and devices described in this book.

OTHER PARTICLES

Some readers may have heard of quarks and other sub-atomic particles. Detailed studies by atomic scientists have discovered that atoms are actually made up of several more sorts of particle. In electronics, however, the only particle we need to know about is the electron.

Electric Current

Some substances let electric charge flow through them. These substances are called **conductors**.

One of the best-known conductors is copper. It conducts so well because the electrons of copper atoms are able to escape easily from the atoms.

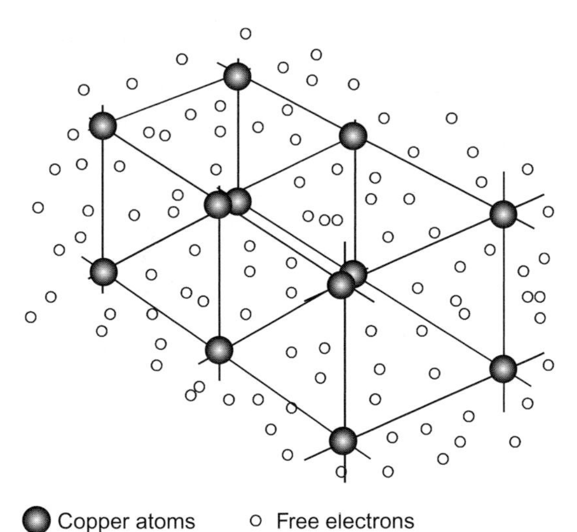

FIGURE 2.1

In a piece of copper, the atoms (large spheres in the drawing above) are arranged in regular rows and columns, called a **lattice**. The electrons that have escaped from the atoms (small circles) are able to wander about freely in the space between the copper atoms.

If we connect a battery to each end of a strip of copper, its negative terminal supplies electrons to the copper. Its positive terminal removes electrons from the other end. They are attracted by the positive (opposite) charge.

Things to do

Test different substances to find out if they are conductors or insulators (non-conductors).

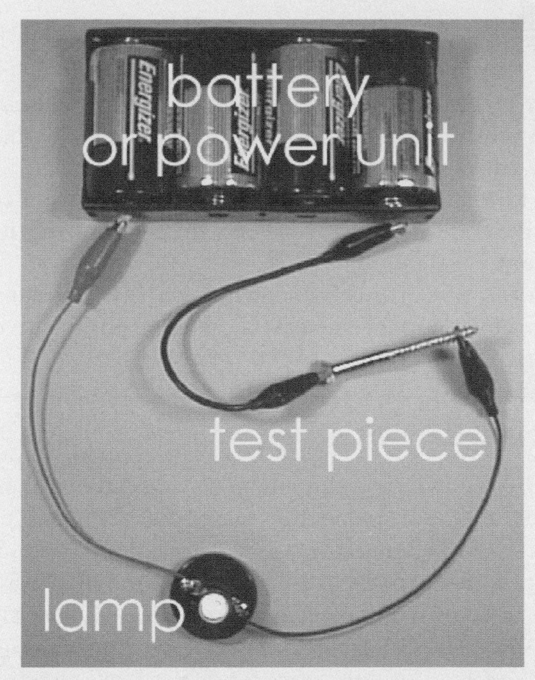

FIGURE 2.2

Try this with different materials, such as: an iron nail or screw (as shown), a piece of brick, a copper strip or wire, a plastic rod, a strip of aluminium kitchen foil, a piece of wood, a 'silver' coin, the 'lead' of a pencil (not really lead, but carbon), a piece of stone, and other materials.

The lamp shines when the material is a conductor. Make lists of conductors and non-conductors.

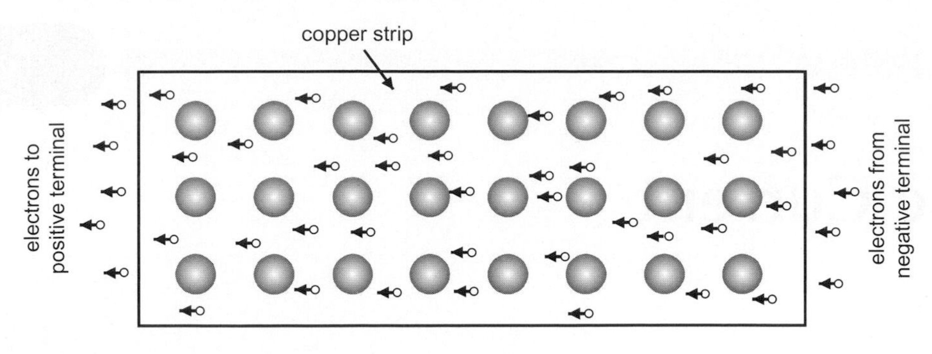

FIGURE 2.3

The flow of electrons along the copper strip is called an **electric current**. The flow is from negative to positive.

DIRECTION OF FLOW

As explained above, an electric current is a flow of negative charge (electrons) from negative to positive.

In electronics, we usually think of the current as flowing from positive to negative. Although this is not what actually happens, most people like to think of it in that way. This idea of a current flowing from positive to negative is known as **conventional current**.

In the rest of this book, when we say 'current' we mean conventional current, flowing from positive to negative.

CONDUCTORS

The best conductors are metals. Copper is the most commonly used conductor because it conducts electric charge better than any metal, except silver. But silver is too expensive to be used. Copper wires are used in almost all electronic equipment. The tracks on a circuit board are also made of copper.

FIGURE 2.4

The next best conductor is aluminium. This is often used in power lines, because of its lightness and cheapness. It is not as strong as copper, so a few strands of steel wire are included when making the cable.

Carbon is a non-metal but it has important uses as a conductor. It does not conduct charge as well as the metals do. Rods of carbon are used for making certain kinds of electric cell. Carbon is also used in making resistors (p. 43).

Solutions of salts in water are reasonably good conductors. Much of the human body consists of such solutions, so the body is a reasonable conductor of electricity. This is why we must be very careful when handling electrical equipment and working with electricity in the laboratory. Even quite a small current through a part of the body can paralyse the nerves and may kill you.

Electricity can also cause unpleasant burns.

INSULATORS

These substances contain few or no free electrons, so they are not able to conduct electric charge. We sometimes call them non-conductors.

Insulators included substances such as:

Many types of plastic, including polyvinyl chloride (PVC), which is used for insulating electrical cables and wires.

Glass and many ceramics.

Dry air.

Paper.

Self test

List these substances in order, from the best conductor to the worst:

aluminium, rubber, copper, carbon, gold, silver

Cells and Batteries

CELLS

Cells are the most compact sources of electric power. They produce electricity as a result of chemical actions inside the cell. When the cell is made, it is packed with chemical compounds that are ready to react together. As current is drawn from the cell the chemical reaction occurs. Current can be obtained from the cell until the chemicals have completed their reaction and no more of the original compounds are left.

There are several different types, based on the chemical reactions that drive them:

Type of Cell	Features	Examples of Uses
Zinc-carbon	Cheapest type. Produces 1.5 V. The voltage falls slowly during the life of the cell. May leak when old and damage the equipment by corroding its metal parts.	Electric torches, handlamps, doorbells, security alarms.
Alkaline	Holds 3 times more charge than a zinc-carbon cell of the same size, but is more expensive. Produces 1.5 V. Can supply a larger current than a zinc-carbon cell. Voltage steady during life of cell; falls sharply at the end (keep a spare handy). Does not leak.	Electronic equipment such as clocks, remote controllers, electronic toys. Also used for the same purposes as zinc-carbon cells and gives superior performance.
Zinc-chloride	Heavy duty cell in standard sizes. Produces 1.5 V. Holds more charge than zinc-carbon cells and is cheaper.	As for zinc-carbon cells.
Silver oxide	Made as disc-shaped 'button' cells. Output = 1.4 V. Delivers a small current for a long time.	Digital watches, small pocket calculators.
Lithium	Made in standard sizes and as 'button' cells. Output = 1.5 V. Produces a small current for a long time (several years) but can produce a large current for a short time.	Memory backup in computers. Pocket calculators, digital cameras, mobile phones. Also used for the same purposes as zinc-carbon cells.

Zinc-carbon, alkaline, zinc-chloride and lithium cells are made in a number of different standard sizes.

The alkaline cell on the right is the AAA size, which is one of the smallest.

The photograph shows the AAA cell about 1.4 times its real size. It produces an electrical force of 1.5 volts (1.5 V, see next Topic).

FIGURE 3.1

RECHARGEABLE CELLS

For a continuous power supply, or occasional large bursts of current, we use rechargeable cells. When the cell is exhausted, we connect it to a charger, powered from the mains supply. It restores the chemicals in the cell to their original state.

Rechargeable cells include:

Type of Cell	Features	Examples of Uses
Nickel-cadmium (NiCad)	Stores less charge than zinc-carbon cells of the same size. Produces 1.2 V. Voltage falls rapidly when nearly discharged. Can deliver high current. Shows the 'memory effect'; it 'remembers' the level to which it has been discharged. After being recharged, it ceases to deliver current when it reaches the remembered level. This reduces its current capacity.	High-current portable equipment such as video cameras, digital cameras, mobile phones.
Nickel metal hydride (NiMH)	Stores about double the charge of a NiCad cell of the same size, but is a little more expensive. Delivers 1.2 V. No memory effect.	As for NiCad cells. Also for electrically powered vehicles.
Lead-acid (Accumulator)	Produces 2 V. Can deliver very high current. Not portable; lead electrodes make the cell heavy. Danger of spilling acid electrolyte.	Car starter motors.

Danger!

NiCad, NiMH, and lead-acid batteries can produce very large currents. They can give a dangerous electric shock. A 12 V lead-acid car battery is capable of providing a bigger current than a 12 V battery made up from zinc-carbon cells. Take care when working with NiCad, NiMH and lead-acid batteries.

BATTERIES

A battery is a number of cells connected together. They are usually connected so that the battery produces an increased output voltage.

For example, the popular PP3 battery produces 9 V. It is made up from six cells, each delivering 1.5 V. It is used in clocks, test meters, and in other low-current equipment.

FIGURE 3.2

This compact battery of 8 alkaline cells is only 28 mm long. It produces a 12 V supply. Small batteries such as this are used where we need a high voltage in a small space.

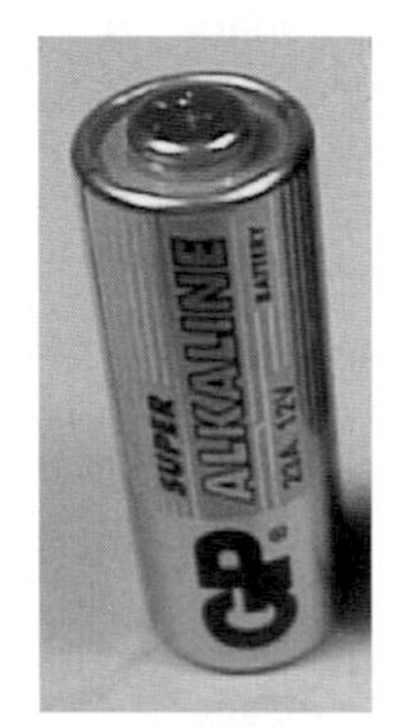

FIGURE 3.3

Examples of this include photographic equipment and key-fob remote controllers.

A battery can be built up from separate cells in a **battery box** (bottom). The plastic boxes have contacts and wires to join the cells together. The box (a) in the photograph has four NiMH size AA cells. Each produces 1.2 V, so the total voltage of the battery is 4.8 V.

Boxes (b) and (c) also hold four AA cells, but differently arranged.

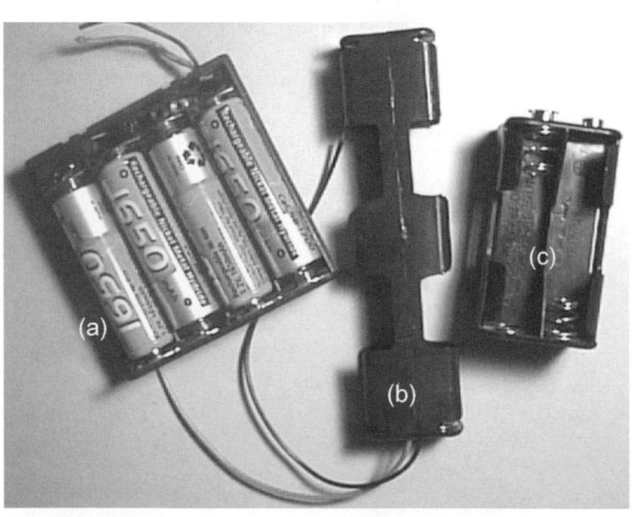

FIGURE 3.4

BATTERY CHARGERS

These are mains-powered and include a transformer to produce low-voltage direct current suitable for charging.

The inexpensive charger in the photo plugs directly into a mains socket. It holds four size AA rechargeable cells. More expensive chargers are able to hold cells of several different sizes.

FIGURE 3.5

Self Test

1. A lead-acid car battery produces 6 V. How many cells does it contain?
2. A torch battery is made up of 3 alkaline cells. What voltage does it produce?
3. What type of cell or battery would you select for powering (a) a mini-computer game, (b) a hearing aid (c) a flying model aeroplane, and (d) a petrol lawn-mower motor?

Current, Voltage and Power

CURRENT

The unit of electric current is the **ampere**. Its symbol is **A**. Few people use the word 'ampere'. They use '**amp**' instead. A typical mains electric lamp needs about one-third of an amp to make it glow brightly. A two-bar room heater needs about 8 amps.

Smaller amounts of current are measured in **milliamps**. A milliamp, symbol **mA**, is one thousandth of an amp. A torch bulb uses 60 mA or less.

An even smaller unit of current is the **microamp**, symbol **μA**. A microamp is one thousandth of a milliamp, or one millionth of an amp. Electronic clocks and watches take only a few microamps. A single AAA cell (p. 9) can power a digital clock for many months.

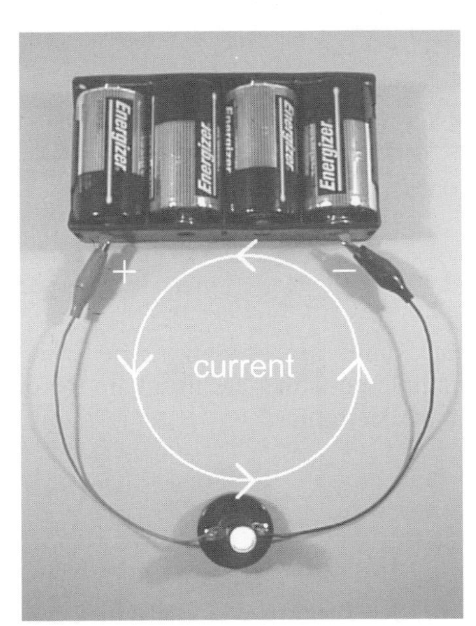

FIGURE 4.1

Current flows round a circuit. It flows out of the positive terminal of the battery, through the lamp and back into the battery through the negative terminal. The same amount of current flows in all parts of the circuit.

VOLTAGE

Voltage is the *electric force* that drives current around an electric circuit. The unit of voltage is the **volt**, symbol **V**. Most cells produce a voltage of about 1.5 V. The voltage of the mains supply is 230 V.

At a power station, the voltage is higher and is measured in **kilovolts**, symbol **kV**. A kilovolt is equal to one thousand volts. On high-power transmission lines the voltage may be as high as 400 kV.

Small voltages are measured in **millivolts**, symbol **mV**. A millivolt is one-thousandth of a volt. Voltages that are smaller still are expressed in **microvolts**, symbol **μV**. A microvolt is one-thousandth of a millivolt or one-millionth of a volt. Electrical signals from a microphone and other sensors are generally measured in millivolts or microvolts.

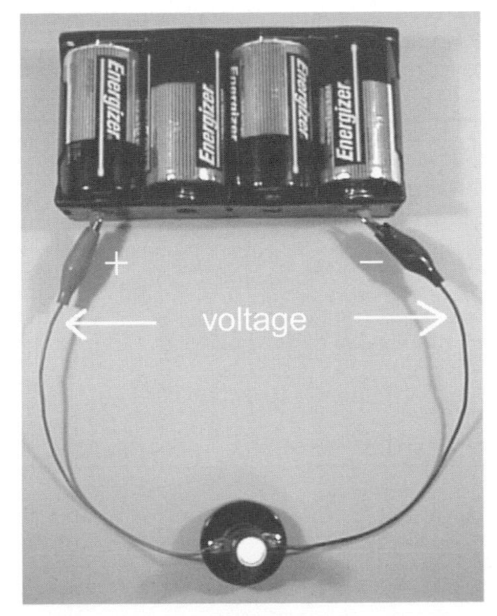

FIGURE 4.2

Current is driven round the circuit by the electric force (the voltage) between the positive and negative terminals of the battery.

POWER

Electric power expresses the rate at which an electrical device is converting energy from one form into another. For example a room heater converts electrical energy into heat energy. The rate at which it does this is its power, expressed in watts. The symbol for watt is **W**.

An average electric lamp runs at 10 W. A typical two-bar heater is rated at 2000 W, or 2 **kilowatts**. The power of an electrical power station (converting energy from coal into electrical energy) is measured in **megawatts**.

It can be shown that the power of a device is proportional to the amount of current flowing through it. It is also proportional to the voltage that is driving the current. The bigger the current and the bigger the driving force, the bigger the power. Writing this as an equation:

$$power = current \times voltage$$

- power is expressed in watts
- current is in amps
- voltage is in volts

Example 1

An electric steam iron runs on the 230 V mains and the current through it is 5.65 A. What is its power?

$$\text{power} = 230 \times 5.65 = 1299.5\ \text{W}$$

This result is close to 1300 W, or 1.3 kW.

Example 2

A 10 W lamp is running on the mains at 230 V. What current is passing through it?

Rearranging the power equation above gives:

$$\text{current} = \text{power}/\text{voltage}$$

Using this version of the equation:

$$\text{current} = 10/230 = 0.0435$$

Current is 0.0435 A, or 43.5.

Self Test

1 A projection lamp runs on a 36 V supply and takes 11 A. What is its power?
2 An electric hair-clipper runs on 230 V, and is rated at 10 W. How much current does it take?
3 A torch uses two alkaline cells and the current through the lamp is 60 mA. What is the power of the torch?
4 A 25 W soldering iron runs on 50 V. How much current does it use?
5 Express **(a)** 34 A in mA, **(b)** 1.2 mA in μA, **(c)** 1.2 mA in A, **(d)** 5505 mA in A, and **(e)** 58 μA in mA.
6 Express **(a)** 4.5 V in mV, **(b)** 11 kV in V, **(c)** 675 mV in V, **(d)** 521 μV in mV, **(e)** 0.55 V in mV, **(f)** 440 μV in V, **(g)** 0.22 mV in V, and **(h)** 3300 V in kV.
7 Express **(a)** 675 W in kW, **(b)** 25 MW in kW, **(c)** 650 mW in W, **(d)** 6 MW in W, **(e)** 4450 kW in MW, **(f)** 2.55 W in mW, **(g)** 79 kW in W, and **(h)** 33 MW in kW.

SUMMING UP

Electrical Quantity	Units of Measurement	Symbols
Current	**amp** (ampere)	**A**
	milliamp	mA
	microamp	μA
Voltage	kilovolt	kV
	volt	**V**
	millivolt	mV
	microvolt	μV
Power	megawatt	MW
	kilowatt	kW
	watt	**W**
	milliwatt	mW

The basic units are in heavy type. The multiple and submultiple units that are listed are the ones most often used in electronics.

Sources of Power

All electronic projects need a source of electrical power. This supplies the charged electrons which make electronic circuits work.

There are four main sources:

- Chemical cells and batteries.
- Solar cells.
- Charged capacitors.
- The mains supply.

CHEMICAL CELLS AND BATTERIES

Portable equipment, such as mobile phones and iPods, needs a portable power source. Topic 3 describes many types of chemical cells and batteries, which are ideal portable power sources for portable equipment. The most portable of these are the small button cells, often used in pocket calculators and hearing aids.

Cells are often required in vehicles such as cars and buses. These can be driven by power from fuel cells.

Mobility often implies being at a great distance from the nearest mains supply. An example is a spacecraft operating on the surface of Mars. Another example is an underwater vessel capable of diving into the deepest trenches of the ocean. Both of these examples have batteries to supply the power they need, when operating far from their base.

Cells are also used as storage for times when the mains power supply is interrupted. An example is the backup battery of a computer which switches on when the mains power fails. The programs and data currently being used are not lost because the backup battery has sufficient power to keep the computer running until the data has been saved.

A security system also needs a backup battery to maintain security while the mains supply is interrupted.

SOLAR CELLS

A solar cell is a slice of silicon coated on both sides with metallic contacts. The silicon has been processed to give it semiconducting (see p. 67) properties. As a result, a voltage is developed between the front and back contacts when light shines on the front surface of the cell.

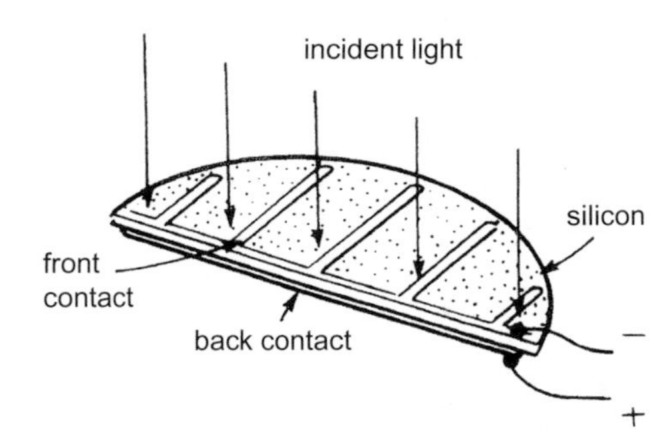

FIGURE 5.1

A typical solar cell develops 0.45 V in full sunlight, and a current of up to 200 mA. For larger voltages and currents, several solar cells are connected together into a battery, mounted on a panel. A battery of over 60 cells can supply 12 V at 500 mA. Latest developments include very large arrays of cells that generate enough power to supply a village.

Solar cells are a useful power source for equipment that is remote from the mains.

Examples: Microwave transmitters on mountaintops, spacecraft and satellites. In regions such as Western Australia, which has plenty of sunshine, there are solar-powered radio-telephones beside out-of-town highways for emergency calls.

CHARGED CAPACITORS

Capacitors are able to store electric charge indefinitely and release it when they are connected to a circuit (see p. 59). Topic 16 explains how they do this. Capacitors with sufficiently high storage capability (1 F or more) are used to provide backup for computer memory. The stored charge is enough to keep a computer running while it saves the data.

MAINS POWER

The mains supply is derived from electromagnetic generators (page 25) which are driven by other sources of energy. The chief sources are:

- **Fossil fuels**, such as coal, gas and oil. The main source of energy for electricity generation in many countries. Unfortunately, they produce greenhouse gases which contribute to global warming. Supplies of fossil fuels will eventually be exhausted.
- **Nuclear energy**. Widely used, but there are risks from radioactivity and difficulty in safely disposing of radioactive by-products.
- **Hydroelectricity**. Used in hilly areas with high rainfall. An important source of power in countries such as the USA and Canada.
- **Wind power**. Suitable for open, windy areas.
- **Wave power**. Experimental generators have been set up in estuaries and on open coastlines.
- **Geothermal power**. Available in volcanic areas such as New Zealand, where the rocks near the surface are hot enough to heat water and produce steam.

The areas in which wind, wave, and thermal power can be generated are of limited extent. This is compensated for by the fact that these sources of energy are inexhaustible and do not contribute to atmospheric pollution or the greenhouse effect.

On the Web

The descriptions in this Topic give only the basic facts. Fill in the details by browsing the World Wide Web. Use *Google, Yahoo!,* or any other browsing software.

Keywords that you could search for include: electricity, power generation, grid, fuel cell, wind power, environment, pollution, greenhouse effect. You will be able to think of many more to try.

Web sites with handy information include:
www.dti.gov.uk
www.aepuk.com
www.natwindpower.co.uk
www.electricityforum.com
www.nucleartourist.com
www.howstuffworks.com

Prepare a project detailing two or three of the energy sources. For each source, write about how it works, where stations are sited and why, efficiency, dangers, safety precautions, disposal of waste, pollution of the environment, whether source is renewable or not, costs of production, specially interesting features.

FIGURE 5.2

IN THE LAB: USING A MULTIMETER

MEASURING CURRENT AND VOLTAGE

A **multimeter** measures several different electrical quantities, usually including voltage and current. There are two kinds of multimeter, analogue (left below) and digital (right).

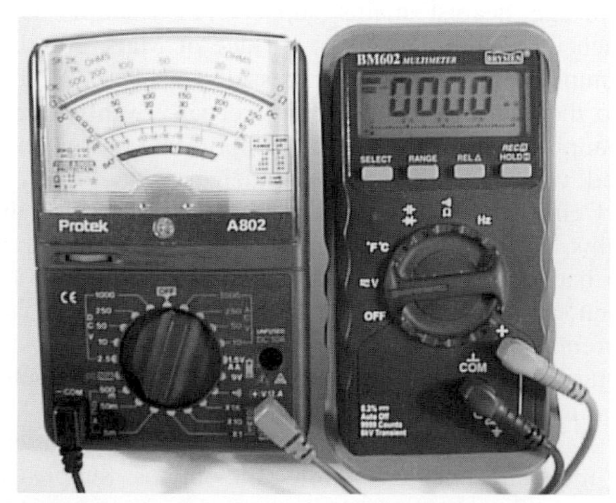

FIGURE 5.3

Both meters have sockets for **probes**:

- Positive, marked '+' and usually red.
- Common (negative), marked COM or '−' and usually black.

Some meters have a second positive socket which **must** be used for high voltages.

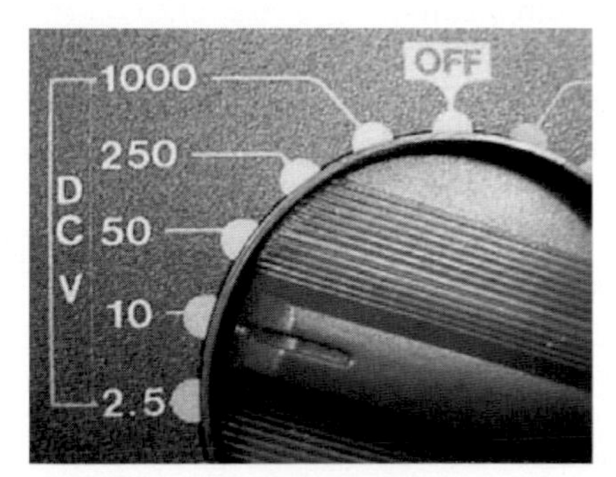

FIGURE 5.4

Both kinds of meter have a knob for selecting which quantity to measure. It often selects the range too. In Figure 5.4, the knob is selecting the 0 V to 10 V range.

With an analogue meter, you must touch the black (common) probe to the more negative point and red probe to the more positive. If you touch them the wrong way round, the needle swings *below* zero. Remove the probes from the test points immediately or you may damage the meter. Digital meters often have **autopolarity**. These work with the probes touching either way and display a minus sign to the left of the reading, if necessary.

With an analogue meter, there is the problem that the reading is wrong if you do not look straight down at the scale. This is called a **parallax error**. To help you avoid this, the meter has an arc of mirror on the scale. You can see the needle reflected in this. When you take a reading, move your head from side to side, until you have lined up the reflection of the needle exactly behind the needle. This makes sure that you are looking straight down on the scale and the reading will be correct.

Digital meters (such as that in the photo) may be **autoranging**. They select the correct range automatically when a measurement is made.

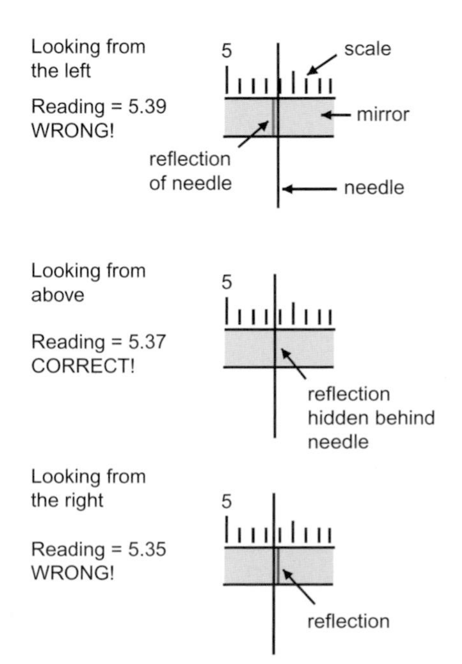

FIGURE 5.5

PLACING THE PROBES

When you measure current, the meter is part of the circuit. The current flows through the meter. So, break the circuit at one point and connect the meter across the break.

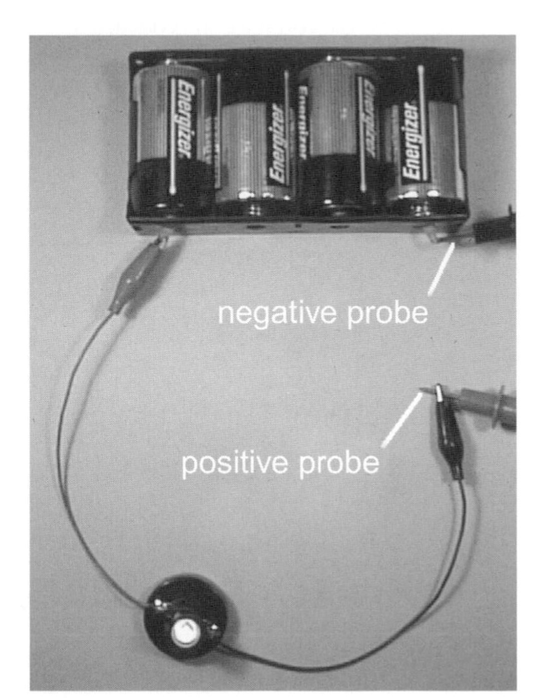

FIGURE 5.6

When you measure voltage, you are measuring the difference of electrical force at two points in the circuit. Do not break the circuit. Touch the probes to the two points to measure the difference.

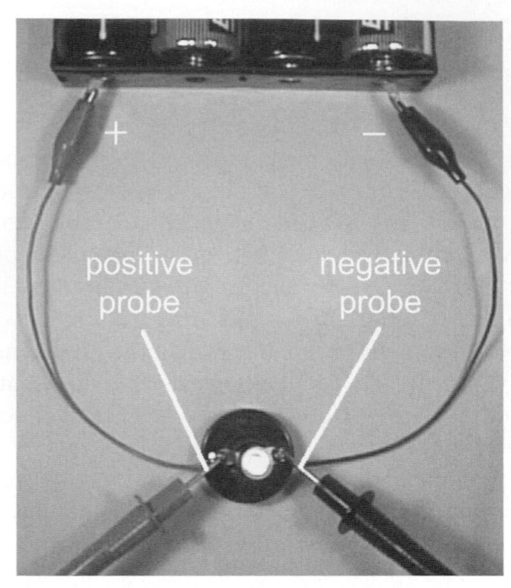

FIGURE 5.7

TEST ROUTINE

1 *Before* you connect the meter to the circuit, set the selector knob to the range you want to use. If the meter has autoranging, you need only select the quantity (current or voltage). If you are not certain which range to choose, select the highest range (for example, 0 V to 1000 V). Then, if the meter shows only a very small reading, you can work down through the ranges to one that gives a readable result.
2 Connect the meter to the circuit, or just touch the probes to two points (remember about polarity – black to negative, red to positive).
3 Take the reading (remember to avoid parallax), and *write it down.*
4 If you need to change the range, disconnect the meter from the circuit first.
5 Disconnect the meter from the circuit when you have taken the reading.
6 Turn the selection knob to the OFF position.

Things to do

Set up circuits like those in the photos, but using cells or batteries of different voltages. You could also use some old batteries that have lost most of their charge. Use various lamps or, instead of the lamp, connect in a small electric motor, an electric bell or a buzzer.

For each circuit:

1 Make a simple drawing of the circuit.
2 Measure the current, and write the result on your drawing.
3 Measure the voltage across the lamp or other device, and write the result on your drawing.

Use your readings of current and voltage to calculate the power of the lamp or device.

Power Equations

POWER AND CURRENT

Power is defined by the equation:

$$\text{power} = \text{current} \times \text{voltage}$$

$$P = IV$$

From Ohm's Law we know that:

$$V = IR$$

Putting this expression for V into the power equation, we get:

$$P = IV = I \times IR = I^2R$$

$$\mathbf{P = I^2R}$$

Example 1

A current of 5 A passes through a 16 Ω resistance. The power is:

$$P = 5^2 \times 16 = 25 \times 16 = 400 \text{ W}$$

Example 2

The current in the previous example is doubled to 10 A. The power now is:

$$P = 10^2 \times 16 = 100 \times 16 = 1600 \text{ W} = 1.6 \text{ kW}$$

Doubling the current has increased the power by four times.

Self Test

1 The resistance of the coil of a DC motor is 1 Ω. The motor is being driven by 600 mA. What is the power?
2 A soldering iron heater coil has a resistance of 3.7 kΩ, and runs at 15 W. What current does it take?
3 An electric room heater runs on the 230 V mains. Its power is 2 kW. What current does it take?

POWER AND VOLTAGE

Another form of the Ohm's Law equation is:

$$I = \frac{V}{R}$$

Putting this expression for I into the power equation, we get:

$$P = IV = \frac{V}{R} \times V = \frac{V^2}{R}$$

$$\mathbf{P = V^2/R}$$

Example 3

A 6 V battery is connected to a 100 Ω resistor. The power is:

$$P = 6^2/100 = 36/100 = 0.36 \text{ W}$$

Example 4

What is the maximum voltage that can be applied across a 220 Ω, 0.25 W resistor without damaging it? Given R and P, we calculate:

$$V = \sqrt{PR} = \sqrt{0.25 \times 220} = \sqrt{55} = 7.4 \text{ V}$$

Example 5

A 10 W lamp runs on a 12 V supply. What is its resistance?

$$R = V^2/10 = 12^2/10 = 144/10 = 14.4 \ \Omega$$

Self Test

1 A 2.2 Ω resistor has a 12 V battery connected across it. What is the power?
2 A 2 kW room heater runs on the 230 V mains. What is its resistance?

Topic 7

Alternating Currents

The voltage at the positive terminal of a battery stays constant until the the cell is exhausted. If we plot a graph of the voltage during a period of time, the graph for a fresh battery is like this:

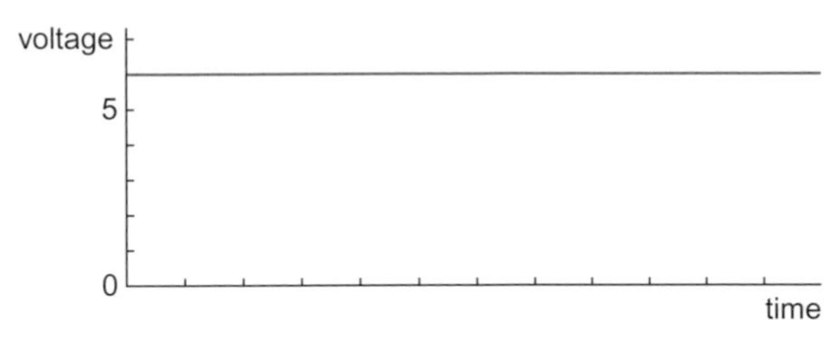

FIGURE 7.1

The graph is a horizontal straight line. It shows that the voltage of the battery is constant at 6 V.

If we connect the battery to a lamp, we can use a multimeter to measure the current flowing through it. Because the voltage is constant, the current that it drives is constant too. The graph of current against time is a straight horizontal line, like the graph above. Constant current like this is called **direct current**. This name is often shortened to **DC**.

DC always flows in the same direction, from positive to negative.

> **Memo**
>
> As usual, we are thinking of current as conventional current (p. 8).

The current that we get from some kinds of generator (including mains generators) is different from this. It repeatedly **changes its direction**.

This kind of current is called **alternating current**. This name is often shortened to **AC**.

ALTERNATING VOLTAGE

The voltage at one terminal of a typical AC generator is shown in this graph:

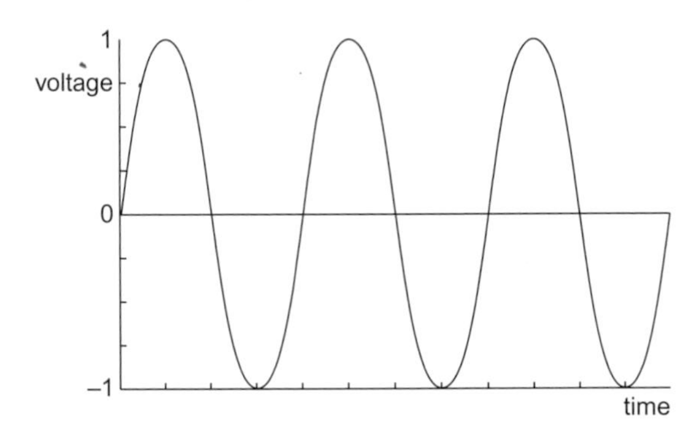

FIGURE 7.2

At one instant, the terminal is 1 V positive of the other. Then the voltage decreases until it is 1 V negative of the other terminal. Then it swings back again to become 1 V positive. This cycle is repeated indefinitely. The effect of this reversing voltage on the current in a circuit is shown below:

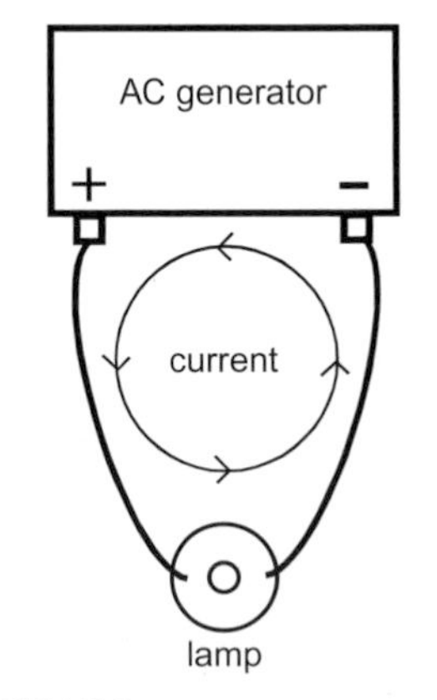

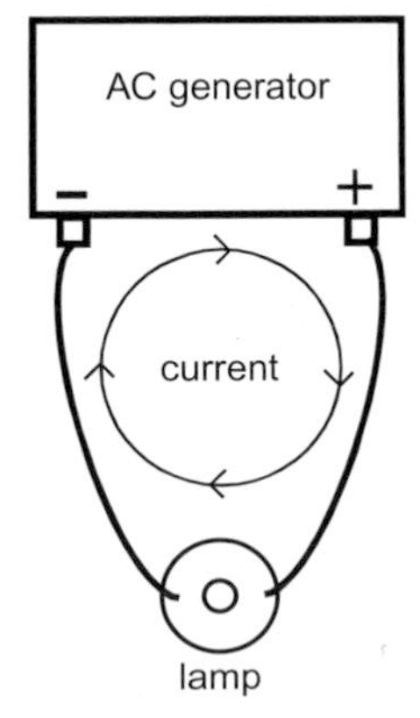

FIGURE 7.3

As the voltages at the terminals swing from positive to negative and back, the current repeatedly changes its direction. It is alternating current.

DESCRIBING AN ALTERNATING CURRENT

A direct current is simple to describe. We just quote its size, in amps. To describe an alternating current we need to state three things:

- **Amplitude:** its maximum size, in amps.
- **Period:** the time for one complete cycle from positive back to positive again.
- **Waveform:** the shape of its graph.

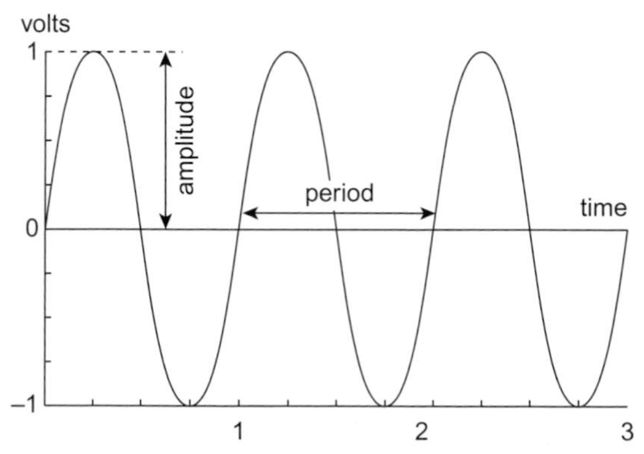

FIGURE 7.4

The AC plotted above has an amplitude $A = 1$ V. It has a period $P = 1$ s.

If we plot a graph of the sine of an angle from 0° to 360°, it has exactly the same shape as one cycle of the AC graph. The AC has a sinewave shape. Not all AC has this shape, but this is the commonest. The mains AC and any AC from mains transformers are sinewaves.

FREQUENCY

The unit of frequency is the **hertz**. Its symbol is **Hz**. The hertz is defined as a frequency of one cycle per second. The AC waveform shown in the graph has a frequency of 1 cycle per second, or 1 Hz.

If the period of a waveform is P seconds, there are $1/P$ cycles in one second. Its frequency is $1/P$ Hz:

$$\text{frequency} = 1/\text{period}$$

Self Test

1 The frequency of the mains is 50 Hz. What is its period?
2 A musical note has a period of 3.9 ms. What is its frequency?
3 Describe this waveform:

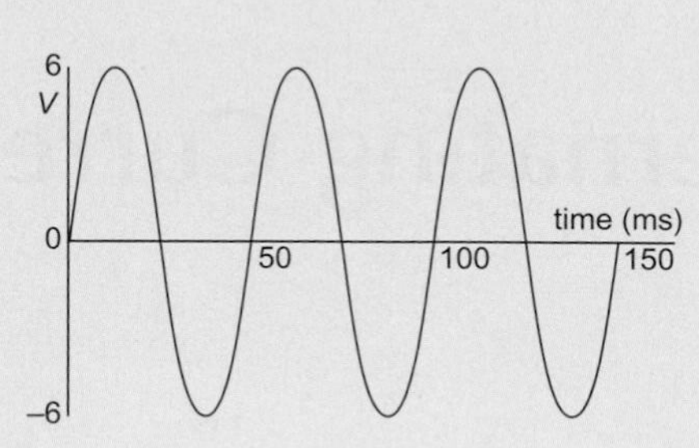

FIGURE 7.5

4 The 'bleep' of a microwave oven has a frequency of 1000 Hz. What is its period?

Things to do

1 Look at a demonstration of the shape of AC waveforms, using an oscilloscope. View the mains AC or the AC from mains transformers. Look at sinewaves of other frequencies and amplitudes from a signal generator.
2 Listen to a demonstration of the sound of sinewaves of various frequencies. This is best demonstrated by connecting an audio amplifier to the output of a signal generator. If possible, view and hear the signals at the same time.
3 Using the same equipment as for item 2 above, try to find out which is the lowest frequency that you can hear. Try to find out which is the highest frequency you can hear.
4 Using a microphone and audio amplifier connected to an oscilloscope, look at the shapes of sound waves produced by various musical instruments. Try to measure the frequency of notes on the musical scale.

ROOT MEAN SQUARE

One way of stating the size of an alternating current or voltage is to quote its **amplitude** (see above). But amplitude tells us the *peak* current or voltage. Sometimes, it is more useful to know the *average*.

Simply taking the average of all values does not work.

The voltage values of the positive half-cycle are exactly cancelled out by the negative values during the next half-cycle. The average value is *zero*!

To avoid this difficulty, we use the fact that the squares of all numbers are positive.

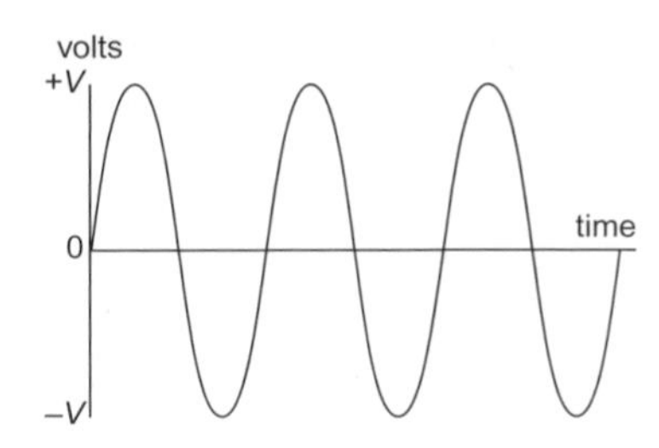

FIGURE 7.6

To find this new kind of average, we use some special maths to square all the thousands of voltage values during a cycle, find the average (or mean) of the squares and finally take its square root. The kind of average we get is called the **root mean square** or **RMS**, for short. The maths is too complicated to describe here, but we can quote its result:

The RMS value of a sinewave is 0.707 times its amplitude

Putting it the other way round:

The amplitude of a sinewave is 1.4 times its RMS value.

Example

When we say that the mains voltage is 230 V, we are quoting its RMS voltage. Its amplitude is 1.4 × 230 = 322 V. The voltage peaks to ±322 V and circuit designers must allow for this when choosing components to use. For instance, capacitors operating at mains voltages must have a working voltage of at least 322 V.

RMS currents must also be considered. For example, when a device is said to operate at 5 A RMS, the current through it will peak at 1.4 × 5 A. That is, it peaks at 7 A.

POWER SUPPLY UNITS

Cells and batteries are ideal for powering hand-held portable equipment such as torches and pocket calculators. Other devices that work on low voltage are powered from the mains, using a low-voltage power supply unit (or **PSU**).

A PSU contains a **transformer** (p. 65) to reduce the voltage from mains voltage to a lower voltage, such as 6 V. It may have a means of switching to a number of standard voltages, such as 3 V, 6 V, 9 V and 12 V. It usually has a **rectifying circuit** (p. 69) to convert the AC output of the transformer into DC. It may also have a **regulator circuit**, so that it produces a steady voltage even when the load on the PSU varies.

Most PSUs produce only small currents, generally less than 1 A, but this is enough for driving most low-current equipment.

PSUs are made in various forms:

- a plastic body that looks like a large mains plug, with three pins for plugging directly into a mains socket. It has a flexible lead for connecting to the equipment.
- a case with two leads, one to connect to a mains plug and the other a low-voltage lead to connect to the equipment.
- a bench PSU is used in laboratories and workshops. It connects to the mains and has terminals that provide the low-voltage DC output. It is usually possible to switch to a range of different voltages and also to vary the output voltage smoothly over a wide range, typically from 0 V to 30 V or more. A limit can be set on the current delivered, to protect circuits powered by the PSU. Meters display the voltage and current being delivered.

QUESTIONS ON ELECTRICITY AND MAGNETISM

(Give the results of calculations to the nearest three significant figures. Answers to questions with short answers are given on the website.)

1. What are the names of the two kinds of electric charge? Which kind of charge is carried on an electron?
2. Why is copper such a good conductor of electricity?
3. List four good conductors and four insulators.
4. What is an electric current?
5. Name the units of: (a) current, (b) voltage, and (c) power. What are their symbols?
6. Write an equation that shows how the three units in question 5 are related to each other.
7. What are the units and symbols for (a) a millionth of a volt, and (b) a million watts?
8. What is the difference between a cell and a battery?
9. List two types of cell, describe their features and give examples of their uses.
10. Name a type of rechargeable cell, describe its features and give examples of its uses.
11. What types of battery could you use for backing up the power supply to a security system? If the system is to run on 24 V, how many cells of this type would you require?
12. In what ways may a 12 V car battery be dangerous?
13. A halogen lamp runs on 12 V with a power of 20 W. How much current does it take?
14. A small radio receiver uses two alkaline cells and takes a current of 12 mA. What is its power?

15 Draw a diagram to illustrate four cycles of AC. Label the amplitude and the period. If the period is 20 ms, what is the frequency?
16 Name the three lines of the mains supply and explain what they do.
17 Explain why a power switch must be placed on the live side of an appliance.
18 Why should an RCCB be used with an appliance such as an electric lawnmower?
19 What conductor is often used in power lines?
20 What is conventional current?
21 Draw a diagram showing three 1.5 V cells connected in series. What is the total voltage?
22 A sine wave produced by an audio generator has an amplitude of 6 V. What is its RMS value?
23 An electric torch has four 1.5 V cells in series. It has a 0.96 W lamp. How much charge passes through the lamp while it is switched on for 10 minutes?
24 A krypton lamp takes 0.7 A when running on a 2.4 V supply. How much work is done while the lamp is switched on for 20 minutes?
25 A compass needle is allowed to come to rest in the Earth's magnetic field. A straight wire is mounted above the needle and parallel with it. A current is switched on, flowing in the wire from North to South. What happens to the needle?
26 Draw a diagram of a magnetic field around a horseshoe magnet, showing the lines of force and their direction.
27 What is a magnetic domain?
28 A current of 2.5 amps flows in a straight wire that is 0.75 m long. The flux density of the local field is 2.5 T. How much is the force on the wire?
29 State Fleming's left-hand rule.
30 Draw a diagram to show the main features of a simple DC motor. Explain how it works.

Mains Electricity

Mains electricity is produced in power stations like the one illustrated in the photo below.

This is one of the two generators at Ironbridge Power Station in Shropshire. Here the generators are driven by turbines. The turbines are turned by steam under pressure. The steam is produced in a coal-fired boiler. The rate of production of electrical energy from the chemical energy in coal is over 200 MW.

In other power stations, the steam may be produced in an oil-fired boiler, or by using heat from a nuclear reactor. In a hydro-electric power station, the turbines are turned directly by the flow of water. Wind is another source of energy for electricity generation.

In a few parts of the World, such as the thermal areas of New Zealand, steam for turbines is generated using energy from the hot rocks just below the Earth's surface.

In the United Kingdom and many other countries, the mains supply to the consumer is close to 230 V. The current is alternating.

MAINS SUPPLY

The electricity supply to a building has three wires:

- Live
- Neutral
- Earth

The **live wire** supplies the current. On this wire, the voltage alternates positive and negative of the neutral wire.

The **neutral wire** returns the current to the power station after it has passed through all the appliances in the building. The voltage on this wire is close to earth voltage, which we take to be 0 V.

FIGURE 8.1

The **earth wire** is connected to the earth. Normally, no current flows in this wire. It is there to provide a path to earth when there is a fault. We explain this later.

To help make sure that wiring is correctly installed, the three wires are colour coded:

- **Live** is brown.
- **Neutral** is blue.
- **Earth** is striped green and yellow.

CABLES

The conductors in mains cable (as in most electrical cables) are made of copper. This is because copper is one of the best conductors. The copper is in the form of several copper strands twisted together. This makes the cable flexible. It can be bent into the right shapes for running it around the building. Cable used for connecting appliances needs to be even more flexible.

The photo shows how a 3-core mains cable is made up. It has three cores of twisted copper strands for live, neutral, and earth. Each core is surrounded by a layer of plastic to insulate it from the other cores.

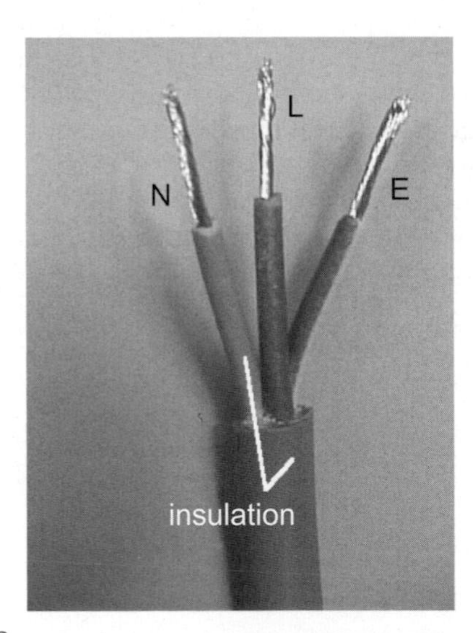

FIGURE 8.2

The three insulated cores are surrounded by an outer plastic layer to hold them together and to give more insulation between the cores and anything that the cable is in contact with.

SWITCHES

Most appliances have a switch for turning them on or off. The switch must always be wired into the live wire of the supply to the appliance. In the top diagram (Figure. 8.3), the switch is open. No current can flow through the appliance, so it is off. It is still

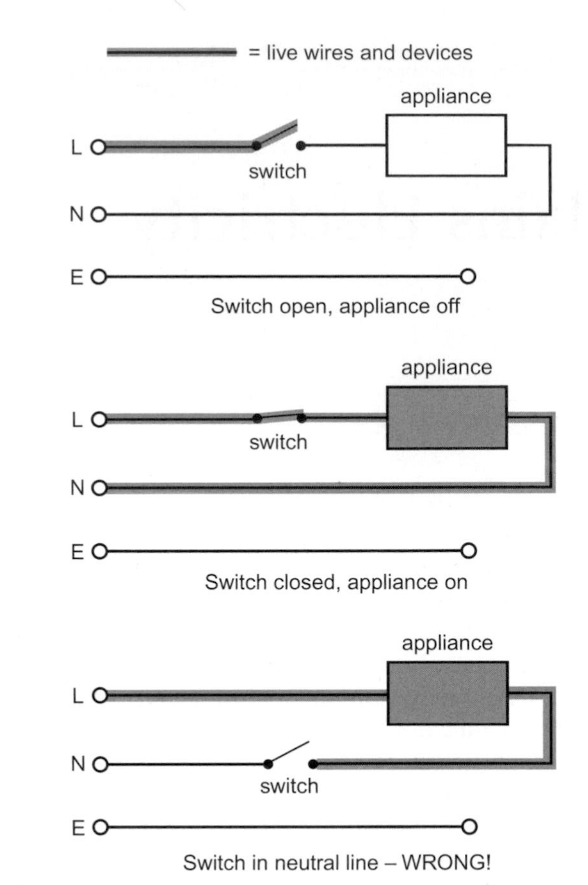

FIGURE 8.3

connected to the neutral line of the mains, but this line is close to earth voltage. If there is a fault in the appliance, there is little danger.

In the middle drawing, the switch is closed and the appliance is on. Current flows through the appliance between the live and neutral lines. There is little danger.

In the bottom drawing, the switch is open and the appliance is off. But it is *still connected to the live wire*. Should there be a fault, the appliance is still live and this is dangerous.

The switching drawing shows the earth wire not connected to anything. Connections for the earth wire are described in the next topic.

Self Test

1. Why is the core of a cable made from copper?
2. What is the reason for making the core from thin strands twisted together?
3. What are the names of the three wires in an electrical supply? What does each do?
4. Why are the cores of a cable covered with plastic?
5. Why is it important for the plastic to be flexible?

Plugs and Fuses

PLUGS AND SOCKETS

Electrical appliances such as electric cookers and immersion heaters need a heavy current. They are usually connected through a switch directly to the mains supply. The supply to ceiling lamps and other fixed appliances is also permanently wired to the mains. A qualified electrician does this.

A moveable appliance is usually plugged in to a socket on the wall. The plug has three pins, live, neutral, and earth. The pins are made of brass. This is a good conductor. It is not quite such a good conductor as copper, but it is stiffer and more hard-wearing. The earth pin is longer than the other two so that it makes contact first when the plug is pushed into the socket. It breaks contact last when the plug is pulled out. This makes sure that the appliance is earthed all the time it is connected to the mains.

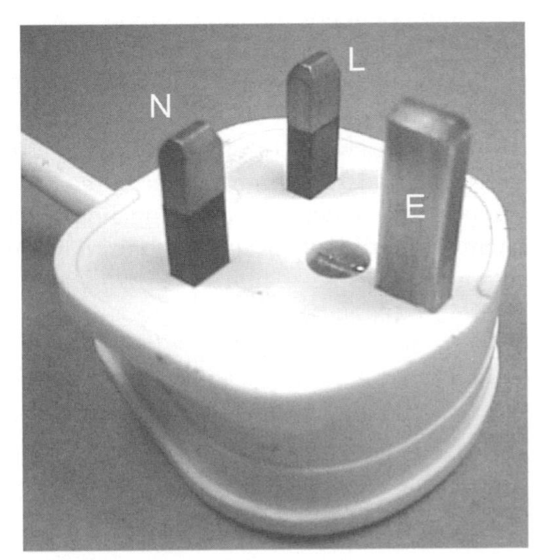

FIGURE 9.1

The body of the plug is made of plastic or hard rubber, which are good insulators.

WIRING A MAINS PLUG

The photograph shows a mains plug with the top removed.

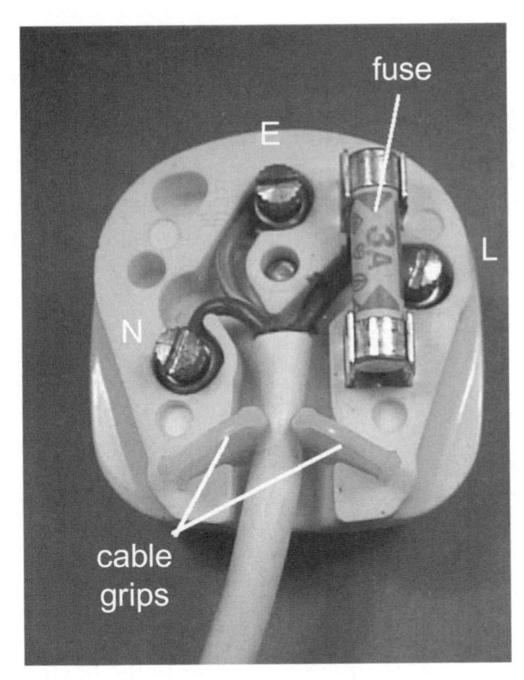

FIGURE 9.2

The steps for wiring the plug are:

1 Unscrew the fixing bolt and remove the top from the plug.
2 Look at the instructions supplied with the plug to find out how much of the outer insulation to remove. Use a wire-stripper to remove it.
3 Check whether any of the three wires needs to be shortened. If so, cut it shorter.
4 Look at the instructions to find out how much insulation to remove from each wire. Use a wire-stripper to remove it.
5 Twist the bare copper strands with your fingers to make them stay together.

6 Usually it is best to thread the cable through the cable grip at this stage.
7 Loosen the terminal screws on each pin. Wrap the bare cores around each screw and tighten each screw.
8 Make sure that the wires run neatly inside their channels.
9 Check that there are no loose strands of copper.
10 Knowing the maximum current that the appliance will require, select a fuse of slightly greater rating. Press it into the fuse clips.
11 **Check again that the three wires are connected to the correct pins.**
12 Replace the top and tighten its retaining screw. If the cord grip has screws, tighten these too.

FUSES

Fuses are used to protect equipment and people against electrical faults. A fuse contains a thin wire made of a special alloy that melts at a fairly low temperature. If the current through the fuse is too high, heat is produced faster than it can be lost from the fuse. The fuse wire gets so hot that it melts and breaks the circuit. This disconnects the supply.

Self Test

Fuses are always placed on the live side of an appliance. Why?

(Hint: see drawing on p. 26)

Fuses are rated to blow if the current exceeds a stated amount. Mains fuses such as the one in the plug (Figure 9.2) are made in 3 A, 5 A and 13 A ratings. Always select the rating according to the device being protected. Use 13 A fuses only with high-current appliances such as room heaters and vacuum cleaners.

DANGER FROM ELECTRICITY

- Touching live metal parts may cause burns or, in serious cases, loss of life. Often, the breathing muscles are paralysed. First aid: turn off the power; then apply artificial respiration.
- A current too heavy for the cable will make the cable hot. This may cause heat damage to the cable and the appliance. It may also lead to fire.
- Sparks, either from static charge or from switches and other devices may ignite inflammable vapours or dusts in the air. This may result in fire or explosion.

CIRCUIT BREAKERS

Some kinds of circuit breaker are used instead of fuses to protect a faulty appliance from drawing too much current. A **thermal circuit breaker** becomes overheated by excess current passing through it. This triggers the breaker switch to open.

A **residual current circuit breaker** (**RCCB**) has a different action. It switches off the mains supply to an appliance when it detects a leakage of current to earth or to another circuit. For example, an exposed metal part of an appliance (such as its case, or a control knob) may become 'live' because of faulty insulation. A person touching that part conducts current to earth and receives a shock. The RCCB measures the current flowing along the live wire and that flowing along the neutral wire. Because some of the current is leaking away (through the person) the live and neutral currents are unequal. The RCCB detects this state and switches off the supply. A typical RCCB detects a current leak as small as 30 mA, and switches off within 20 ms. This greatly reduces the chances of serious electric shock.

RCCBs do not switch off if the leakage is direct from live to neutral. This is because the currents are *equal*. A person accidentally touching both live and neutral wires or terminals at the same time receives a shock.

Electricity and Magnetism

A current flowing in a wire produces a magnetic field around the wire. We can detect this by placing a compass below the wire, as in the photographs.

FIGURE 10.1

FIGURE 10.2

The wire is horizontal, running in a north-south direction. When the current is off (Figure 10.1), the compass needle aligns itself with the Earth's magnetic field, and comes to lie parallel with the wire. When the current is switched on (Figure 10.2) and flowing through the wire from the bottom to the top of the picture, it generates a strong magnetic field, which deflects the compass needle.

Figure 10.3 (right) represents the magnetic field as an array of lines of force. These lines follow the direction in which a compass needle points when it is placed in the field. The lines of force are concentric circles centred on the wire.

On the right, the drawing shows how to find the direction of the field. Grip the wire with your right hand, your thumb pointing in the direction of the current. Your fingers, gripping the wire, show the direction of the field.

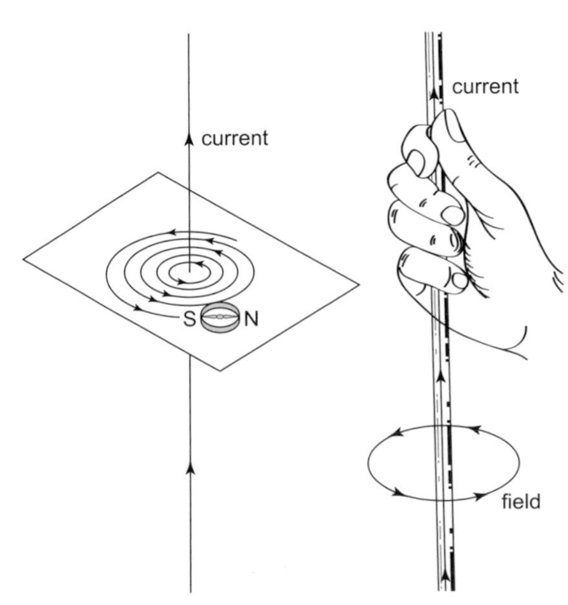

FIGURE 10.3

MAGNETIC POLES

Every magnetic object has two poles, North seeking (N) and South seeking (S). The poles of a compass needle are at the pointed ends. If the needle is free to swing in the Earth's magnetic field, it comes to rest with the poles pointing toward the Earth's magnetic North and South poles, respectively. By definition, the lines of force run from the South pole to the North pole inside the needle. They run from North to South in the region around the needle.

FIGURE 10.5

THE FIELD AROUND A COIL

When the current is on, each turn of a coil produces a magnetic field. The fields due to the individual turns of the coil combine to form a field as shown at (a) and (b). The field of the long, narrow coil is similar to the field around a bar magnet.

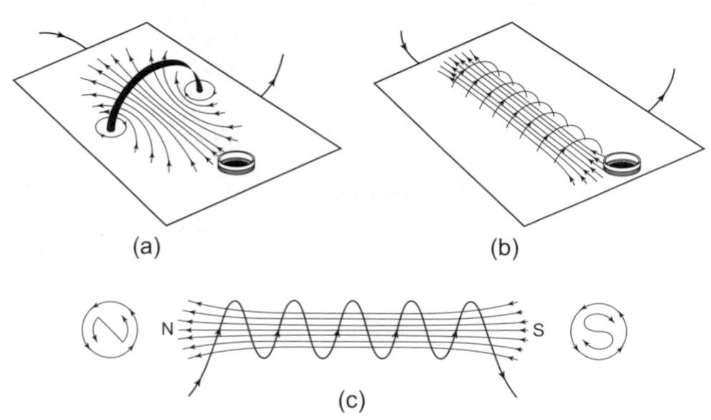

FIGURE 10.4

Drawing (c) illustrates a way of remembering which is the North end and which is the South end of the coil.

PLOTTING A MAGNETIC FIELD

Place a bar magnet on a sheet of paper and draw round its outline so that you can remember where the magnet was placed.

Place a small compass up against the north pole of the magnet. Mark a dot on the paper to show where the ends of the needle come to rest. These two dots are shown in Figure 10.5. A line joining these dots would show the direction of the field.

Move the compass so that the needle end which was nearest the magnet is now over the second dot. Plot a third dot beside the other end of the needle.

Proceed in this way, plotting a line of force, which runs out from the magnet and curves round toward the south pole of the bar magnet. The photo below shows several lines of force.

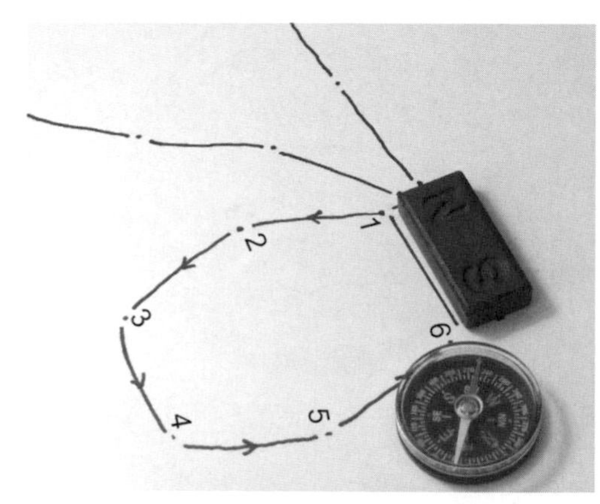

FIGURE 10.6

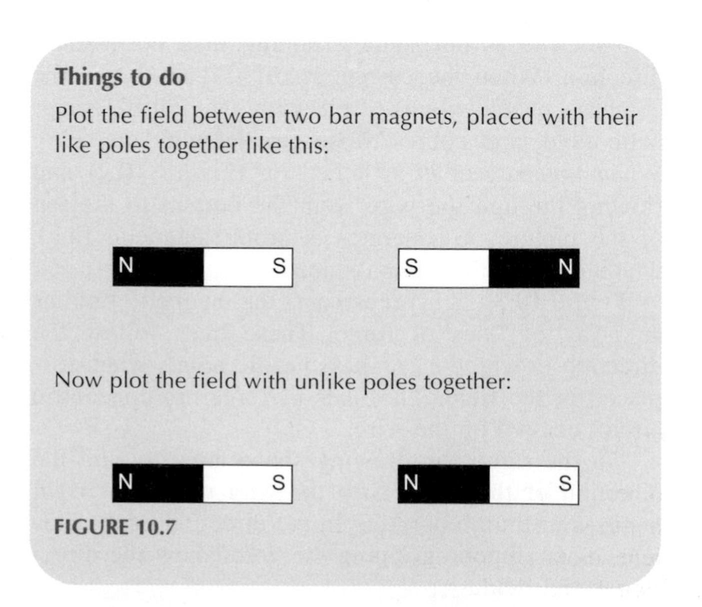

Things to do

Plot the field between two bar magnets, placed with their like poles together like this:

Now plot the field with unlike poles together:

FIGURE 10.7

THE FORCE ON A WIRE

When current flows through a wire that is in a magnetic field, the force F on the wire is

$$F = BIL$$

Where F is the force in newtons, B is the magnetic flux density in tesla, I is the current in amps and L is the length of the wire in metres. B is calculated from:

$$B = v \times \frac{1}{c^2} \times E$$

The magnetic field B is produced by a unit charge of 1 coulomb, moving at v metres per second in an electric field of E volts per metre. The constant c is the velocity of light in m/s, which equals 3×10^8 m/s.

One way of defining electrical field strength is to consider two points in the field, D metres apart with the potential difference of V between them. The electric field strength is defined as:

$$E = V/\mathrm{D} \text{ volts per metre}$$

This equation expresses the fact that the electric field strength depends on the potential difference over a given distance.

Given the electric field strength, we calculate B and then calculate F.

ELECTRIC MOTORS

When a wire is carrying a **current**, and that wire is on a magnetic **field**, there is a force that makes the wire **move**.

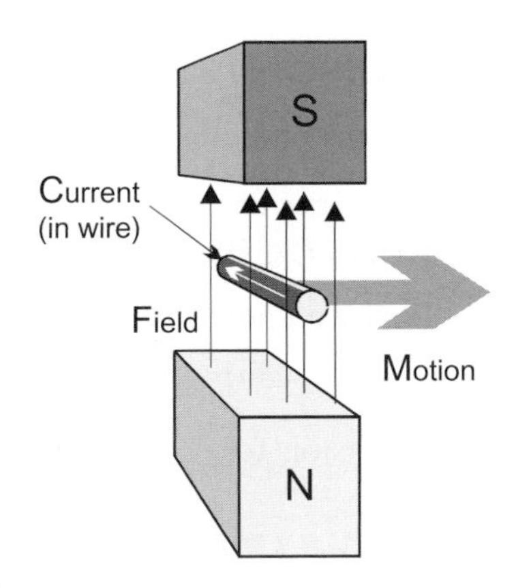

FIGURE 10.8

FIELDS AND MOTION

When a charged object moves in a magnetic field it is subject to a force. The direction of the force depends on **Fleming's left-hand rule**, as shown in Figure 10.9.

Hold your thumb, first finger and second finger at right angles, as in the drawing below. Point your first finger in the direction of the electric field and your second finger in the direction of the current. Your thumb points in the direction of the motion.

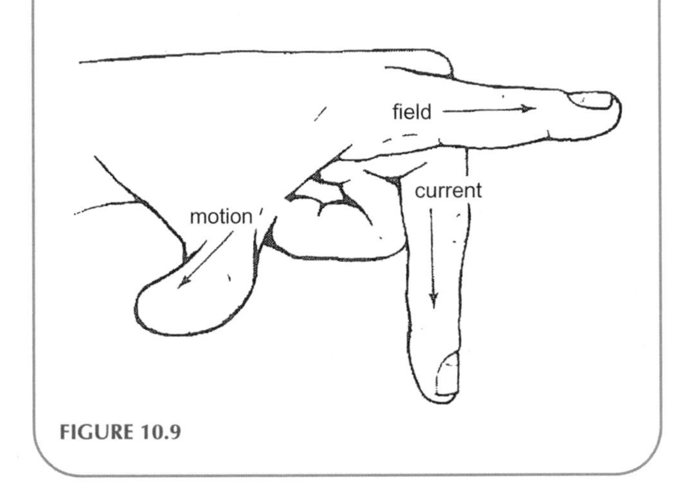

FIGURE 10.9

HOW A SIMPLE DC MOTOR WORKS

The drawing below represents a simple DC motor. There is a permanent magnet to provide the field. A coil mounted on an axle spins between the poles. The coil is shown with a single turn. In a real motor, it would have several hundred turns.

The ends of the coil are connected to a **commutator**, consisting of two metal half-rings. There are two springy metal **brushes** in contact with the half-rings.

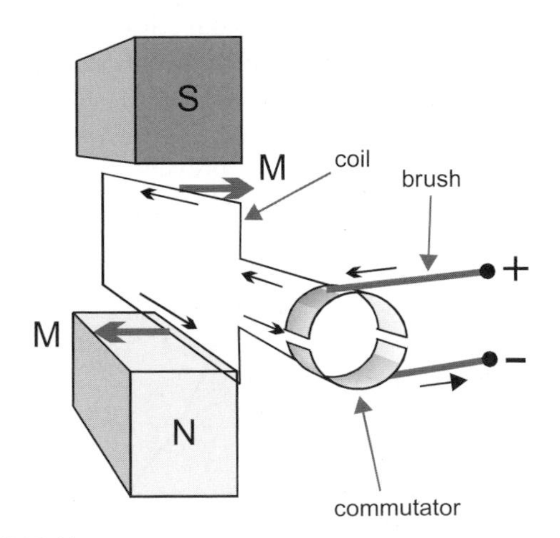

FIGURE 10.10

When a DC voltage is applied to the terminals, current flows along the upper brush to the commutator, around the coil to the other half-ring of the commutator and back along the lower brush. The current is flowing *away* from the commutator in the upper section of the coil. The field and current are as in the Figure 10.8. Applying Fleming's Left-hand Rule, the upper section of the coil is forced to move to the right. Applying the rule to the lower section, where the current is flowing *toward* the commutator, that section is forced to move to the left. The two forces cause the coil to rotate in an anti-clockwise direction.

The coil spins round until the gaps between the half-circles come under the brushes. For an instant, the current stops. Inertia makes the coil continue to spin until the brushes contact the half circles again. But the section that is now the upper one still carries current away from the commutator. The section that is now the lower one still carries current toward the commutator. So the coil continues to turn in a clockwise direction.

In a real motor, the coil has a core of layers of iron (an **armature**) to help guide the lines of magnetic force through the centre of the coil.

GENERATORS

A DC motor can also be used as a generator. If we turn the coil, by hand, or perhaps by using a petrol engine or a steam driven turbine, the motion of the coil through the field generates a current.

The direction of the current is found by using **Fleming's Right-hand Rule** (Figure 10.11). Hold your right hand with the thumb and first two finger at right-angles to each other. If the thu**M**b points in the direction of **M**otion, and your **F**irst **F**inger points in the direction of the **F**ield, your Se**C**ond finger is pointing in the direction of the **C**urrent. Applying this rule we see that the current flows in the opposite direction to that in the motor. This is because the induced current is trying to prevent the coil from rotating. To turn the coil we have to supply extra energy, which appears as generated electrical power.

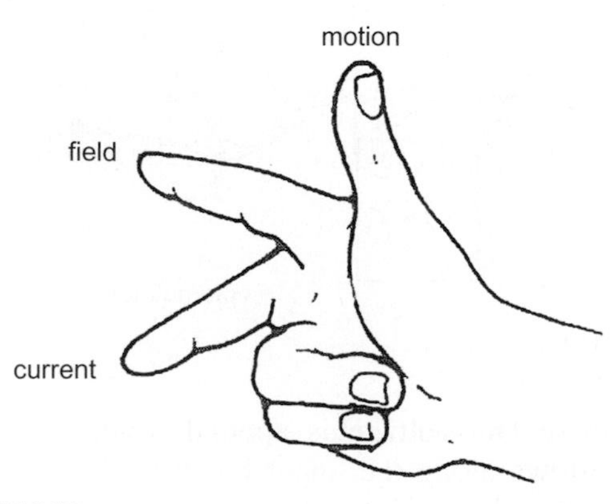

FIGURE 10.11

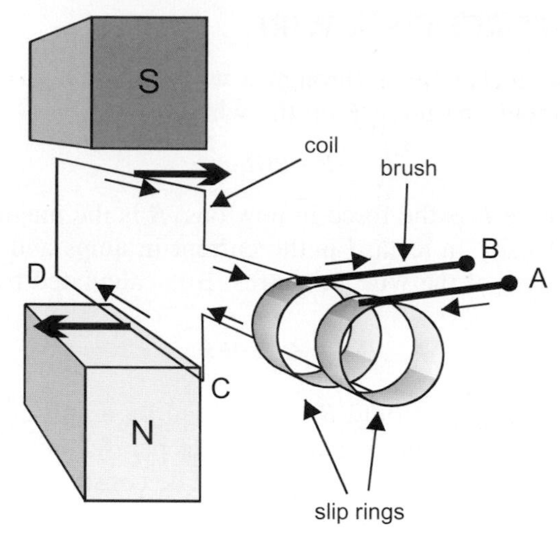

FIGURE 10.12

An AC generator is similar to a motor, but has two slip rings instead of a commutator. These are connected to the ends of the coil. As the coil is rotated, each brush remains in contact with the same ring (Figure 10.12).

In the drawing, the part of the coil labelled CD is moving past the north pole. Current flows in through terminal A (negative) and out through terminal B (positive). As the coil is rotated, CD moves upward and then moves past the South pole. The current in it reverses. Now terminal A is positive and terminal B is negative. The current is alternating. It goes through one cycle for each rotation of the coil.

POWER AND CURRENT

Power is defined by the equation:

$$\text{power} = \text{current} \times \text{voltage}$$
$$P = IV$$

From Ohm's Law we know that:

$$V = IR$$

Putting this expression for V into the power equation, we get:

$$P = IV = I \times IR = I^2R$$
$$\mathbf{P = I^2R}$$

Example 1

A current of 5A passes through a 16Ω resistance. The power is:

$$P = 5^2 \times 16 = 25 \times 16 = 400\ \text{W}$$

Example 2

The current in the previous example is doubled to 10A. The power now is:

$$P = 10^2 \times 16 = 100 \times 16 = 1600\ \text{W} = 1.6\ \text{kW}$$

Self Test

1 The resistance of the coil of a DC motor is 10 Ω. The motor is being driven by 600 mA. What is the power?
2 A soldering iron heater coil has a resistance of 3.7 kΩ, and runs at 15 W. What current does it take?

POWER AND VOLTAGE

Another form of the Ohm's Law equation is:

$$I = \frac{V}{R}$$

Putting this expression for I into the power equation, we get:

$$P = IV = \frac{V}{R} \times V = \frac{V^2}{R}$$

$$\mathbf{P = V^2/R}$$

Example 1

A 6 V battery is connected to a 100 Ω resistor. The power is:

$$P = 6^2/100 = 36/100 = 0.36\ \text{W}$$

Example 2

What is the maximum voltage that can be applied across a 220 Ω, 0.25 W resistor without damaging it? Given R and P, we calculate:

$$V = \sqrt{PR} = \sqrt{0.25 \times 220} = \sqrt{55} = 7.4\ \text{V}$$

Self Test

1 A 2.2 Ω resistor has a 12 V battery connected across it. What is the power?
2 A 2 kW room heater runs on the 230 V mains. What is its resistance?

MAINS POWER DISTRIBUTION

Mains power is generated in power stations, as outlined in Topic 8.

It is distributed to homes, offices, shops, factories and other places in which it is used by the National Grid. The National Grid is a network of cables spreading to all parts of the country. Power stations supply electricity to the Grid, the amount they supply depending on local demand at the time.

The cables of the Grid have thick copper or aluminium conductors, so their resistance is low. However, the cables are many kilometres long and their resistance can not be ignored. As electricity passes along the cables there is a loss of power. Electrical energy is converted to heat energy, which escapes into the surroundings.

The power loss is proportional to the square of the current. To keep the loss as small as possible, the current must be as small as possible. For a given power level, to make current as small as possible, the voltage must be as large as possible.

Power is distributed through high-voltage power lines. These may run underground or may be above ground and supported by pylons (Figure 10.13). Underground power lines are necessary in built-up areas, but are more expensive to install. There are also complications in routing the power lines through areas where water mains, gas mains and sewers are already in position. It is relatively cheaper to string cables from pylons, which is a very important factor for lines many kilometres long in country areas. Unfortunately, most people consider rows of pylons to be unsightly.

FIGURE 10.13

MULTIPLE CHOICE QUESTIONS

(Answers are on the website)

1 Of these, the best electricity conductor is:
   A nylon.
   B copper.

C aluminium.
D carbon.

2 In the mains supply, the wire that supplies the current is called the:
A neutral.
B earth.
C ground.
D live.

3 The nucleus of an atom of carbon consists of:
A one proton.
B several protons.
C protons and electrons.
D electrons and neutrons.

4 A typical residual current circuit breaker limits leakage to earth to:
A 20 mA.
B 13 A.
C 30 mA.
D 1 A.

5 A 12 V battery is used to light a small lamp. The current flowing through the lamp is 500 mA. The power of the lamp is:
A 6 W.
B 24 A.
C 0 .24 mW
D 500 mA.

6 A nickel metal hydride cell:
A produces 1 V.
B shows the memory effect.
C stores less energy than a NiCad cell.
D is rechargeable.

7 Three lithium cells in series produce:
A 4.5 V.
B 3.6 V.
C 6.0 V.
D 1.5 V.

8 The switch for a mains powered circuit should be in the live lead because:
A this is a legal requirement.
B it isolates the circuit from the live mains.
C putting it there is a convention.
D the switch controls mains voltages.

9 The longest pin of a 13 amp plug is:
A earth.
B live.
C neutral.
D all the same length.

10 The body of the mains plug is made from plastic or hard rubber because they are:
A flexible.
B insulators.
C cheap.
D lightweight.

SI Units

The International System (SI) of units defines seven **base units**. These include:

- the **metre** (symbol, m), the unit of distance,
- the **kilogram** (symbol, kg), the unit of mass,
- the **second** (symbol, s), the unit of time, and
- the **ampere** (symbol A), the unit of electric current.

USING THE BASE UNITS

On its own, each base unit describes a particular quantity — distance, mass, time and current. We also combine these quantities to describe other quantities. If a model plane flies a distance of 1 metre in 1 second, we say its **velocity** is 1 metre per second. Velocities are measured in metre per second (in symbols, m/s).

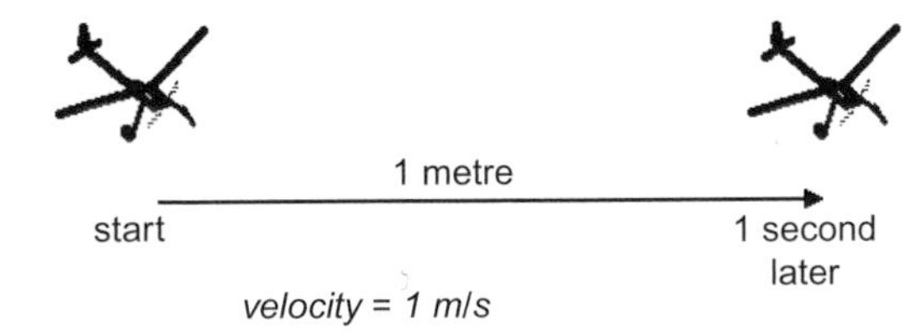

FIGURE 11.1

Note the division symbol (/). To measure velocity, we *divide* the distance by the time. For example, if the plane flies 3 m in 2 s, its velocity is 3/2 m/s. Completing the division gives:

$$\text{velocity} = 1.5\ \text{m/s}$$

From velocity we obtain **acceleration**. This is the amount by which the plane's velocity increases per second. Suppose that it is flying at a velocity of 1.5 m/s and then, 1 second later, it is flying at a velocity of 2.5 m/s.

Its velocity has increased by 1 m/s in a period of 1 s. Its acceleration is 1 metre per second per second (in symbols m/s/s or m/s^2).

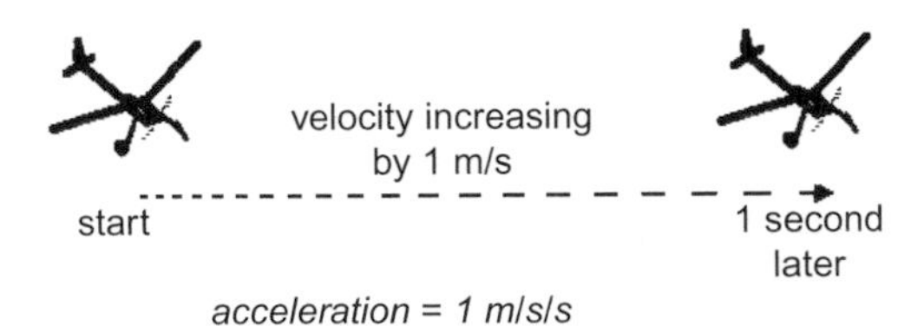

FIGURE 11.2

We have seen two examples of combining units by dividing a measurement made in one unit by a measurement made in the same or a different unit. We can also *multiply* measurements made in different units. An example of this occurs when we define the unit of **force**.

A force is applied to an object of mass 1 kilogram and the object accelerates at 1 m/s^2 in the direction of the force. In the figure, the object is a model sailing-boat and the force is that of the wind on its sails. We assume that this drives the boat in the same direction as the wind.

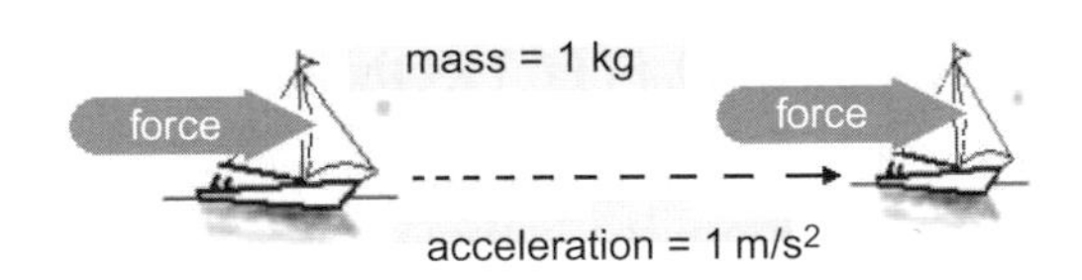

FIGURE 11.3

The force is calculated by multiplying the mass of the boat by its acceleration. The figure gives the unit of force as kg m/s^2. But measuring the force in 'kilogram metres per second per second', although it demonstrates that force is related to the SI base units, is rather too lengthy. So we give the unit of force a special name: the newton (symbol, N).

SI includes many other units, but all of these, like the newton, can be obtained from the base units by multiplication and division, without using any numerical factors. These are called derived units.

DERIVED UNITS

Some of the derived units are based on the newton. An example is the unit of work or energy, the **joule** (symbol, J). The joule is defined as the work done (or energy used) when a unit force of 1 N is applied to an object and that object moves a unit distance of 1 metre in the direction of the force.

In the case of the model sailing boat, if it moves a distance of 1 m, and the wind force is 1 N, the work done is 1 J.

This is another instance of obtaining a unit by multiplication. To calculate the work done we multiply the force by the distance moved and express the result in newton metres (N m), usually referred to as joules.

The unit of power is the **watt** (symbol, W). This is the rate of doing work, in joules per second. If a person does 1 J of work in 1 s, the rate of working is 1 J/s, or 1 W. In the case of the sailing boat, if it took 1 s for the wind to accelerate the 1 kg boat at 1 m/s^2 and during that time it travelled 1 m, the power of the wind would be 1 watt.

We have already met the watt as unit of electrical power (p. 14). Now we are describing it in terms of wind and sailing boats! The explanation is that the watt is the unit for the rate of doing *any* sort of work, whether it be an electric heater warming a room, or a donkey pulling a cart.

These examples illustrate how the system consists of units that are related to each other and can all be related back to the base units.

ELECTRICAL AND ELECTRONIC UNITS

The base units of SI include one electrical and electronic unit, the **ampere**. In practical electricity and electronics, we hardly ever refer to it by its full name. Usually we call it the **amp**.

The ampere is defined by the force (measured in newtons) between two wires carrying an electric current. The force is caused by the magnetic fields surrounding the wires (p. 29), causing them to be repelled from each other. Once again we are referred back to the SI base units. Having defined the amp, referring to SI base units, we are now able to define other electrical and electronic units.

The first of these is the unit of electrical charge, the **coulomb** (symbol, C). In the diagram (see top right), if the switch is closed for exactly 1 second, and the current is 1 A, the amount of electrical charge flowing through the resistor is 1 coulomb.

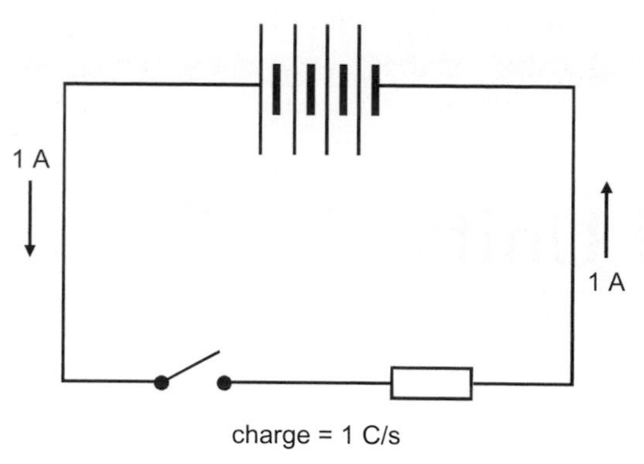

FIGURE 11.4

Next, we define the unit of electrical potential, the **volt** (symbol V). This is the more usual name for the unit, joule per coulomb (J/C).

Now we come to the interesting unit, the unit of electrical power, the **watt**. In the left-hand column we stated that it is both mechanical and electrical. How can this be?

On this page we define the watt in mechanical units:

$$watt = joule/second$$

On p.14, we defined the watt in electrical units:

$$watt = volt \times amp$$

But now we have defined the volt as:

$$joule/coulomb$$

and the amp as:

$$coulomb \times second$$

So we can write:

$$\begin{aligned} watt &= volt \times amp \\ &= (joule/coulomb) \times (coulomb \times second) \\ &= joule \times second \end{aligned}$$

The 'coulomb' terms have cancelled out, bringing us back to the mechanical definition of the watt. Mechanical power and electrical power are just different forms of the same thing — power.

Three important electrical units remain to be defined:

- **Resistance:** unit, the ohm (symbol, Ω), definition: *ohm = volts/amp* (p. 41)
- **Capacitance:** unit, the farad (symbol, F), definition: *farad = volt/coulomb* (p. 57)
- **Inductance:** unit, the henry (symbol, H), definition: *henry* = (−1/*volt*) × (*amp/second*) (p. 64).

There is one other SI unit that is often used in electronics. This is the hertz (symbol Hz), the unit of **frequency**. Frequency is the number of oscillations per second, so 1 Hz is 'one per second'.

MULTIPLES

Suppose that we want to write down a very large quantity, such as the average distance between the Earth and the Sun. Unfortunately the SI base unit, the metre, is too small for convenience. The distance is approximately:

150 000 000 000 m

There are too many zeroes and it is easy to write too few or too many. One way of overcoming this problem is to write the number in exponential form.

This means expressing it as a number multiplied by a power of 10:

$$1.5 \times 10^{11}$$

SI provides another way. Multiples of units are given prefixes:

Multiple	*Prefix*	*Symbol*	*Example*
10^3	kilo	k	kilovolt
10^6	mega	M	megawatt
10^9	giga	G	gigahertz
10^{12}	tera	T	teraohm

So the distance from Sun to Earth is 150 Gm.

FRACTIONS OF UNITS

There are also many instances where the SI unit is far too large. The wavelength of the light from a blue light-emitting diode (p. 71) is:

0.00000047 m

As before, there are too many zeroes. To express such small quantities we can write them in exponential form (with negative indices) or use these SI prefixes:

Fraction	*Prefix*	*Symbol*	*Example*
10^{-3}	milli	m	mm
10^{-6}	micro	μ	μs
10^{-9}	nano	n	nH
10^{-12}	pico	p	pF
10^{-15}	femto	f	fW
10^{-18}	atto	a	aV

So the wavelength of the blue light is 470 nm.

On the Web

1. Search the Web for the biggest, smallest, heaviest, longest, shortest, and fastest things you can find. Express their sizes in SI units.
2. Look for other SI units as keywords, such as candela, mole, kelvin and pascal.

Part 2

Electronic Components

Resistance

We begin by investigating what happens when we use different voltages to make a current flow through a conductor. We use a poor conductor, such as carbon, so that the current will not be too large to measure with an ordinary multimeter.

Things to do

A suitable piece of carbon for this investigation is a short length (about 3 cm) of Artists' charcoal.

1. Connect a 6 V battery, a multimeter and the piece of carbon in a circuit, as shown in the photograph. Use short leads with crocodile clips at each end.
2. Leave the positive (red) probe of the meter free. By touching this against different metal contacts in the battery box, you can obtain voltages of 1.5 V, 3 V, 4.5 V and 6 V. Alternatively, use a PSU with switchable output instead of the battery.

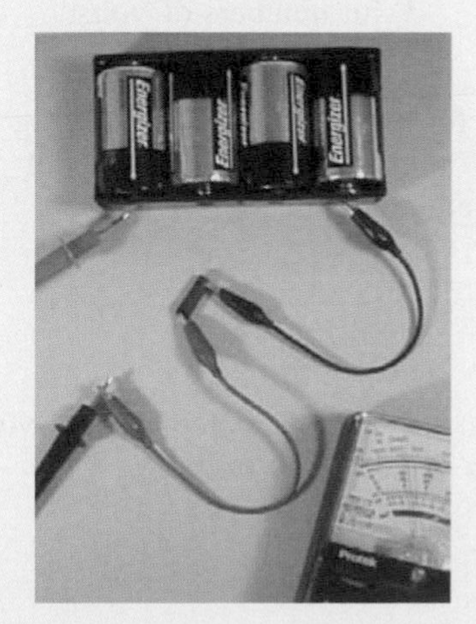

FIGURE 12.1

3. Set the multimeter to its largest current range (500 mA or 1 A). If the readings are too small to read accurately, you can switch to a lower current range.
4. Take the first reading with the red probe at the 6 V point. Measure the current, using the meter. Record these results in a table:

Voltage (volts)	Current (amps)	Voltage/Current
6		
4.5		
3		
1.5		

5. Repeat with the voltage set to 4.5 V, 3 V and 1.5 V.

What do you notice about the current as we reduce the voltage? To investigate this further, divide each voltage by the corresponding current and write the results in the third column. What do you notice about the values in the third column?

OHM'S LAW

The results of the investigation above show that:

The current passing through the carbon is proportional to the voltage difference between its ends.

This was first discovered, using lengths of wire, by Gregor Ohm, so it is called **Ohm's Law**. It applies to all conductors. We can state it as an equation:

$$\frac{voltage}{current} = constant$$

You may have noticed in your class that different people, working with different pieces of carbon, have found different constants. The value of the constant is a property of a particular piece of carbon. We call it the **resistance** of the piece of carbon. Now we can write the equation like this:

$$\frac{voltage}{current} = resistance$$

UNITS OF RESISTANCE

If the voltage is expressed in volts and the current in amps, the unit of resistance is the **Ohm**, symbol Ω. This symbol is the Greek capital letter *omega.*

Self Test

1. When a voltage of 6 V is applied to a piece of carbon, the current is 120 mA. What is the resistance of the carbon?
2. When a voltage of 230 V is applied to a lamp, the current is 260 mA. What is the resistance of the lamp?

Larger units of resistance are the kilohm (kΩ) and megohm (MΩ).

$$1\ \text{k}\Omega = 1000\ \Omega$$
$$1\ \text{M}\Omega = 1000\ \text{k}\Omega$$

OHM'S LAW EQUATIONS

The equation given opposite can be written in three forms:

$$resistance = \frac{voltage}{current}$$
$$current = \frac{voltage}{resistance}$$
$$voltage = current \times resistance$$

If we are given any two of the quantities, we can calculate the third. These three equations are used more than any in electronics, so you will need to remember them. An easy way of remembering is to memorise this diagram:

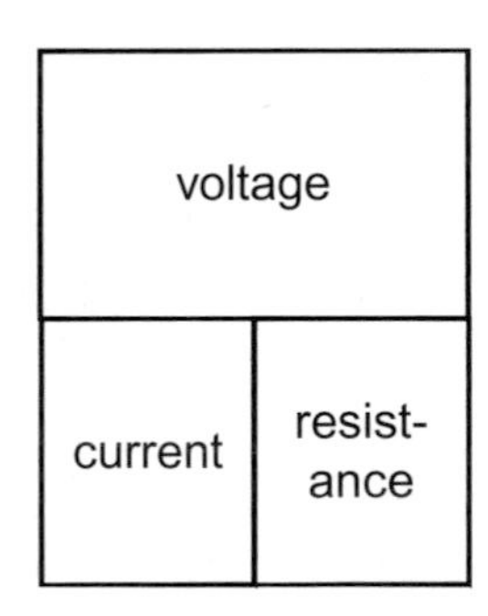

FIGURE 12.2

To use the diagram, cover the quantity that you want to calculate. The diagram shows the remaining quantities as they appear in the equation.

Example

To calculate current:

Current equals 'voltage over resistance'.

Try it for the other two quantities.

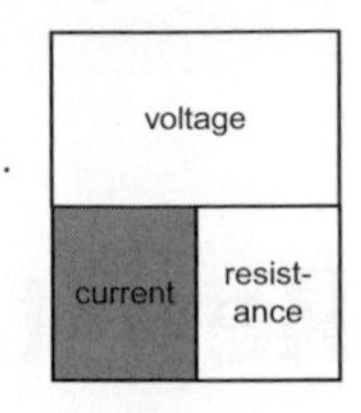

Self Test

1. When 25 V is applied to a length of cable, the current through it is 10 A. What is the resistance of the cable?
2. A coil of wire has a resistance of 600 Ω and the voltage across it is 4.5 V. How much current flows through the cable?
3. If a wire has a resistance of 2.4 Ω, and a current of 3.5 A is flowing through it, what is the voltage difference between its two ends?
4. A 2.3 kW heater runs on the 230 V mains. What is its resistance?

SYMBOLS FOR QUANTITIES

You have already used symbols for *units*, such as A, V, and W. It is useful to have symbols for *quantities*, too. This makes it quicker to write out equations. The symbols for quantities are:

I for numbers of amps
V for numbers of volts
R for numbers of ohms

Symbols for quantities are in slanting letters (italics). Using these symbols instead of words, the Ohm's Law equations become:

$$R = V/I$$
$$I = V/R$$
$$V = IR$$

Try not to confuse V, which means 'volts' with *V*, which means 'numbers of volts'.

Resistors

In most circuits, we join the components together with copper wires. This is because copper is a good conductor of electricity. It has very low resistance.

Some connections may need greater resistance than that of a copper wire. This is when we use resistors.

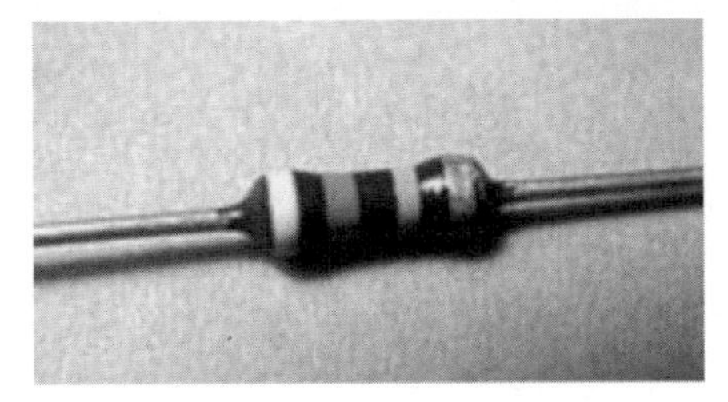

FIGURE 13.1

The photo shows a typical **fixed resistor**. Resistors of this kind are sold in a range of different resistances, from less than 1 Ω and up to 10 MΩ.

Below are two different symbols used for resistors in circuit diagrams. We use the rectangle symbol in this book. The zigzag symbol is less often used nowadays.

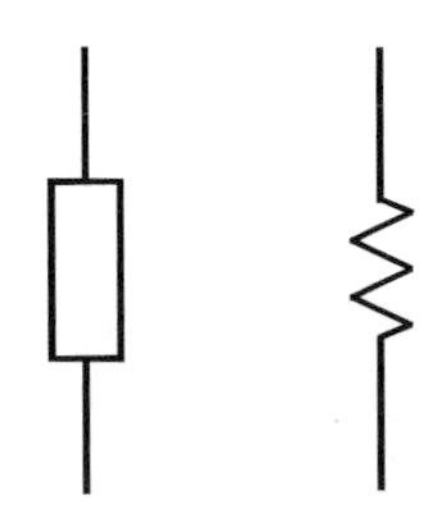

FIGURE 13.2

PREFERRED VALUES

Resistors are made in a range of values (in ohms):

1.0 1.1 1.2 1.3 1.5 1.6 1.8 2.0 2.2 2.4 2.7 3.0
3.3 3.6 3.9 4.3 4.7 5.1 5.6 6.2 6.8 7.5 8.2 9.1

After these 24 values, the sequence repeats in multiples of ten:

10 11 12 13 ... up to ... 82 91,
then 100 110 120 ... up to 820 910,
then 1 k 1.1 k 1.2 k ... up to 8.2 k 9.1 k,
('k' means kilohms)
then 10 k 11 k 12 k ... up to 82 k 91 k,
then 100 k 110 k 120 k ... up to 910 k 1M.

This is the **E24 series**. The **E12 series** comprises the alternate values of the E24 series: 1.0, 1.2, ... , 6.8, and 8.2, with multiples.

RESISTOR COLOUR CODE

Three coloured bands are used to tell us the resistance of a fixed resistor. The bands are close together at one end of the resistor. The colour of each band represents a number.

Reading from the end, the meaning of the bands are:

First band	**First** digit of resistance
Second band	**Second** digit of resistance
Third band	**Multiplier** — a power of 10, or the number of zeroes to follow the two digits.

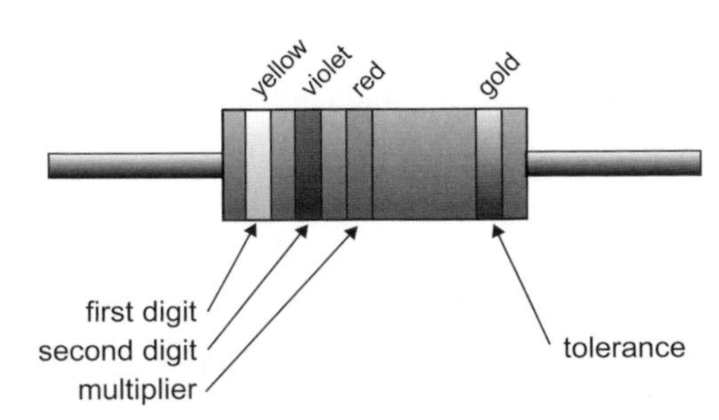

FIGURE 13.3

The table shows the meanings of the colours.

Colour	Number
Black	0
Brown	1
Red	2
Orange	3
Yellow	4
Green	5
Blue	6
Violet	7
Grey	8
White	9

Example 1

The bands are yellow, violet, red.

Yellow means '4'
Violet means '7'
Red means '2'

Write '4', then '7', then follow with two zeroes. This gives:

4700 Ω

We normally write this as:

4.7 kΩ

Example 2

The bands are white, brown, yellow.

White means '9'
Brown means '1'
Yellow means '4'

Write '9' then '1', then follow with four zeroes. This gives:

910 000 Ω or 910 kΩ

There is a similar system with four bands for marking high-precision resistors.

Self Test

1 State the resistances of resistors that have bands of these colours:
(a) orange, white and brown.
(b) green, blue and yellow.
(c) brown, black and green.
(d) brown, black and black.
(e) red, red, and red.

2 What colours are the bands on resistors of these values?
(a) 33 Ω
(b) 200 kΩ
(c) 750 Ω
(d) 43 kΩ
(e) 820 Ω.

BREADBOARDS

A breadboard makes it easy and quick to build circuits. It is a plastic block with rows of sockets. The sockets in each row are connected electrically, as indicated in the photo below.

If you plug two or more component wires into the same row, current can flow from one to the other.

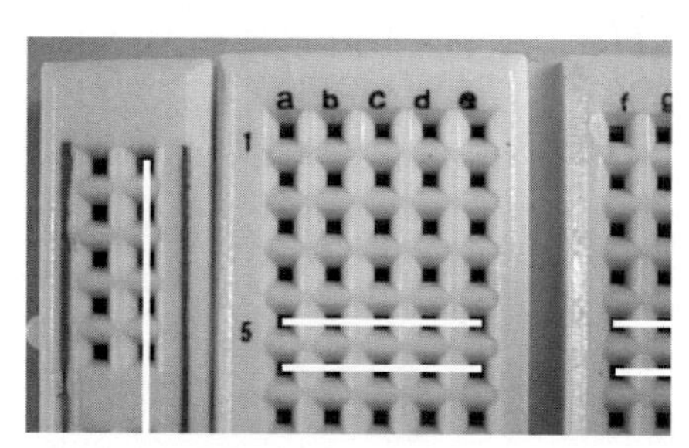

FIGURE 13.4

Things to do

You need:

- 10 fixed resistors of different values.
- A multimeter.
- A breadboard.

1 Find out the value of one of the resistors by reading its colour code.
2 Use the multimeter to check the resistance. First select a suitable range on the meter.
3 Plug the wire leads of the resistor into *different rows* on the breadboard. Touch one probe to each of the resistor leads.

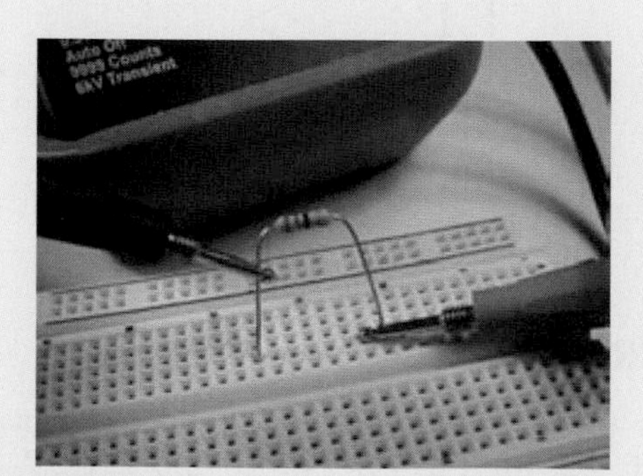

FIGURE 13.5

(continued)

4 Read the resistance on the meter. Does it agree with the colour code?
5 Repeat with the other resistors.

RESISTANCE ON AN ANALOGUE METER

The resistance scale runs from right to left and is not linear. This means that the markings are more crowded toward the higher (left) end of the scale.

The reading in the photo (on the top scale) is 72 Ω.

When using an analogue meter, adjust the zero setting occasionally. To do this, touch the two probes together. Then set the 'Ohms adjust' knob or wheel until the needle lines up with zero (on the *right* of the scale).

FIGURE 13.6

More about Resistors

If we want to be able to vary the resistance of part of a circuit, we use a **variable resistor**. One type of variable resistor is called a **potentiometer**. This type is a often used for volume controls in audio equipment. It is more often called a 'pot', for short.

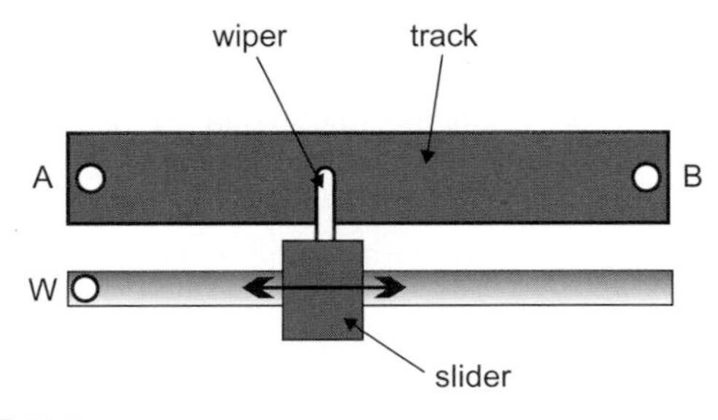

FIGURE 14.1

Above is a slider pot, which has a track made from a film of carbon. In the more expensive, hard-wearing and accurate types, the track is a layer of conductive ceramic. The ends of the track are connected to two terminals.

The third terminal is connected to the **wiper**. This is a springy metal strip that presses firmly against the track and makes electrical contact with it. It is attached to a sliding knob, used to move the wiper along the track. As it moves, the distance between one end of the track (say, A) and the wiper is changed. This changes the electrical resistance between A and the wiper. The resistance can have any value between zero and the resistance of the whole track.

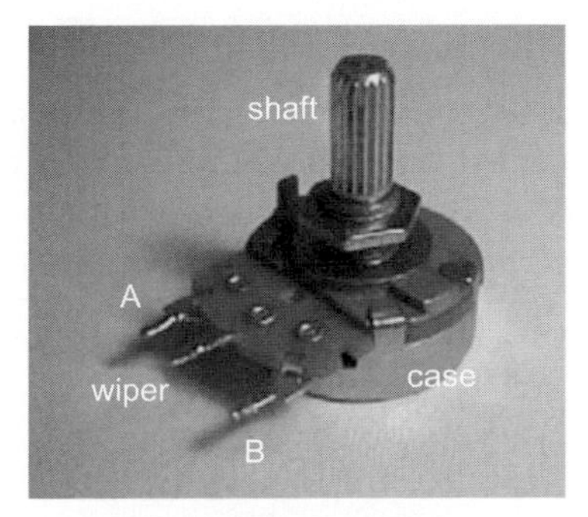

FIGURE 14.2

Slider pots are often used on audio equipment for setting the frequency response.

A common form of variable pot is the **rotary pot**. This has a curved track, along which the wiper moves as the shaft is turned (Figure 14.2)

Often the track covers an angle of about 270°.

Rotary pots are often used as volume controls in audio equipment and also for controlling the brightness of lamps, the speed of motors and many other purposes. The shaft can be fitted with a knob, if preferred.

Sometimes we want to adjust a resistance when we are setting up a circuit. After 'pre-setting' or 'trimming' the circuit, we may not need to change the resistance again.

We use a smaller rotary pot, known as a **preset pot**, or as a **trimmer**, or **trimpot**, which is adjusted with a screwdriver (Figure 14.3).

Special symbols are used for pots (left) and presets (right) (Figure 14.4). Zigzag versions are also used.

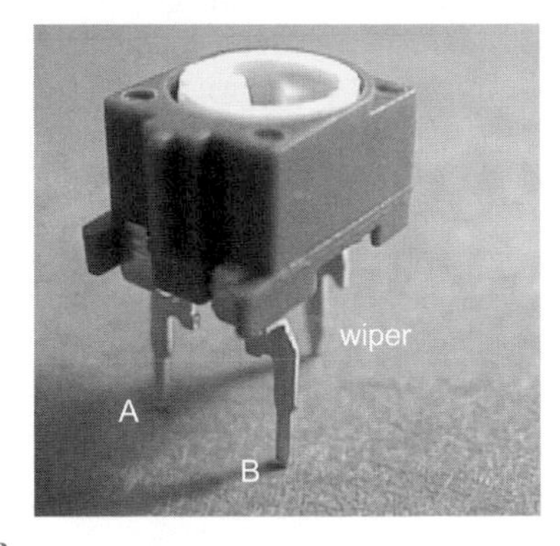

FIGURE 14.3

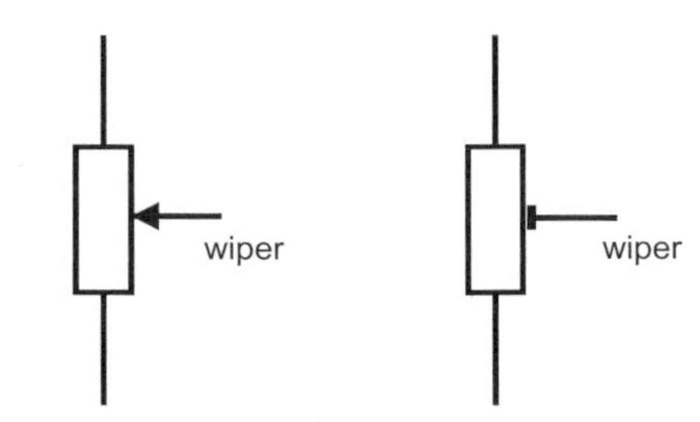

FIGURE 14.4

Electronics: A First Course.

Design Tip

Pots need careful use. If the wiper is near one end of the track, the resistance between the wiper and the near end is small. As a result, the current through that part of the track is too high and burns out the track. If possible, wire a small fixed resistor in series with the wiper or track to limit the current to a safe amount.

POWER RATINGS

Many fixed resistors are intended to conduct electricity at a maximum power of a quarter of a watt (250 mW). This power must not be exceeded.

Example 1

The current through a 250 mW resistor is 10 mA and the voltage across it is 5 V. The power is $0.01 \times 5 = 0.05$ W = 50 mW. Because the resistor is rated at 250 mW, it can safely run at 50 mW. It becomes only slightly warm as a result of the current passing through it.

Example 2

If the current through a 250 mW resistor is 50 mA and voltage is 5 V, the power is 250 mW. This is as much as the resistor can safely stand. It will get hot but it is not damaged. If either the current or voltage is increased above these amounts, the resistor gets too hot. It may scorch or burn and possibly crack into pieces. Even if it is not totally destroyed, its resistance changes permanently as a result of overheating.

Resistors are made with higher power ratings, such as 0.5 A, 1 A, 5 A. Some are able to run at several hundred watts. These are larger than the typical low-power resistors. Those of the highest ratings usually consist of a coil of thin wire wound on a ceramic core.

Typical pots are rated to run at a maximum of 0.2 W to 0.5 W. Pots of higher rating, up to 3 W, usually consist of a coil of wire wrapped on a ceramic core.

TOLERANCE

There is usually a fourth coloured band on a resistor, at the end opposite to the three bands. This band indicates the tolerance or precision of the resistor. This tells us how far the actual resistance may differ from the **nominal resistance**, as shown by the colour code.

Example 3

A 470 Ω resistor has a gold tolerance band. This means a tolerance of ±5%. Calculate 5% of 470 Ω, which is $470 \times 5/100 = 23.5$ Ω.
The actual resistance of the resistor may be somewhere between:

$$470 - 23.5 = 446.5\ \Omega$$
and
$$470 + 23.5 = 493.5\ \Omega$$

Colour	Tolerance
Red	±2%
Gold	±5%
Silver	±10%
No band	±20%

Example 4

A 220 kΩ resistor has no tolerance band. Its tolerance is ±20%. Calculate 20% of 220 kΩ, which is $220 \times 20/100 = 44$ kΩ.

The actual value of the resistor may be somewhere between:

$$220 - 44 = 176\ \text{k}\Omega$$
and
$$220 + 44 = 264\ \text{k}\Omega$$

THE REASON FOR E24

It would be very expensive to make and stock all possible values of resistor from 1 Ω up to 1 MΩ. Also, because 5% tolerance is good enough in most circuits, the limited number of resistances of the E24 series (p. 43) covers the whole range of values. Take these four nominal values as examples:

Nominal	Lowest (−5%)	Highest (+5%)
390	370.5	409.5
430	408.5	451.5
470	446.5	493.5
510	484.5	535.5

The range of each value *slightly* overlaps each of its neighbours. At 5% tolerance, there is no point in making resistors with values in between these E24 values.

Resistor Networks

When two or more resistors are joined together we create a **resistor network**. This Topic describes several types of network and their properties.

RESISTORS IN SERIES

If two or more resistors are joined end-to-end so that current flows through each one in turn, we say they are joined **in series**. Compare this with cells joined in series. We find the **effective resistance** of the series by **adding** together their resistances.

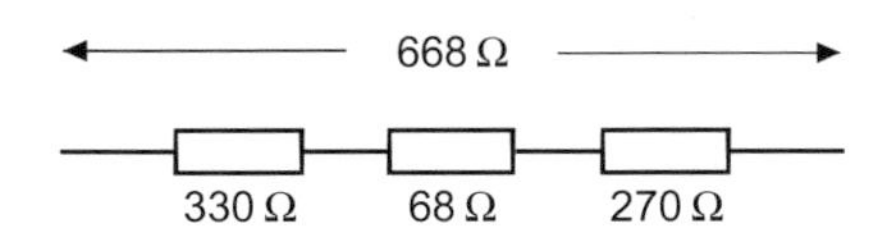

FIGURE 15.1

Example

In the series of three resistors above, the effective resistance is 330 + 68 + 270 = 668 Ω. As a check on the calculation, note that the effective resistance is always bigger than the biggest of the individual resistances.

Self Test

1 What are the effective resistances of each of the following groups of resistors when wired in series?
 (a) 220 Ω and 3.3 kΩ.
 (b) 5.6 Ω, 39 Ω, and 16 Ω.
 (c) 13 kΩ , 1 MΩ and 390 kΩ.
2 A student needs a 470 Ω resistor, but the lab is out of stock. How could this be made up from resistors of other E24 values?
3 Make up a 130 kΩ resistor from other E24 values.

Things to do

You need:

- 10 fixed resistors of different values and various tolerances.
- A multimeter.
- A breadboard.

1 Read the colour codes to find the resistance of each resistor. Then check your reading by measuring the resistances with a multimeter (p. 16).
2 Take any two or three resistors and plug them into the breadboard so they are in series. Calculate the effective resistance of the series. Then check your calculations by using the multimeter to measure the resistance of the series. Repeat for four more groups of resistors.
3 From its colour bands, find the resistance and tolerance of a resistor. Work out what its lowest and highest resistances could be. Check with a multimeter that the actual resistance is within these limits.

CURRENT RULES

1 Three or more wires in a network meet at one point. It is not possible for charge to build up at the junction. It is not possible for charge to decrease at the junction. Therefore:

The total current arriving at a junction equals the total current leaving it.

Example

The total current arriving is 2.1 + 2.4 = 4.5 A

The total current leaving is 1.5 + 3.0 = 4.5 A

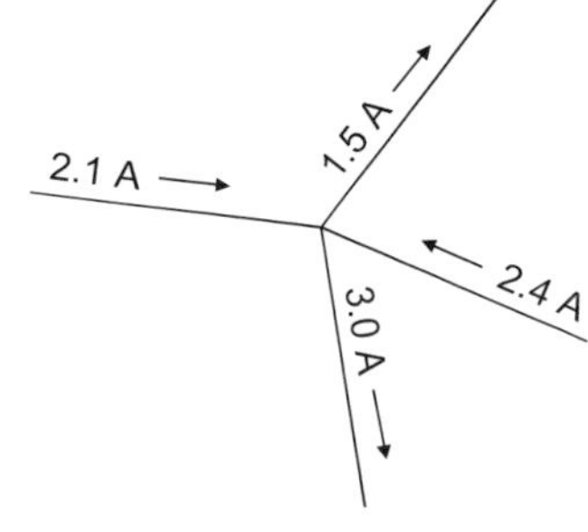

FIGURE 15.2

2 In a series circuit (one in which all the components are in series), there is no point at which charge can enter or leave the circuit. Therefore:

The current is the same everywhere in a series circuit.

Example

The cell and three resistors are in series. The same current (= i amps) flows in all parts of the circuit.

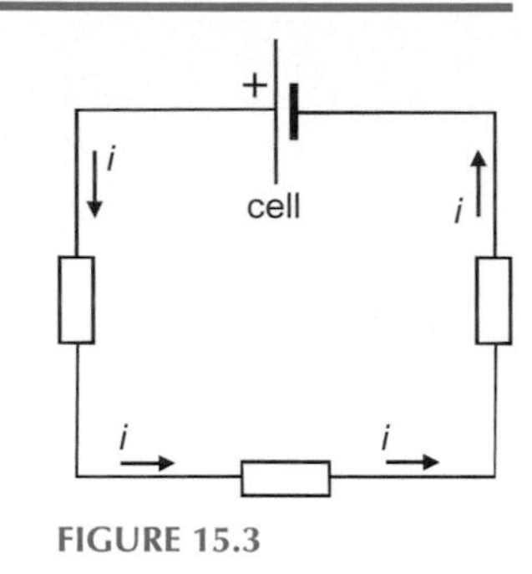

FIGURE 15.3

VOLTAGE RULES

1 Going round a series circuit in the direction of the current, there is a voltage drop across each of the resistors. There is a voltage rise across the cell. The size of each voltage drop depends on Ohm's Law. The voltage rule is that:

The sum of the voltage drops in a series circuit equals the voltage across the circuit.

Example

The voltage drops are v_1, v_2, and v_3. Their total is equal to the voltage *rise* v_T across the cell.

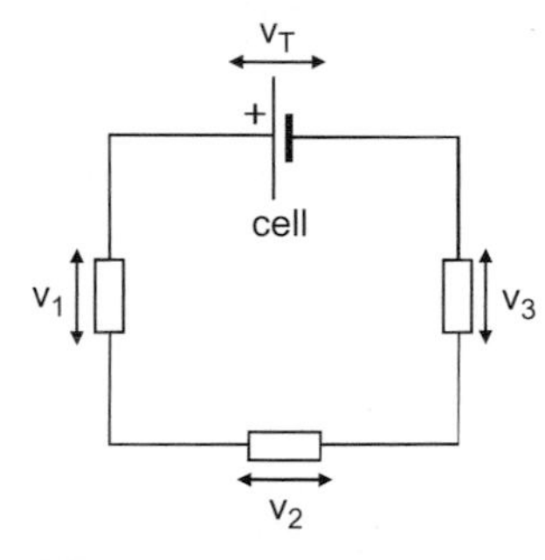

FIGURE 15.4

2 The drawing (see top right) shows a circuit in which the resistors are joined **in parallel**.

One end of each resistor is connected to the positive terminal of the cell. The other end of each resistor is connected to the negative terminal. This means that:

In a parallel circuit, there is the same voltage across each component.

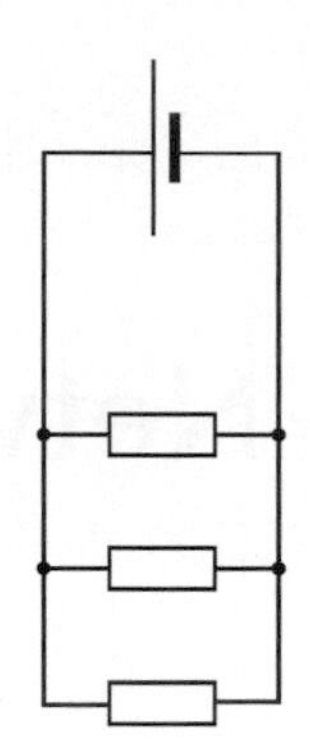

FIGURE 15.5

VOLTAGE DIVIDER

This kind of network is also called a **potential divider**. The input to a voltage divider is a voltage v_{IN}. This drives a current i through the two resistors. Because the two resistors are in series, the same current flows through both (Current Rule 2).

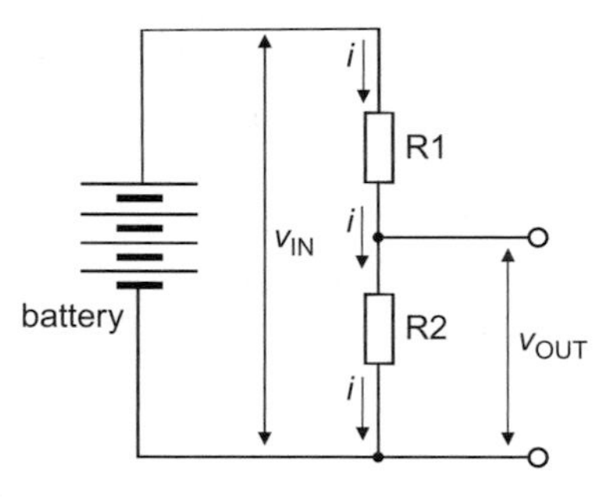

FIGURE 15.6

The effective resistance of the two resistors in series is R_1+R_2. The voltage across them is v_{IN}. By Ohm's Law, the current is:

$$i = \frac{v_{IN}}{R_1 + R_2}$$

Using Ohm's Law again, the voltage across R2 is:

$$v_{OUT} = v \times R_2$$

Substituting the value of i from the previous equation:

$$v_{OUT} = v_{IN} \times \frac{R_2}{R_1 + R_2}$$

This is the equation for calculating the output voltage of a voltage divider. By choosing two suitable resistors, we can obtain any output in the range from 0 V to v_{IN}.

Example

In a voltage divider, $v_{IN} = 6$ V, $R_1 = 220\ \Omega$ and $R_2 = 390\ \Omega$. Calculate v_{OUT}.

$$v_{OUT} = 6 \times \frac{390}{390 + 220} = 3.84 \text{ V}$$

There are questions about the current rules, the voltage rules and about voltage dividers on pp 54–5.

RESISTANCES IN PARALLEL

The effective resistance of two or more resistances in parallel is calculated from:

$$\frac{1}{R} = \frac{1}{R_1} + \frac{1}{R_2} + \frac{1}{R_3} + \ldots$$

There are as many terms on the right as there are resistances in parallel.

Example

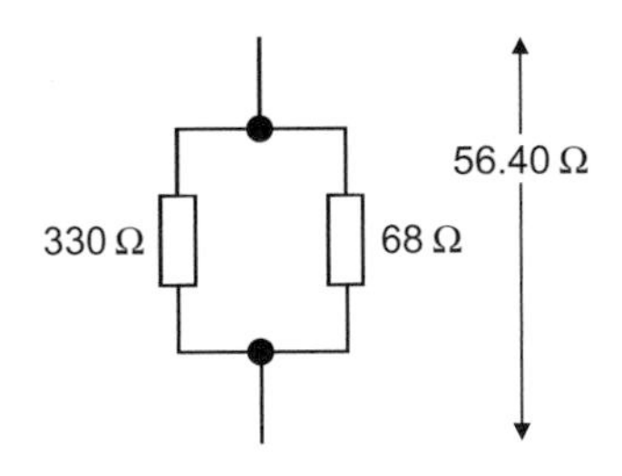

FIGURE 15.7

The effective resistance of this network is found by calculating:

$$\frac{1}{R} = \frac{1}{330} + \frac{1}{68} = 0.00303 + 0.01471 = 0.1774$$

$$\Rightarrow R = \frac{1}{0.01774} = 56.40\ \Omega$$

Check: the effective resistance is always *less than* the smallest of the resistances in parallel.

Self Test

Find the effective resistance of each of the following sets of resistors, when wired in parallel:

(a) 22 Ω and 270 Ω
(b) 120 Ω and 10 kΩ
(c) 27 Ω, 47 Ω, and 15 Ω
(d) 390 kΩ, 18 kΩ, and 91 kΩ
(e) 100 Ω, and 100 Ω.

Things to do

Repeat the tests you made on p. 49, but with the resistors in parallel.

MIXED NETWORKS

A mixed network has some resistances in series and others in parallel. Look for groups of resistances in the network, which are all in series or all in parallel. Re-draw the network, replacing each group with its equivalent resistance. Gradually simplify the network to a single resistance.

Example 1

The 56 Ω and 33 Ω resistances in parallel can be replaced with 20.8 Ω. This gives two resistances in series. Replace these with 67.8 Ω.

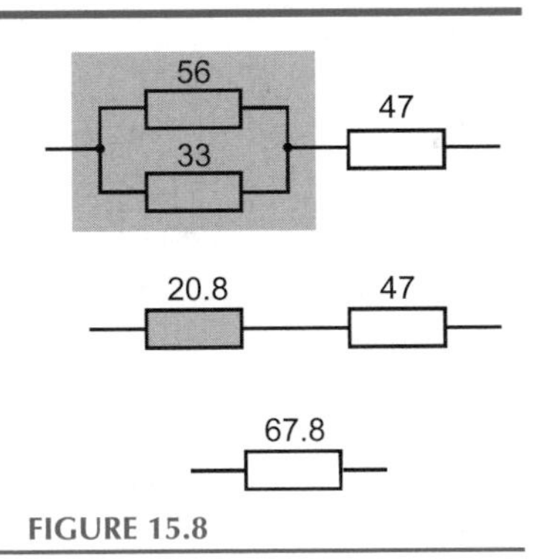

FIGURE 15.8

Example 2

There are two sets of resistances in series. Simplify them to 550 Ω and 473 Ω. These are in parallel, so may be replaced with 254 Ω.

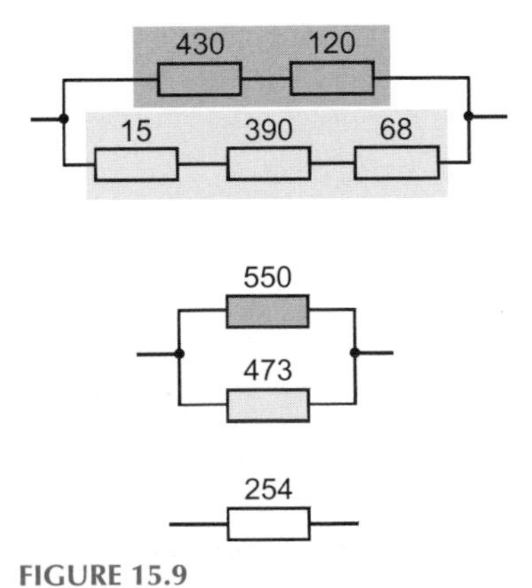

FIGURE 15.9

Self Test

1 Find the effective resistances of these networks:

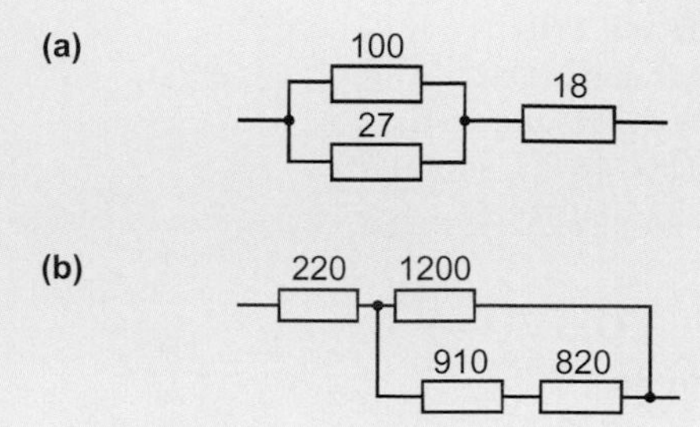

FIGURE 15.10

2 What is the effective resistance if the 1200 Ω of the network above is **(a)** removed, and **(b)** replaced with a copper wire?

RESISTOR PRINTED CODE

This is sometimes used instead of the resistor colour code. It is also known as the **BS1852 code**. The code uses one of three letters to show the unit and the position of the decimal point. The letters are:

R = ohms, **K** = kilohms, **M** = megohms.

Examples

Code	Value	Code	Value
8R2	8.2 Ω	82 K	82 kΩ
82R	82 Ω	820 K	820 kΩ
820R	820 Ω	8M2	8.2 MΩ
8K2	8.2 kΩ	–	–

A second letter at the end of the code indicates tolerance:

G = ± 2%	K = ± 10%
J = ± 5%	**L** = ± 20%

Examples

9K1J means a 9.1 kΩ resistor with ±5% tolerance.
12KK means a 12 kΩ resistor with ±10% tolerance.

Self Test

1 What are the values of resistors marked **(a)** 33K, **(b)** 4M7, **(c)** 2R2, **(d)** 1R8J, and **(e)** 27KK?
2 Write the codes for resistors with the following ratings: **(a)** 47 Ω, **(b)** 100 kΩ, **(c)** 9.1 kΩ, **(d)** 3.9 kΩ ±5%, and **(e)** 750 kΩ ±10%.

MULTITURN TRIMMERS

These are used when we need to be able to adjust a variable resistor with high precision. The wiper is moved by a screw mechanism. It takes several turns of the screwdriver to move the wiper the full length of the track.

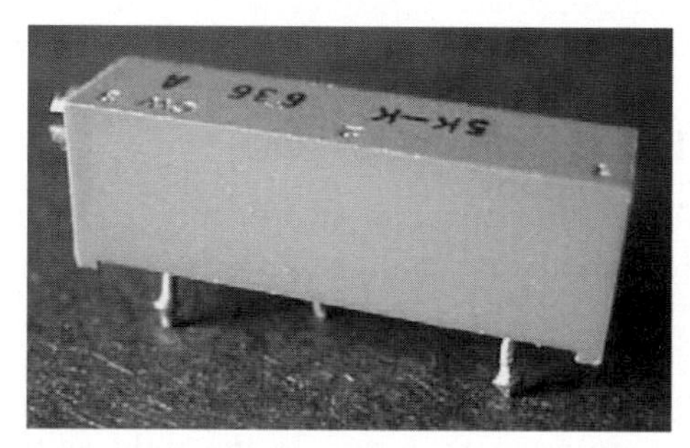

FIGURE 15.11

The screw head can be seen on the left. It requires 10 turns of the screwdriver to move the wiper from one end of the track to the other. Even more precise trimmers need 18 or 25 turns.

Self Test

Can you read the value and tolerance of the trimmer illustrated above?

ENERGY TRANSFER

A charge carrier (such as an electron) loses energy as it moves through a conductor. The energy is transferred to the atoms of the conductor. This makes them vibrate — the conductor becomes hot. Electrical energy is converted to heat. We use this action to heat electric kettles and room heaters. In electronic circuits the resistors, and other components, become warm or hot. This is usually a disadvantage because energy is being wasted. In an electric lamp, the conductor is a thin wire that becomes white hot and emits visible light.

The rate of energy conversion is measured in watts.

MORE ABOUT VOLTAGE DIVIDERS

Things to do

You need:

- A breadboard.
- A power supply (6 V battery or PSU).
- A 22 kΩ resistor, a 10 kΩ resistor, and one other.
- A multimeter.

1 With the 22 kΩ resistor and one other, build a voltage divider to give an output of 3.6 V, as shown in the drawing. You have to work out what the value of the other resistor (R2) should be. Do not put the 10 kΩ resistor in the circuit yet.

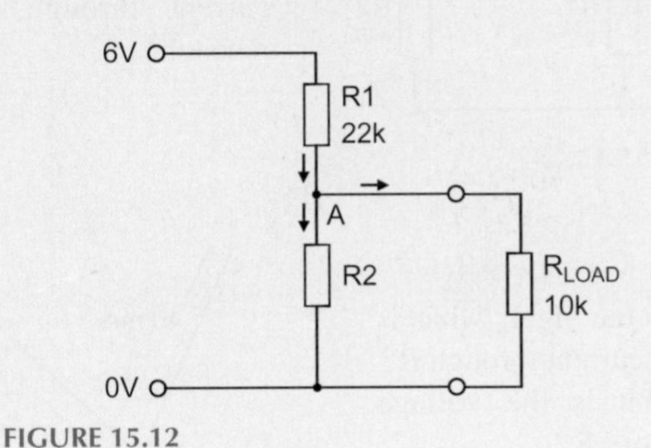

FIGURE 15.12

2 Use the multimeter to check that the divider is working. Its input voltage should be 6 V and its output should be 3.6 V.

3 Connect the load resistor (10 kΩ) to the divider. Measure the output voltage again. What has happened to it?

The results of the investigation above are explained by thinking of the currents at point A. According to current rule 1 the currents leaving point A to flow through R2 and the load, must be equal to the current arriving through R1. The current splits and more goes through the load than through R2. This is because the load has a lower resistance than R2.

Because the current through R2 is reduced when the load is connected, the voltage across R2 is decreased (Ohm's Law). The output of the divider is less than we first calculated. Try working out the voltages and current in the network to check your results.

CIRCUIT DIAGRAMS

There are two points that you should notice:

- **Cell or battery symbol:** These are not often used. Instead, the diagram shows a pair of terminals. One, labelled 0V, is the negative terminal of the power supply. The other is labelled with the voltage of the positive terminal of the supply. The supply *may* be a cell or battery, but it is more likely that you will obtain the supply from a bench PSU.
- **Ohms symbol:** This is omitted. Instead we use a shorter form, using the resistor printed code.

Design Point

We need to avoid a serious voltage drop caused by the load taking too much current from the divider. A simple rule is that the current flowing through the divider from positive supply to the 0V line must be at least **10 times** the current being taken by the load. There will still be a drop, but it will not be unduly large.

VARIABLE VOLTAGE DIVIDER

If a variable output voltage is needed, use a divider based on a variable resistor.

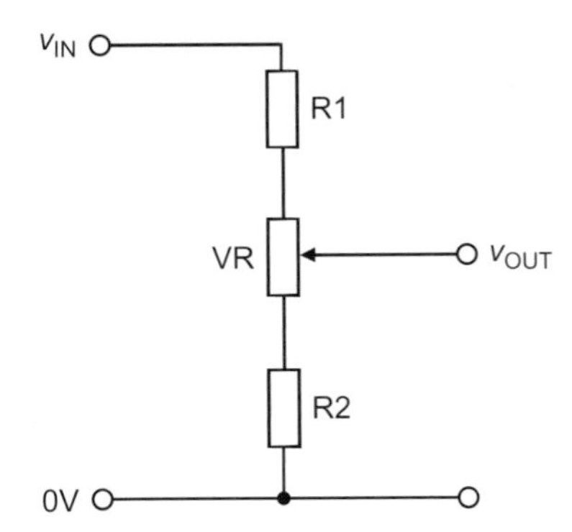

FIGURE 15.13

The values of R1 and R2 decide the upper and lower limits of v_{OUT}. Without R1 and R2, the divider output ranges from 0V to v_{IN}.

MEASURING RESISTANCE

This is another way of measuring resistance. It uses two separate meters, an ammeter for current (I) and a voltmeter for voltage (V).

R is the resistance being measured. The ammeter measures the current *through* R. The voltmeter

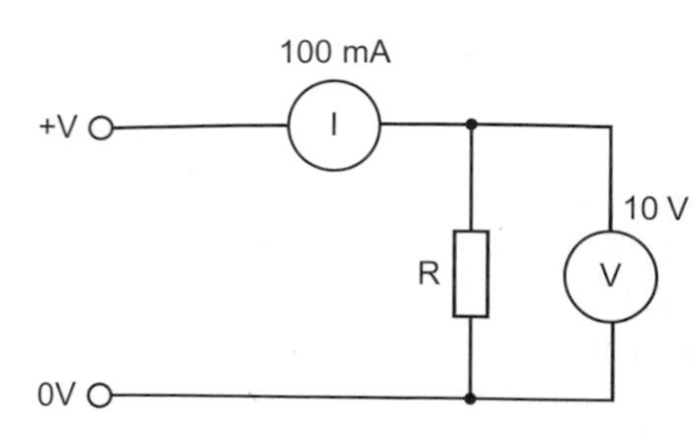

FIGURE 15.14

measures the voltage *across* R. The values beside each meter in the diagram are their **full scale deflection**. This is the largest current or voltage that the meter can measure.

The supply *V*+ is variable up to 10 V. Use a PSU or connect different numbers of cells. Try 4 or 5 different voltages. Measure *I* and *V* at each and calculate $R = V/I$ at each.

There is an error in this technique. Some of the current that goes through the ammeter goes *through the voltmeter*, not through R. So the ammeter reading is too big. But a voltmeter takes very little current compared with that through R, so the error is small.

QUESTIONS ON RESISTANCE

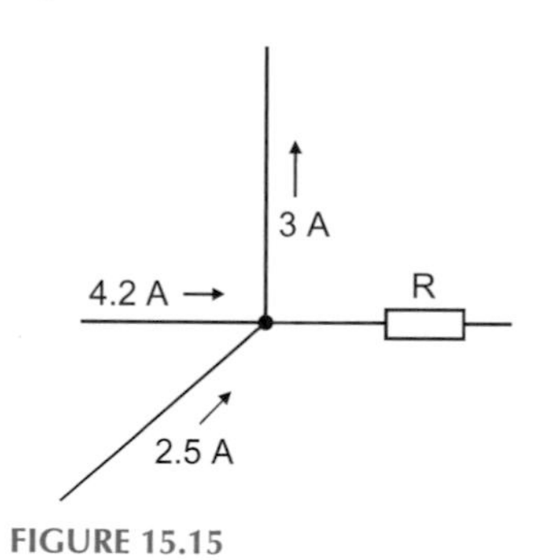

FIGURE 15.15

1 How much current flows through R? What is its direction?
2 If the 4.2 A current is reduced to 0.7 A, what is the current through R? What is its direction?

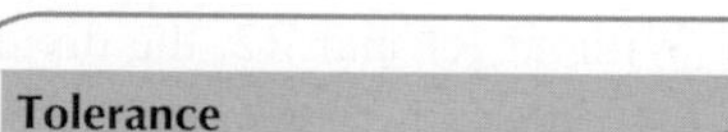

Tolerance

In all questions, assume that the resistors have the *exact* values stated. Tolerance does not have to be allowed for.

3 In the network at top right, which resistor has the biggest voltage across it?
4 What is the current through the network? What is the voltage drop across each resistor?

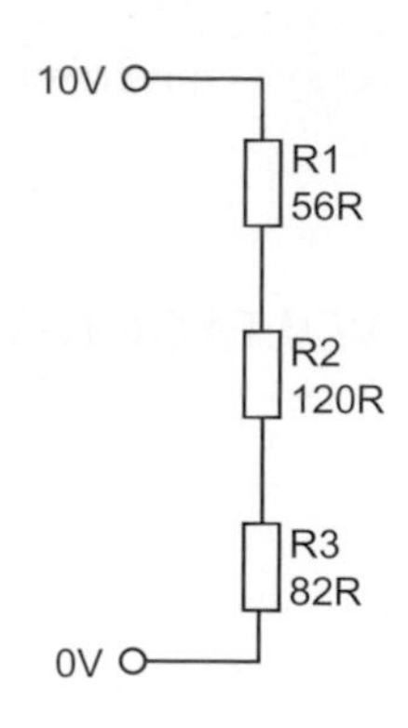

FIGURE 15.16

5 In the network, what supply voltage produces a current of 30 mA?

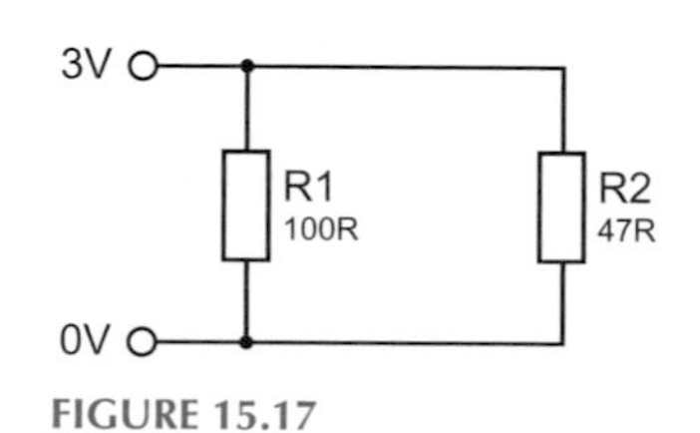

FIGURE 15.17

6 On the left, what is the voltage across, and the current through, each resistor?

7 On the right, what is the current through R?
8 What is the voltage across R?

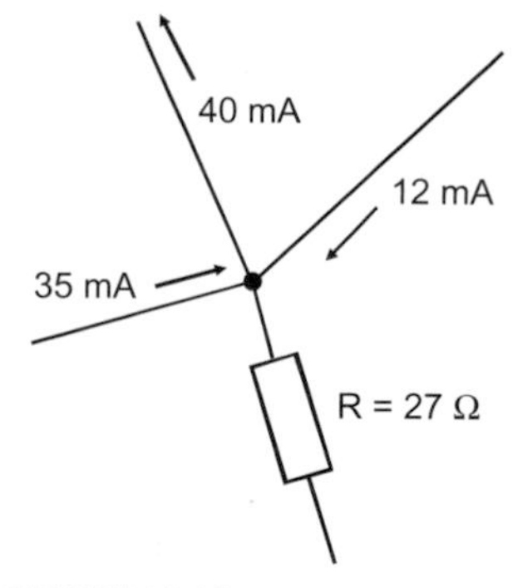

FIGURE 15.18

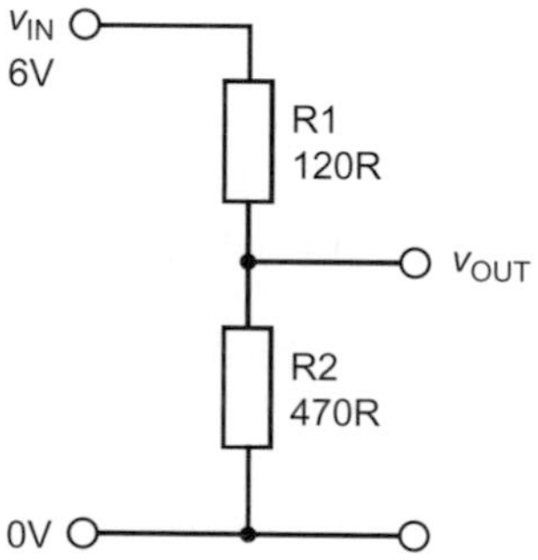

FIGURE 15.19

9 What is the output voltage of the divider on the left?
10 What is its output voltage if the input voltage is raised to 15 V?
11 What is the output in the diagram (left) if R1 is increased to 680 Ω?
12 What input is required in the diagram to give an output of 5 V?

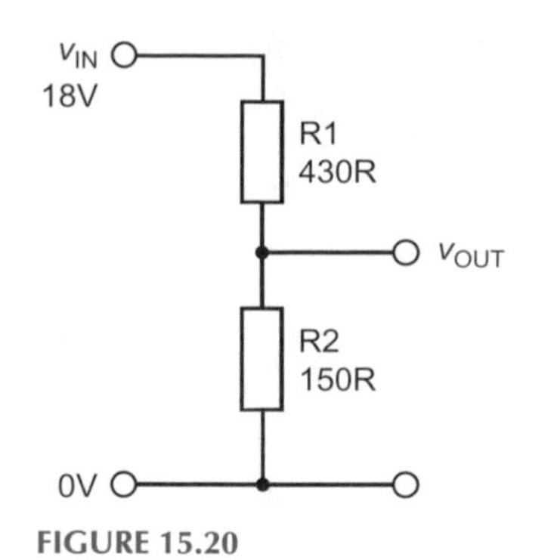

FIGURE 15.20

13 What is the output of the divider on the left?

14 What is the output if the input is reduced to 3 V?

15 What is the output if the resistors are exchanged?

16 Design a divider to produce an output of 4.5 V, given an input of 9 V.

17 Design a divider to produce an output of 4.8 V, given an input of 12 V.

18 State the equation that describes Ohm's Law. If a voltage of 5 V is applied to a resistor and the current through the resistor is 61 mA, what is the resistance of the resistor?

19 If a 2.2 kW electric kettle runs on the 230 V mains, what is the resistance of its heating element?

20 A coil has a resistance of 200 Ω. What voltage is required to produce a current of 400 mA through it?

21 What value of resistor is required so that a current of 65 mA passes through it when 5 V is applied across it? What is the nearest value in the E24 series? What is the colour code for the nearest resistor, assuming it has 5% tolerance?

22 Describe three types of variable resistor and give a practical use for each.

23 What is the maximum current that may be passed by a 0.5 W resistor, when the voltage across it is 12 V?

24 The tolerance band on a resistor is silver. What does this mean? If the three colour code bands on this resistor are all red, what are the lowest and highest values that the resistor can have?

25 Two resistors, 270 Ω and 1 kΩ, are connected in parallel. What is their effective resistance?

26 Three resistors have values of 10 kΩ, 22 kΩ, and 18 kΩ. What is their effective resistance when wired in parallel? Which resistor carries the largest current when a voltage is applied across the network?

27 Three resistors, 470 Ω, 560 Ω and 220 Ω are wired in parallel. What is the current through the network when a 6 V battery is connected across it? What is the voltage across the 560 Ω resistor?

28 Find the effective resistance of each of the networks shown in Figure 15.21

29 A circuit is set up for measuring resistance as on p. 51. With the test resistance R connected, the milliammeter reads 176 mA and the voltmeter reads 12 V. What is the value of R?

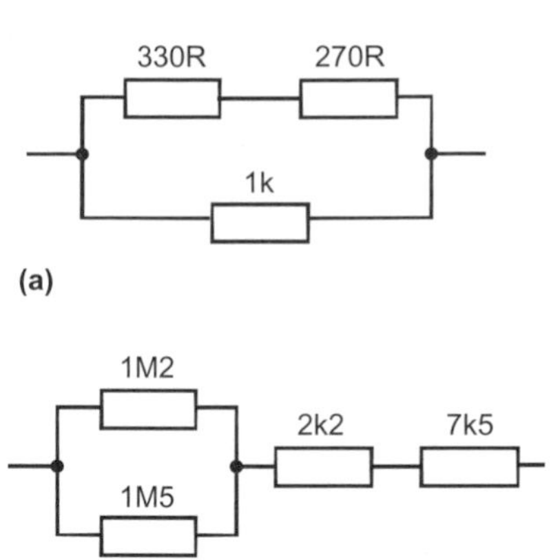

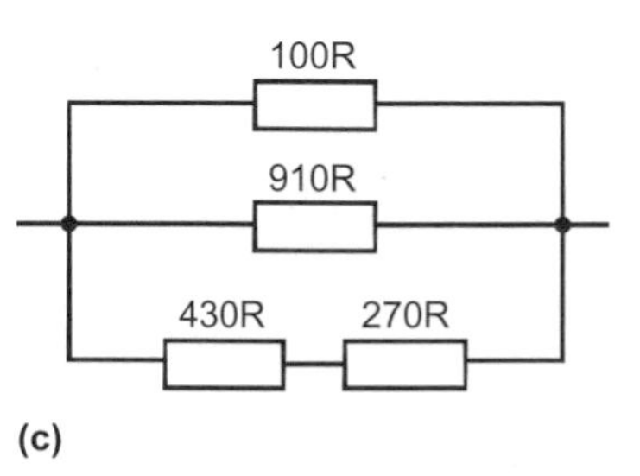

FIGURE 15.21

30 What are the resistor printed codes equivalent to the following colour codes?
- **(a)** red, violet, brown; red tolerance.
- **(b)** blue, red, yellow; gold tolerance.
- **(c)** brown, blue, red; red tolerance.
- **(d)** red, red, green; no tolerance band.

31 What are the resistor colour codes equivalent to the following printed codes?
- **(a)** 1M8L.
- **(b)** 56RJ.
- **(c)** 750KK.
- **(d)** 2K4G.

32 (a) In the circuit in Figure 15.22 what are the lowest and highest obtainable values of V_{OUT}? (b) If a load of 50 Ω is connected between the output terminal and the 0 V line, what is the highest obtainable value of V_{OUT}?

33 What is the maximum practicable current that can be drawn from the voltage divider above? To what values should the resistors be changed, to allow up to 40 mA to be drawn from the divider, keeping the same minimum and maximum voltages?

34 A voltage divider is made from a single variable resistor, value 1 kΩ, and connected to a 15 V supply. The wiper is set so that the output voltage is 9 V. Then a 470 Ω load is connected to the output. What is the new value of the output voltage and how much current flows through the load?

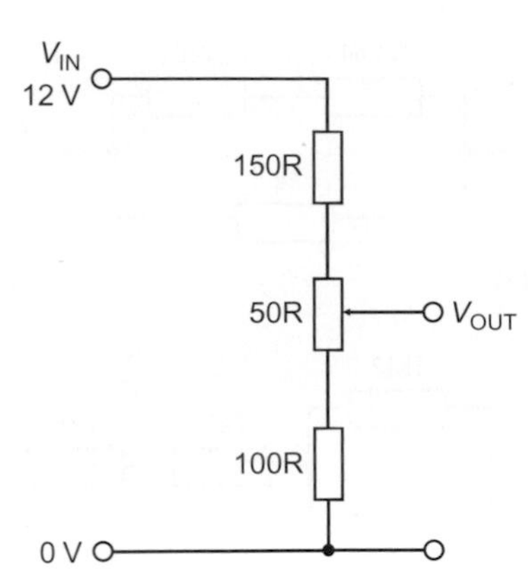

FIGURE 15.22

35 Describe how to use an ammeter (or a milliammeter) and a voltmeter to measure resistance. State why there is an error in this technique.

36 Explain why a conductor becomes warm or hot when an electric current passes through it. Name two useful applications of this effect.

MULTIPLE CHOICE QUESTIONS

1 Which one of these values is NOT an E24 value?
A 21 Ω.
B 910 Ω.
C 5.6 Ω.
D 62 Ω.

2 Which of these values is an E12 value?
A 110 Ω.
B 7.5 kΩ.
C 4.7 MΩ.
D 240 Ω.

3 The multiplier band in the three-band colour code is the:
A tolerance band.
B first band.
C black band.
D third band.

4 If the tolerance band is red, the resistor has a tolerance of:
A ±2%.
B ±5%.
C ±10%.
D ±20%.

5 The effective resistance of two or more resistors wired in series equals:
A the resistance of the largest resistor.
B the sum of the resistances.
C the average of the resistances.
D the resistance of the smallest resistor.

6 Three resistors, 10 Ω, 100 Ω and 220 Ω, are joined in parallel. A battery is connected across the network. The resistor that is passing the largest current is:
A 10 Ω.
B 100 Ω.
C 220 Ω.
D All pass the same current.

7 In the circuit of question 6, the resistor that has the biggest voltage across it is:
A 10 Ω.
B 100 Ω.
C 220 Ω.
D All have the same voltage.

Capacitors

A simple capacitor consists of two metal plates with a layer of insulator between them.

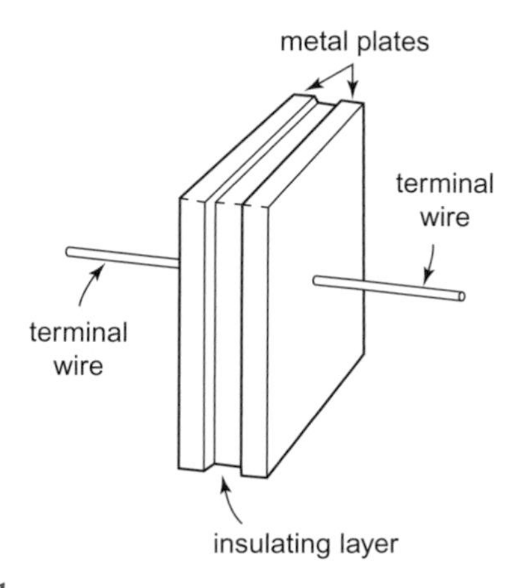

FIGURE 16.1

The insulating layer may be a thin sheet of plastic, but some types of capacitor have a layer of air instead.

If a capacitor is connected to a source of DC electric power, electrons accumulate on the plate that is connected to the negative supply terminal. These repel electrons from the opposite plate. The repelled electrons flow toward the positive terminal.

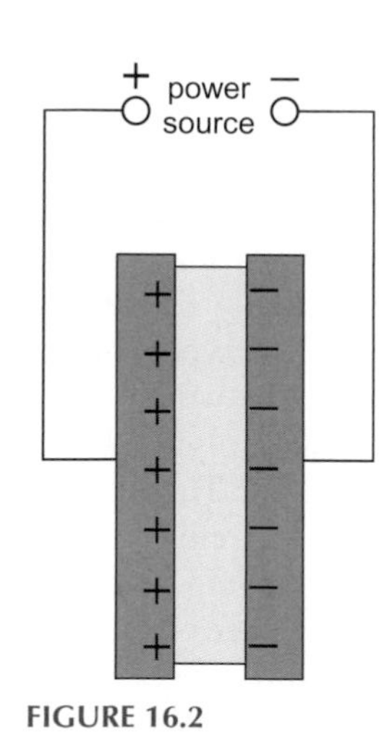

FIGURE 16.2

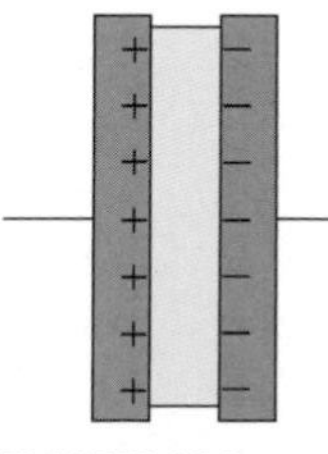

FIGURE 16.3

A capacitor connected like this to a power supply instantly becomes charged. The voltage between its plates equals that of the supply.

If the capacitor is removed from the supply, it remains charged.

Because of the insulating layer, current can not flow through the capacitor. The capacitor remains charged indefinitely. For this reason, capacitors are useful for **storing charge**.

CAPACITANCE

The ability of a capacitor to store charge is called its **capacitance**, symbol **C**.

The unit of capacitance is the **farad**, symbol **F**.

The farad is defined as the amount of charge stored (in coulombs) per volt:

$$capacitance = \frac{charge}{voltage}$$

Example

The amount of charge stored on a capacitor is 6 coulombs. The voltage between its plates is 2 V. What is its capacitance?

capacitance = 6/2 = 3 F

Self Test

1 A capacitor receives a charge of 2.5 C and the voltage across it is 10 V. What is its capacitance?
2 A 2 F capacitor has 5 V across it. How much charge is it storing?

Capacitors rated in farads are used for backing up the power supply to computer memory. However, most electronic circuits need much smaller capacitances. The units in which most capacitors are rated are:

- **microfarad**, millionths of a farad, symbol **μF**.
- **nanofarad**, thousandth of a microfarad, symbol **nF**.
- **picofarad**, thousandth of a nanofarad, symbol **pF**.

Self Test

Express in nanofarads: **(a)** 1000 pF, **(b)** 2.2 μF, **(c)** 1 F, **(d)** 47 pF, **(e)** 56 μF.

TYPES OF CAPACITOR

There are many types of capacitor, of which only the most often used are described here:

Polyester: The insulating material is polyester, which gives a relatively high capacitance. The plates are made of metal foil, or a metallised film is deposited on the insulator. The 'sandwich' of plates and insulator is rolled to make it more compact and the roll is coated with insulating plastic. Polyester capacitors (the three on the left below) are general-purpose capacitors and widely used.

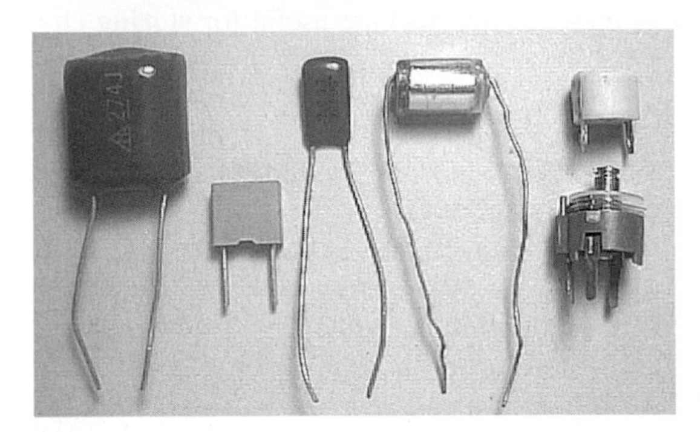

FIGURE 16.4

Polystyrene: (Figure 16.4, fourth from left) These are constructed in a way similar to polyester capacitors. Using polystyrene as an insulator results in lower capacitances than with polyester. However, they can be made with much closer tolerance, so are suitable for tuning circuits and filters.

Variable: These have two sets of plates, the alternate plates being electrically connected. One set is fixed. The other set can be turned so that we can vary the amount by which the sets overlap (below). This varies the capacitance.

The larger capacitors used for tuning radio receivers have air as the insulator. Plastic film is used in small trimmer capacitors (Figure 16.4, right). Some trimmers work by having a screw adjustment that compresses or loosens the plates and film to vary the capacitance.

ELECTROLYTIC CAPACITORS

These are used to store large amounts of charge. Their capacity is usually 1 μF or more and may be as much as 10 000 μF.

The two commonest types of electrolytics are **aluminium electrolytics** (Figure 16.5, left and centre) and **tantalum bead capacitors** (right).

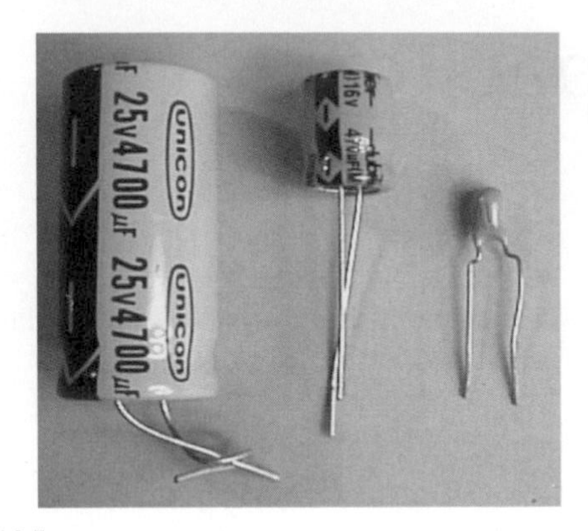

FIGURE 16.5

Electrolytics can store large amounts of charge for hours. When building or testing circuits, there is danger from electric shock if you touch the terminal wires without giving the capacitor time to discharge. When storing large electrolytics, twist their terminal wires together (Figure 16.5, left) so that they can not retain a charge.

Electrolytics are **polarised**, which means that they have positive and negative terminals. They *must* be connected the right way round. In the photo, there are markings on the case to indicate the negative terminal. If an aluminium electrolytic capacitor is connected the wrong way round, gas is formed inside and it will explode. Tantalum bead capacitors can be destroyed in a few seconds by wrong connection.

The insulation between the plates in electrolytics (especially the aluminium type) is not as high as that in other types of capacitor. A few microamps leak across between the plates.

Electrolytics have wide tolerance, normally ±20% or wider. They can not be used in precision filters or timing circuits.

Tantalum bead capacitors are made with smaller capacitances than for aluminium electrolytics. However, they are generally smaller in size so are useful where space is limited.

Charging Capacitors

STORING CHARGE

The ability of a capacitor to store charge is one of its most important properties.

Things to do

Set up the circuit shown in the diagram, using a breadboard. The wire from the positive terminal of the electrolytic capacitor is a **flying lead**. It is a piece of wire about 10 cm long. One end is plugged into the same row of contacts as the positive wire of the capacitor. The other can be plugged into the board to connect either to (A) the positive supply or (B) the lamp.

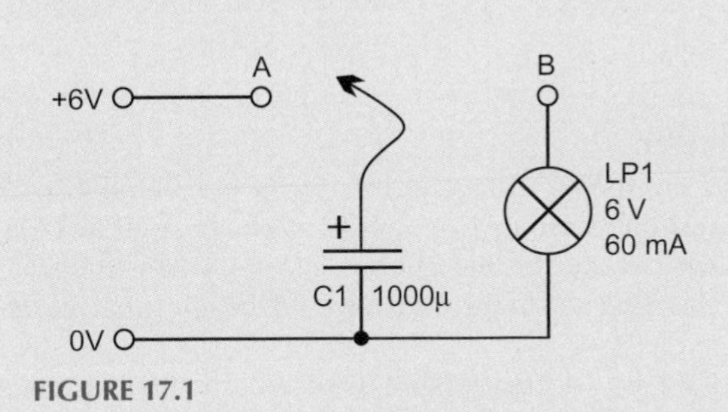

FIGURE 17.1

1 Connect the flying lead C to the positive power line (A) to charge the capacitor.
2 Quickly connect C to the lamp at B, to discharge the capacitor. Did you see the lamp flash? If not, try again.
3 Connect the lead to A again. Remove it from the socket, but wait 10 seconds before connecting it to the lamp. Does the lamp flash?
4 Repeat **3**, but wait for longer times before connecting to the lamp. How long does the capacitor hold enough charge to flash the lamp?

This investigation demonstrates that capacitors can hold charge for a long time.

CHARGING AND DISCHARGING

The rate of flow of charge depends on the voltage across the capacitor.

Things to do

This circuit has a switch for charging and discharging. It has a resistor to make the current smaller. Because of this, charging and discharging take longer. It gives you time to see what happens. The meters can be a separate voltmeter and ammeter, or a pair of multimeters.

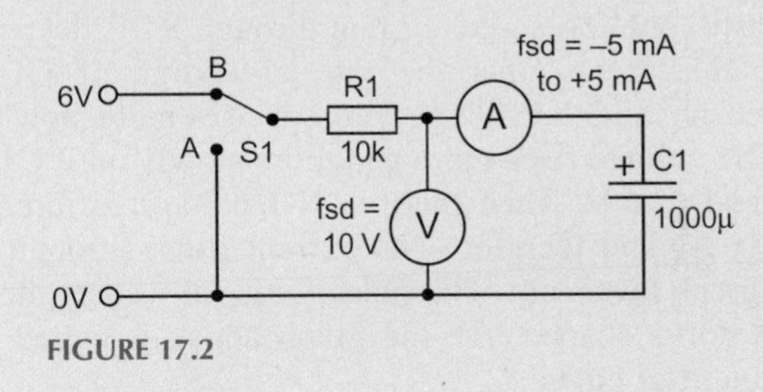

FIGURE 17.2

1 Set the switch to A to discharge the capacitor if it is already charged.
2 Set the switch to B and watch the meters as the capacitor is charged. You and a partner could watch one meter each.
3 Set the switch to A and watch the meters as the capacitor is discharged.
4 Repeat **2** and **3** until you can answer these questions:
   - When does the biggest current flow into the capacitor?
   - When does the biggest current flow out of the capacitor?
   - When does no current flow into or out of the capacitor?
   - When does the voltage across the capacitor change most quickly?

In the investigation, the flow is too fast to see exactly what happens. The voltage change can be more clearly seen by connecting an oscilloscope in place of the voltmeter. The display looks something like this:

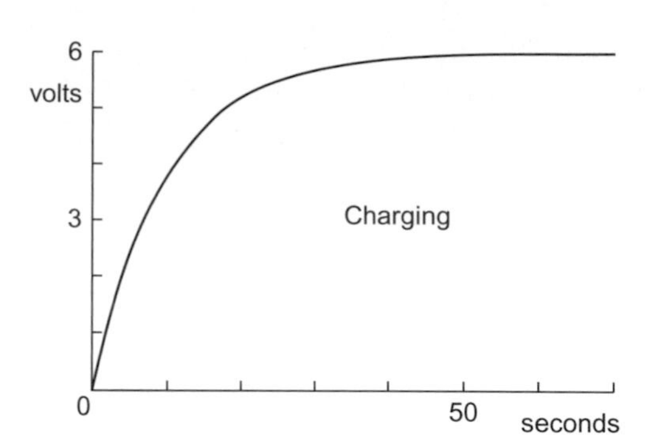

FIGURE 17.3

When the switch is turned to A, the voltage on the supply side of the resistor is 6 V and that on the capacitor side is 0 V. By Ohm's Law, the current through the resistor is 6/10 000=600 μA. Charging begins and the voltage across the capacitor (see graph) rises steeply. The voltage on the supply end of R1 stays at 6 V, but the voltage at its other end is increasing. The voltage difference across R1 is *decreasing*. Ohm's Law still applies, so the current through R1 is decreasing. This means that the rate of charge of C1 is decreasing and the voltage across it rises more slowly.

The voltage rises more and more slowly until C1 is charged to 6 V. Then there is NO voltage difference across R1 and therefore NO current flows through it. The graph levels out. The capacitor is fully charged.

A curve shaped like the graph above is called an **exponential curve**.

The reverse happens when the capacitor is discharged. At first, there is a voltage difference of 6 V across R1, so 600 μA flows out of the capacitor, through R1 to the 0 V line. The voltage becomes less as the capacitor discharges. The voltage drops more and more slowly. When it reaches zero, the capacitor is uncharged.

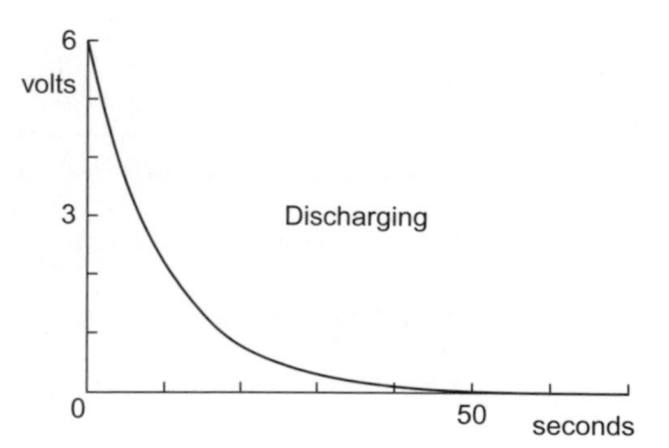

FIGURE 17.4

CAPACITORS IN PARALLEL

Wiring two or more capacitors in parallel is the equivalent of adding together the areas of their plates. For this reason, the effective capacitance is the sum of the individual capacitances.

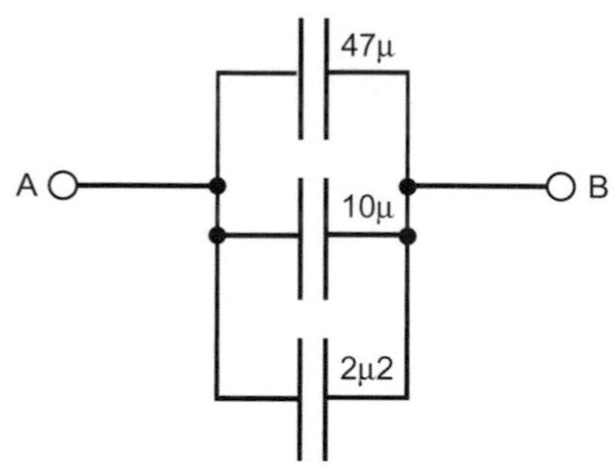

FIGURE 17.5

Example

In the diagram above, the effective capacitance is:

$$C = 47 + 10 + 2.2 = 59.2\ \mu F$$

The effective series capacitance is always greater than the greatest of the individual capacitances.

Self Test

1 If the supply voltage in the circuit in column 2, p. 59 was increased to 10 V and R1 was increased to 18 kΩ, what would be the current into C1 when the switch was first set to B? What would be the final current into C1?

2 If C1 was decreased to 470 μF, will the time to charge the capacitor be longer? Or the same? Or shorter?

Design Tip — Capacitor Values

Capacitors are made in a range of values that is similar to the range of preferred resistor values (p. 43). Capacitors have wider tolerances than resistors so there is no point in having 24 basic values in the range. Instead, capacitors values are the E12 series:

1.0 1.2 1.5 1.8 2.2 2.7
3.3 3.9 4.7 5.6 6.8 8.2

The values are repeated in multiples of 10.

Design Tip — Marking Capacitors

Values are often marked on the case (see photo, top right, p. 58). There is no room for this on small capacitors so the value is coded. The code has two digits. The first two digits are the first two digits of the capacitance, in picofarads. The third digit is the number of zeroes following the two digits.

Example

The code '223' means '22' followed by three zeroes. This gives 22 000 pF, which is equal to 22 nF.

Tolerance is coded by an extra letter, as for the resistor printed code (p. 52).

Design Tip — Selecting Capacitors

For **high capacitance** (1 μF or more), use aluminium electrolytic capacitors. These may have axial or radial lead wires. Axial leads are used one at each end of the capacitor. This can be useful if you need to 'jump a gap' on the circuit board. More often, especially on PCBs, you will need radial leads (both leads at the same end).

If space is short, use tantalum bead capacitors, but these are more expensive.

For **medium capacitance** (10 nF to 1 μF) normally use polyester or ceramic capacitors (not suitable for audio circuits). For better temperature stability, use polycarbonate.

For **low capacitance** (under 10 nF) use polystyrene or ceramic capacitors.

Capacitors have a **working voltage**, usually marked on the case. The capacitor is destroyed if this is exceeded.

Polyester, polystyrene, polycarbonate and ceramic capacitors usually have working voltages of 100 V or more, so you will usually have no problems. Electrolytics have lower working voltages. If it is important to reduce leakage, choose an electrolytic with higher working voltage than you need (say, 63 V). However, it will be more expensive and larger than one with a lower working voltage (10 V or 25 V).

QUESTIONS ON CAPACITORS

1. State the values and tolerances of capacitors marked as follows: (a) 473 J, (b) 394 K, (c) 102 J.
2. What is the effective capacitance of 470 nF, 150 nF and 1.2 μF connected in parallel?
3. What are the main features of an aluminium electrolytic capacitor? When would you use such a capacitor?
4. Draw a curve to represent the change in voltage across a capacitor while it is being charged from 0 V to the supply voltage. What name is given to this curve? Explain why it has this shape.
5. In question 4, how could you make the charging time longer?
6. What is the insulating material in the most widely used type of capacitor?
7. State two ways of varying the capacitance of a variable capacitor.

Multiple choice questions on capacitors are on p. 66.

Inductors

It can be shown that, when a current flows in a wire, a magnetic field is produced around the wire. If the wire is formed into a coil, the magnetic field resembles that of a bar magnet.

The magnetic field is represented by **lines of force** that show the direction of the field in and around the coil.

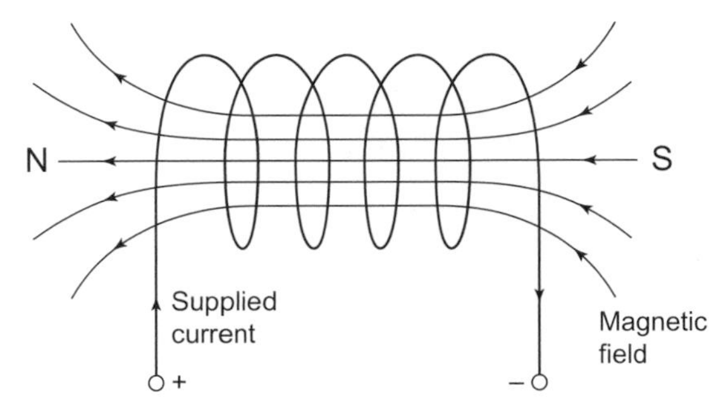

FIGURE 18.1

INDUCTION

A current through a coil produces a magnetic field, as described above. The reverse action also occurs. In the drawing below, a bar magnet moving toward a coil **induces** a current in the coil:

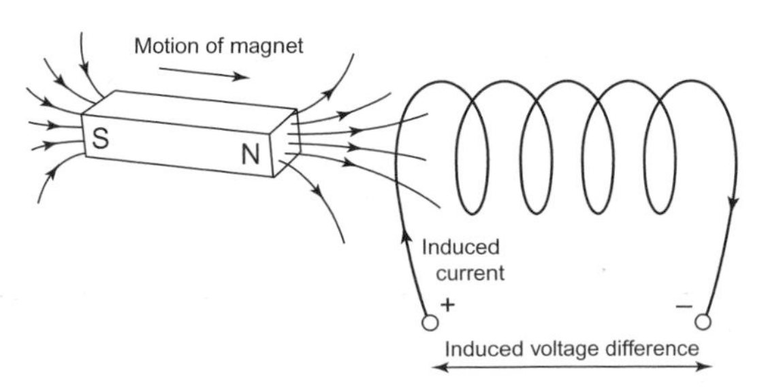

FIGURE 18.2

The current flows and the voltage difference is produced only when the magnet is *moving*. If we hold the magnet still, the current stops. If we move the magnet *away from* the coil, the current flows in the opposite direction.

Note that the current is flowing in the same direction in the two drawings. This means that the induced current is producing a magnetic field with its North pole at the end nearer the magnet. This induced magnetic field is trying to *repel* the magnet (like poles repel). It is trying to stop it from moving toward the coil.

The induced magnetic field opposes the motion of the magnet. When we move the magnet away, the current reverses and the field reverses too. Now there is a South pole near the North pole of the magnet. Unlike poles attract. There is a force trying to prevent us from moving the magnet away.

Whichever way we move the magnet, there is a force to *oppose* the motion. We have to do extra muscular work to move the magnet. The extra energy we use appears as the voltage difference between the ends of the coiled wire.

Things to do

Connect a coil of wire to a multimeter switched to a low voltage range. Take a bar magnet and move it towards one end of the coil. Try it with the North pole nearer the coil, then with the South pole nearer the coil. Hold the magnet still. Move the magnet away from the coil. Move the magnet slowly. Move the magnet quickly. What do you notice about the voltage?

SELF INDUCTION

When the current through a coil changes, the magnetic field through the coil changes. This changing field acts just like a magnet being moved around near the coil — it induces another current in the coil. The direction of the current opposes the change in current through the coil. This effect, in which a coil induces current *in itself*, is called **self induction**.

Self induction is important in the action of chokes (see below). If there is a very rapid change of current, such as when the current is switched off, self induction produces a large current that damages components (p. 97).

TYPES OF INDUCTOR

An inductor is a component in which the main cause of its action is its inductance. Usually an inductor is a coil of wire. It often has a **core**, consisting of a rod of magnetic material. There are two main types of inductor:

Chokes: '*Things to do*' on p. 63 demonstrated that the faster you move the bar magnet, the larger the induced voltage. In other words, the voltage induced in a coil is proportional to the *rate of change* of the magnetic field.

A high-frequency signal changes from positive to negative and back again very quickly — much faster than a low-frequency signal. So, when an alternating signal is passing through a coil, the rate of change of the magnetic field is greater for high-frequency signals than for low-frequency signals.

This means that a high-frequency signal self-induces a higher voltage to oppose the signal. As a result of this effect, low-frequency signals pass through the coil much more easily than high-frequency signals.

A choke is a coil which makes use of this action to block high-frequency signals from passing through from one part of a circuit to another. High-frequency signals are partly blocked while low-frequency signals or DC voltage levels are able to pass through.

Large chokes look like transformers (Figure 18.3), but have only one coil. Small chokes consist of beads or collars made of ferrite, threaded on to the wire that is carrying the high-frequency signals. Ferrite is an iron-containing material, so it acts as a core to contain the lines of magnetic force around the wire. Sometimes a choke is made by winding the wire around a ferrite ring.

A large choke may be used to prevent mains-frequency 'hum' from getting through from the power supply to the amplifying circuits of an amplifier. The photo below shows a small ferrite ring choke on the lead connecting a digital camera to a computer for downloading the pictures.

Tuning coils: Used in radio transmittters and receivers to tune a circuit to a particular radio frequency (Figure 18.4).

The coil is wound on a plastic former. It may have a core of ferrite or iron dust ceramic that can be screwed in or out of the coil to tune it.

Two or more coils can be wound on one former to make a transformer.

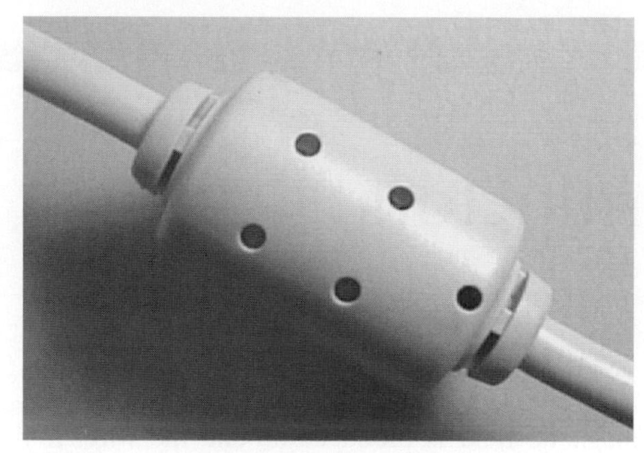

FIGURE 18.3

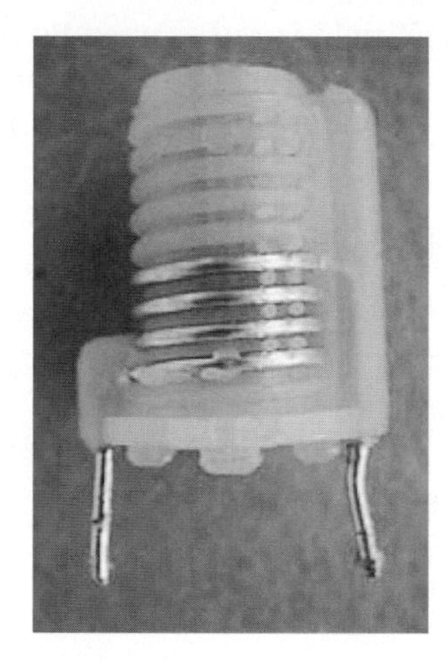

FIGURE 18.4

INDUCTANCE

The unit of inductance and self-inductance is the henry, symbol L. It is defined by this equation:

$$\textit{voltage} = -L \times \textit{rate of change of current}$$

Putting it into words, if the current through an inductor is changing at the rate of 1 A per second, the inductance of the coil is numerically equal to the voltage induced in the coil. The negative sign indicates that the induced voltage opposes the change in current.

Example 1:

The current through an inductor changes from 2 A to 0.5 A in 0.3 s. The rate of change is $(2 - 0.5)/0.3 = 1.5/0.3 = -5$ A/s. If the induced voltage is +50 V (opposing the fall in current), the inductance of the coil is:

$$L = -50/-5 = 10\ \text{H}$$

Example 2:

The current through an inductor changes by 2 A in a period of 1 ms. The rate of change is 2/0.001 = 2000 A/s. If the inductance is 0.5 H, the induced voltage is:

$$V = -0.5 \times 2000 = -1000\text{ V}$$

Self Test

1 A current increasing at the rate of 100 mA per second is passing through an inductor of 0.25 H. What is the voltage induced across the coil?
2 A voltage of −200 V is induced in an inductor of 400 mH (a millihenry is one thousandth of a henry). What is the rate of change of current through the coil?

TRANSFORMERS

A transformer consists of two coils wound on a core. The core is made of layers of iron.

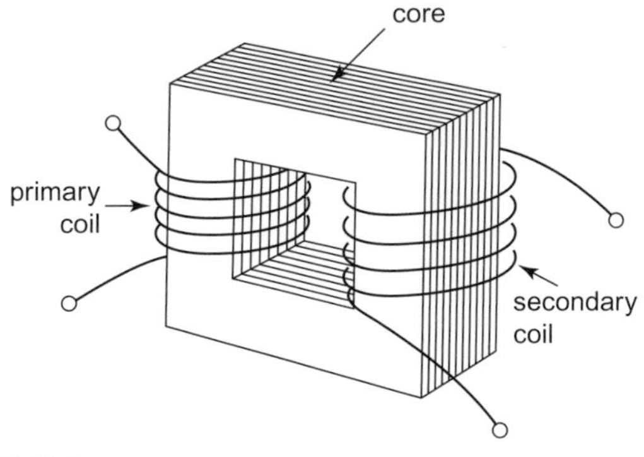

FIGURE 18.5

The coils normally have many more turns than are shown in the drawing.

When a current is passed through the primary coil, it produces a magnetic field. The core is to provide a path for the lines of magnetic force so that they almost all pass through the secondary coil. Induction occurs only when there is a *change* of magnetic field. So a transformer does not work with DC. When AC flows through the primary coil there is an alternating magnetic field. This induces an alternating current in the secondary coil.

Frequency: The frequency of the induced AC equals that of the inducing AC.

Amplitude: If V_P is the amplitude of the voltage in the primary coil, and V_S is the amplitude of the voltage in the secondary coil, then:

$$\frac{V_S}{V_P} = \frac{\text{secondary turns}}{\text{primary turns}}$$

Example

A transformer has 50 turns in the primary coil and 200 turns in the secondary coil. The amplitude of the primary AC is 9 V. What is the amplitude of the secondary AC?

Rearranging the equation above gives:

$$V_S = V_P \times \frac{\text{secondary turns}}{\text{primary turns}}$$

$$V_s = 9 \times \tfrac{200}{50} = 36\text{ V}$$

The amplitude of the secondary current is 36 V. These calculations assume that the transformer is 100% efficient.

Self Test

1 A transformer has 25 primary turns and 1200 secondary turns. The amplitude of the primary AC is 5 V. What is the amplitude of the secondary AC?
2 A transformer is used to transform 20 V AC to 120 V AC. It has 50 primary turns. How many secondary turns does it have?

QUESTIONS ON INDUCTORS

1 A bar magnet has its south pole nearer to one end of a coil. The bar magnet is being moved away from the coil. Draw a diagram to illustrate this and show the direction of the current induced in the coil.
2 Describe one type of choke and what it does.
3 A transformer has 6 V AC supplied to its primary coil and is to deliver 24 V AC from its secondary coil. The primary coil has 400 turns. How many turns has the secondary coil? If the frequency of the 6 V supply is 100 Hz, what is the frequency of the 24 V output from the transformer?
4 A transformer has 24 V AC applied to its primary coil. The primary coil has 500 turns and the secondary coil has 40 turns. What is the voltage across the secondary coil?
5 State Fleming's Left-hand Rule.
6 In the drawing of a DC motor shown in Figure 18.6, state which terminal, A or B, is made positive in order to make the coil rotate in the direction shown. Name the parts labelled C and D.
7 What is the name of the rule that relates the directions of motion, current and field in a generator?
8 Why does an AC generator produce an alternating voltage? If the coil is turned at 120 revolutions

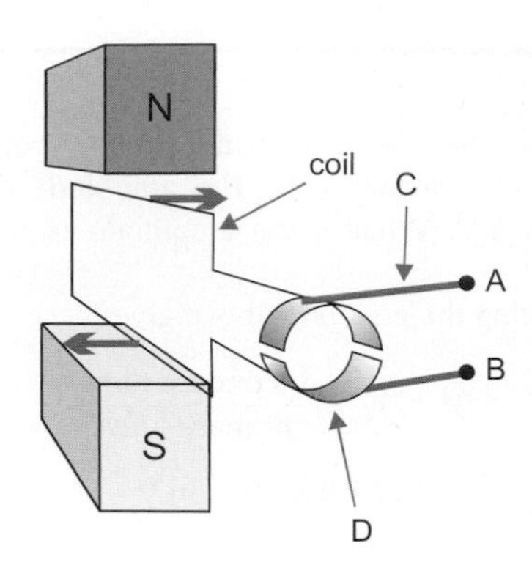

FIGURE 18.6

a minute, what is the frequency of the AC produced?

9 If a 43 Ω resistor is rated at 0.5 W, what is the maximum voltage that can be placed across it?

10 A pocket calculator uses power at an average rate of 8 mW. It is powered by two lithium cells delivering a total of 3 V. What is its effective resistance?

11 A current of 3 A flows through a low-voltage projector lamp when it is running at 35 W. What is the resistance of its filament when the lamp is running at this rate?

12 Why is mains power distributed at high voltage?

13 Why is mains power distributed as an alternating current?

14 When it is connected to the 240 V AC mains, the secondary output of a transformer is 48 V AC. If the primary coil has 1000 turns, how many turns has the secondary coil?

MULTIPLE CHOICE QUESTIONS ON CAPACITORS AND INDUCTORS

1 Answer 'True' or 'False' :
- **a)** Electrolytic capacitors store large amounts of charge.
- **b)** Polystyrene capacitors have high tolerance.
- **c)** Polystyrene capacitors have a high leakage current.
- **d)** A capacitor does not work properly if the applied voltage is less than its working voltage.
- **e)** Electrolytic capacitors are polarised.
- **f)** A charged capacitor loses its charge as soon as it is disconnected from the supply.
- **g)** The unit of capacitance is the farad.
- **h)** A capacitor must have a layer of plastic between its plates.
- **i)** A trimmer is a small variable capacitor.
- **j)** Electrons can flow from one plate of a capacitor to the other plate.
- **k)** All inductors have an iron or ferrite core.

2 Two capacitors wired in parallel have values 15 μF and 2.2 μF. Their effective capacitance is:
- **A** 15 μF.
- **B** 2.2 μF.
- **C** 17.2 μF.
- **D** 12.8 μF.

3 When a capacitor is discharged through a resistor, the current through the resistor is:
- **A** greatest at the beginning.
- **B** constant.
- **C** greatest when fully discharged.
- **D** greatest when half discharged.

Diodes

A diode is made of silicon. Silicon is neither an insulator nor a conductor. It is a **semiconductor**. This means that its properties are different from ordinary conductors such as copper.

Small amounts of substances are added to the silicon to give it the very special properties of a diode. In this Topic we find out what these properties are.

A diode is contained in a small capsule made of glass or plastic. It has two terminal wires. One of these is called the **anode**. The other is called the **cathode**.

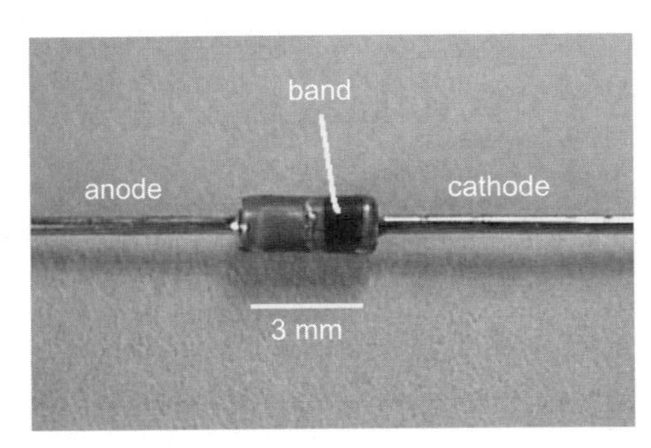

FIGURE 19.1

Usually there is a band marked on the diode to show which wire is the cathode.

Things to do

You need:

- A 6 V battery or a PSU.
- A 6 V lamp in a socket, with connecting wires.
- A diode. Type 1N4148 is suitable, but any ordinary silicon diode will do.
- A breadboard.

1. Examine the diode. Look for the band and identify the cathode wire.
2. Set up the circuit shown at top right on this page, but not the battery or PSU.
3. Check that the diode is connected the right way round. Its cathode wire goes to one terminal of the lamp.

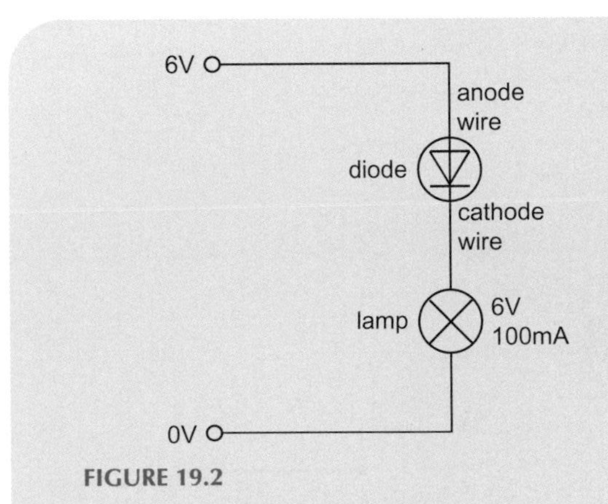

FIGURE 19.2

4. Connect the battery and see what happens to the lamp. What can you say about the diode?
5. Disconnect the battery. Remove the diode and replace it the other way round. It now has its anode wire connected to the lamp.
6. Connect the battery and see what happens to the lamp. What can you say about the diode now?

CONDUCTION THROUGH A DIODE

The experiment above demonstrates that:

A diode conducts in only one direction
and
Conduction is from anode to cathode.

As we shall see, these properties are very useful.

When a diode is connected as in the diagram above, with its anode to positive, we say that it is **forward biased**. A diode conducts only when it is forward biased.

When a diode is connected in the reverse direction, with its cathode to positive, we say that it is **reverse biased**. A diode does not conduct when it is reverse biased.

VOLTAGE DROP

In a voltage divider circuit (p. 50), the output voltage is a proportion of the input voltage. The values of the two resistors decide what the proportion is. We will see what happens if we replace one of the resistors with a diode. The diode is forward biased so that current can flow through it.

Things to do

You need:

- 6 V battery or PSU.
- 220 Ω resistor.
- diode.
- multimeter or 2 V voltmeter.
- breadboard.

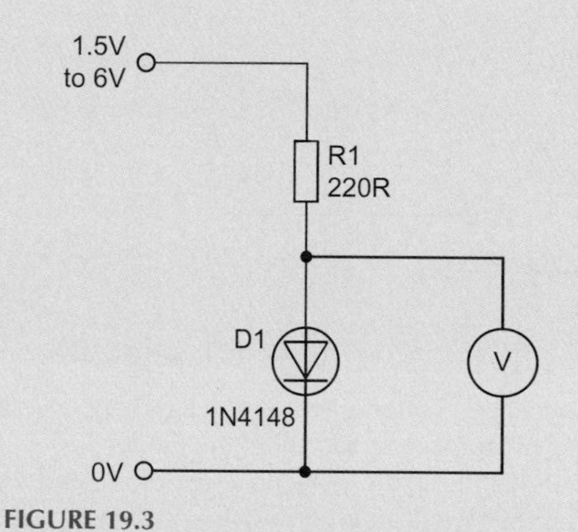

FIGURE 19.3

Input Voltage	Output Voltage
6	
4.5	
3	
1.5	

1 Set the input voltage to 6 V. Measure the voltage across the diode. Record the result in a table.

2 Repeat (1) with the voltage set to 4.5 V, 3 V, and 1.5 V.

3 What do you notice about the output voltage as you change the input voltage?

4 Is output proportional to input?

5 Does this circuit work like a voltage divider?

The results of the investigation show that a diode does not behave like a resistor. It does not obey Ohm's Law. The output voltage (that is, the voltage across the diode) varies only slightly with input voltage. It stays very close to 0.7 V.

We can sum up the results of the investigation by saying that:

A forward biased diode has a voltage drop of about 0.7 V.

This voltage drop is called the **forward voltage drop**.

TESTING DIODES

Many multimeters have a **diode test** function. It measures the forward voltage drop of the diode.

Things to do

You need:

- several diodes of different types, including some faulty ones.
- multimeter that has the diode test function.

1 Turn the range selector knob to the diode test function.

2 Connect the test probes of the meter to the diode, with the negative (black) probe to the cathode and the positive (red) probe to the anode.

3 With a good diode the reading is about 0.7 V (700 mV). Diodes vary slightly, so it may range between 400 mV and 900 mV. Anything outside this range indicates a faulty diode.

4 Reverse the connections to the diode. There should be a 'zero' or equivalent reading (depending on the meter) to indicate that the diode is not conducting when reverse biased.

Rectifier Diodes

One of the most important uses of diodes comes from their ability to conduct in only one direction. See what happens in this circuit:

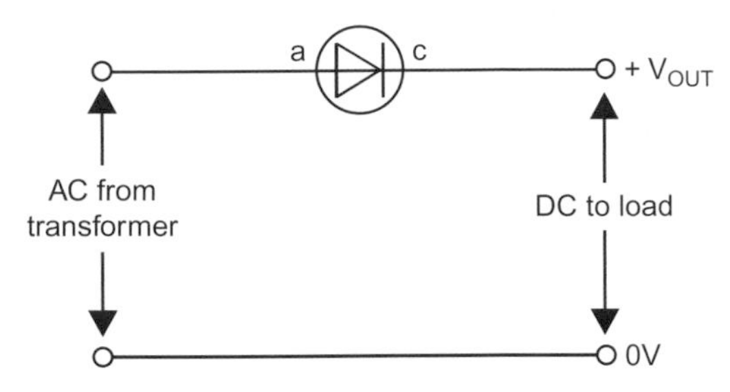

FIGURE 20.1

The supply is alternating current, from a transformer.

In the diagram below, there is a load connected to the circuit. The diagram shows the path of the current during the positive half-cycles of the AC. The diode is foward biased, so it conducts. Current flows through the diode to the load and returns along the 0 V line.

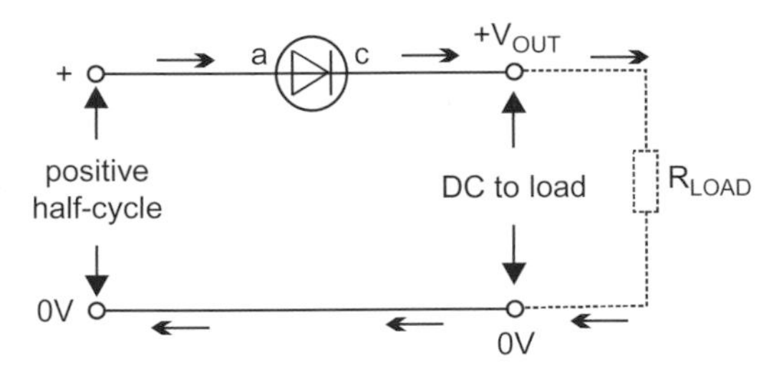

FIGURE 20.2

The diode does not conduct during the negative half-cycles, as shown below:

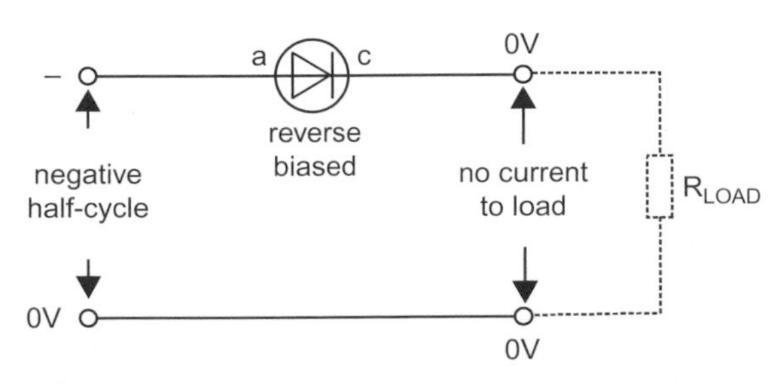

FIGURE 20.3

The waveform of the current through the load is plotted below. Although the voltage is pulsing, it is always positive. It is the equivalent of DC.

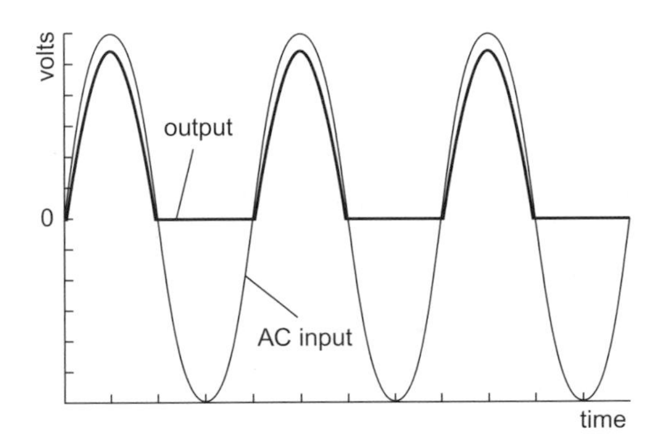

FIGURE 20.4

Comparing the graphs of the AC input and pulsed DC output, we see that:

- There is no output during negative half-cycles. Half the input power is wasted.
- The output amplitude is less than the input amplitude. This is because of the forward voltage drop across the diode.

A circuit that converts AC to DC is called a **rectifier**. Because this rectifier produces current only from the positive cycle, it is called a **half-wave rectifier**.

FULL-WAVE RECTIFIER

The circuit below rectifies AC by using a **bridge** of four diodes:

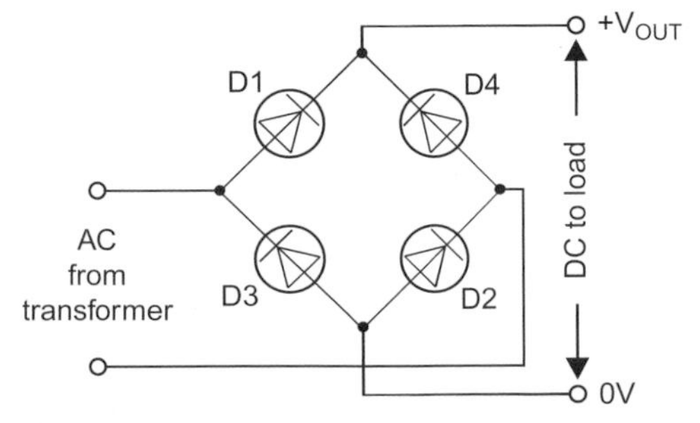

FIGURE 20.5

During the positive half-cycle, diodes D1 and D2 are forward biased, so they conduct. Diodes D3 and D4 are reverse biased and do not conduct. Current flows through the load shown in the next diagram.

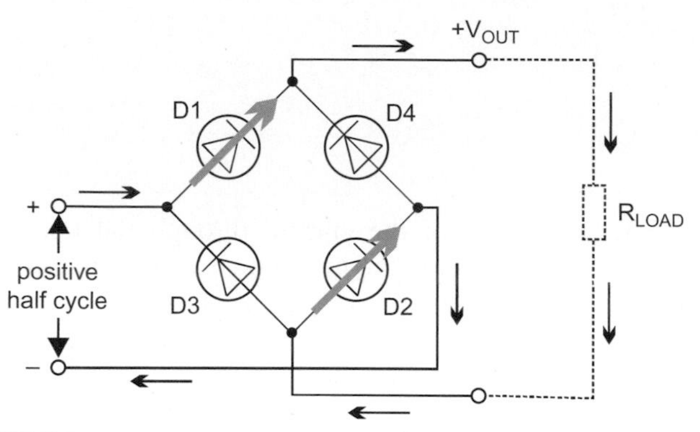

FIGURE 20.6

During the negative half-cycle (see below), diodes D1 and D2 are reverse biased so they do not conduct. Diodes D3 and D4 are forward biased and conduct.

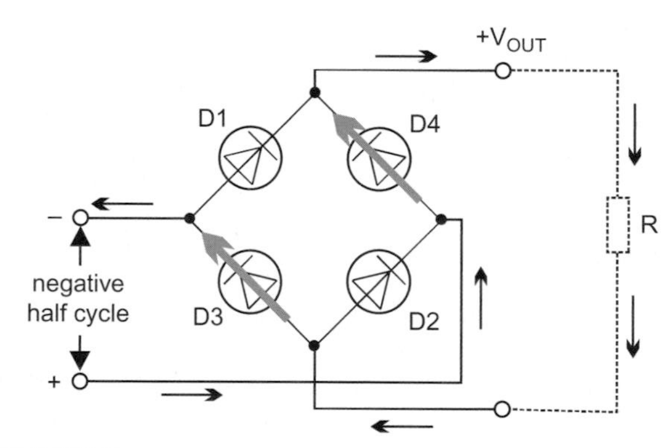

FIGURE 20.7

The result is that current continues to flow through the load, *in the same direction* as before. The graphs of input and output are:

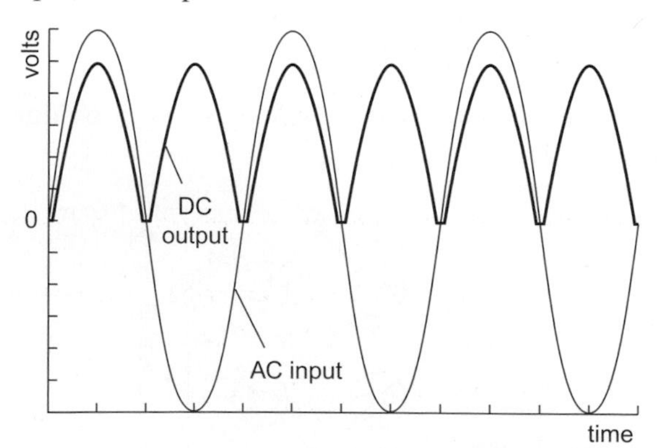

FIGURE 20.8

The rectifier produces output during both half-cycles, so it is 100% efficient. It is called a **full-wave rectifier**. In each half-cycle, the current flows through *two* diodes, so the output amplitude is *two* voltage drops (about 1.4 V) less than the input amplitude.

Rectifiers are used in PSUs and other power units to produce DC from the low-voltage output of a mains transformer.

SMOOTHING

The pulsed DC from a rectifier is unsuitable for powering circuits until it has been smoothed. This is done by connecting a large-value capacitor across the DC output.

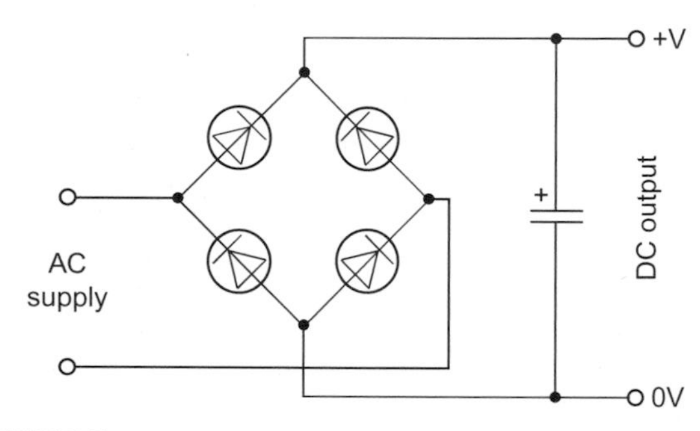

FIGURE 20.9

The capacitor is usually an aluminium electrolytic one and has a capacity of 1000 μF or more. Repeated pulses of DC soon charge the capacitor up to the peak voltage. When current is being drawn from the circuit the voltage starts to fall after each peak, but it is returned to peak level by the next pulse. The result is DC with a slight **ripple**.

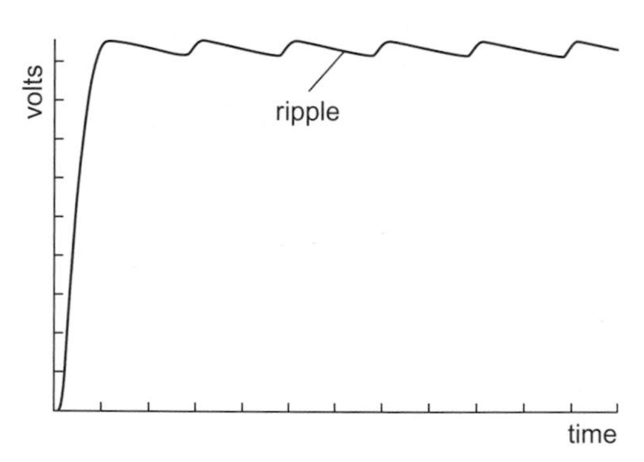

FIGURE 20.10

If the capacitor is large enough and the current drawn by the load is not too large, the voltage is almost as smooth as a pure DC.

Things to do

Use a low-voltage mains transformer and an oscilloscope to view the output from half-wave and full-wave rectifiers, unsmoothed and smoothed.

Light Emitting Diodes

Light emitting diodes, usually known as **LEDs**, give off light when a current flows through them. The original LEDs were red, but now there are orange, yellow, green, blue and white LEDs. There are also infrared LEDs, which produce infrared instead of visible light.

The most recent LEDs are exceedingly bright. They are used for torches, reading lights, car indicator lamps, and street lights. The most powerful take up to 5 A, yet need only 3.5 V to operate them. They are very energy efficient.

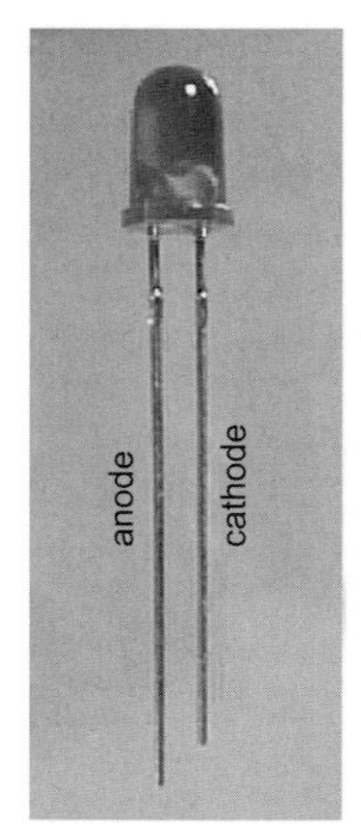

FIGURE 21.1

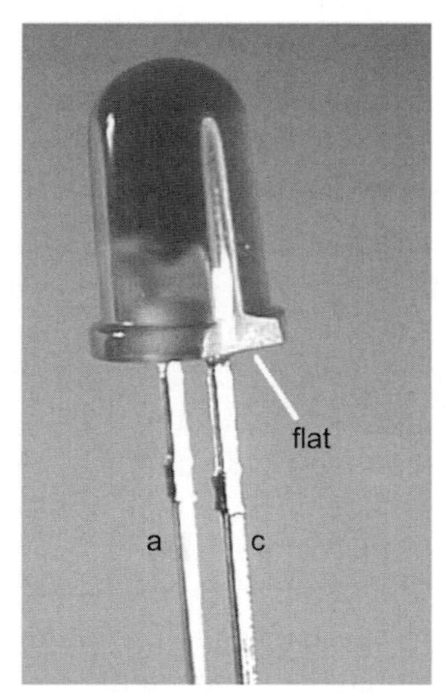

FIGURE 21.2

A typical LED is in a domed plastic package, with a rim. There are two terminal wires at the bottom. Usually, though not always, the cathode wire is shorter than the anode wire.

Another way to tell cathode from anode is to look at the rim (if the LED has one). The rim is flat on the side nearer the cathode wire.

An LED needs about 20 mA to light it to full brightness, though as little as 5 mA still produces a clearly seen glow. The forward voltage drop (p. 68) of an LED averages about 1.5 V, so a 2 V supply will light most types to their maximum brightness. When lit by a higher voltage, the LED may be burnt out if the forward voltage across it exceeds 2 V. It is essential to wire a **current limiting resistor** in series with it.

CURRENT LIMITING RESISTOR

A suitable value for a current limiting resistor is calculated as follows.

The supply voltage is V_S volts. The current that we want to flow through the LED is i amps. Assume that the forward voltage drop will be 2 V.

The voltage drop across the resistor must be $V_S - 2$.

By Ohm's Law, this voltage drop equals iR. Therefore:

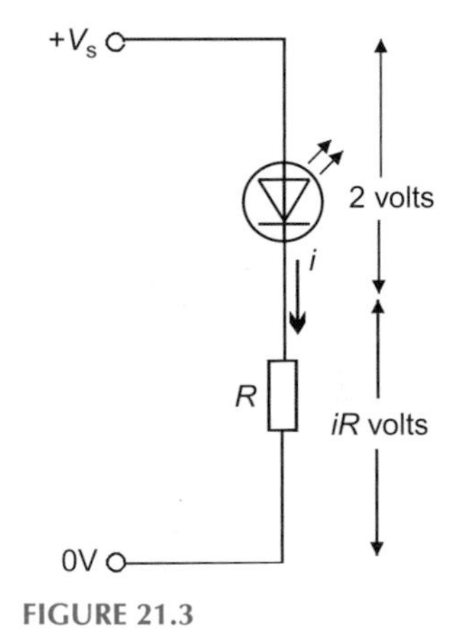

FIGURE 21.3

$$V_S - 2 = iR$$

Rearranging terms gives:

$$R = \frac{V_S - 2}{i}$$

Example

An LED is lit from a 9 V supply and takes a current of 15 mA. The value of the series resistor is:

$$R = \frac{V_s - 2}{i} = \frac{9 - 2}{0.015} = 466\ \Omega$$

Use the next higher resistance in the E24 series, which is 470 Ω.

Self Test

Calculate the value of the series resistor required for an LED that:

(a) runs at 25 mA on a 12 V supply.
(b) runs at 10 mA on a 15 V supply.

SHAPES AND SIZES

Some LEDs are used as indicator lamps — for example, to indicate that the power to an appliance is switched on. They are also used for informative or decorative displays. They are made in a variety of shapes, including circular, rectangular, and triangular LEDs

Arrays of shaped LEDs are used for displays. The commonest of these is the seven-segment display, used for displaying numerals and letters.

FIGURE 21.4

One or more rows of these are used for displaying messages.

LEDs are made in a range of sizes. The smallest are about 1 mm in diameter, used as indicators on panels where space is short. At the other extreme the jumbo LEDs, 10 mm in diameter, are useful where a readily noticeable warning lamp is needed.

FIGURE 21.5

LEDs are ideal as indicators because they require very small currents when compared with filament lamps. This makes them very suitable in battery-powered equipment, in which a filament lamp would soon exhaust the supply. There is also the factor that filament lamps have a limited life. Sooner or later the filament burns out. LEDs last almost for ever.

REVERSE BIAS

An LED is not able to withstand a reverse bias of more than a few volts. Most can survive a reverse bias of 5 V, but no more. This is quite different from most diodes, which can withstand reverse bias of hundreds of volts.

Since the circuits in which LEDs are used often have a supply voltage of 6 V or more, it is important to make sure that the diode is wired into the circuit the right way round.

BICOLOUR LEDs

An LED that can change colour is useful in some applications. For instance the LED could indicate 'all systems GO' when it is green and 'fault condition' when it is red. On a digital camera, there may be a 'record/play' indicator that is red when the camera is in recording mode and changes to green when it is in play mode. LEDs that can display two colours are called **bicolour LEDs**.

A bicolour LED has two LEDs of different colours inside one package. There are two ways in which the LEDs may be connected.

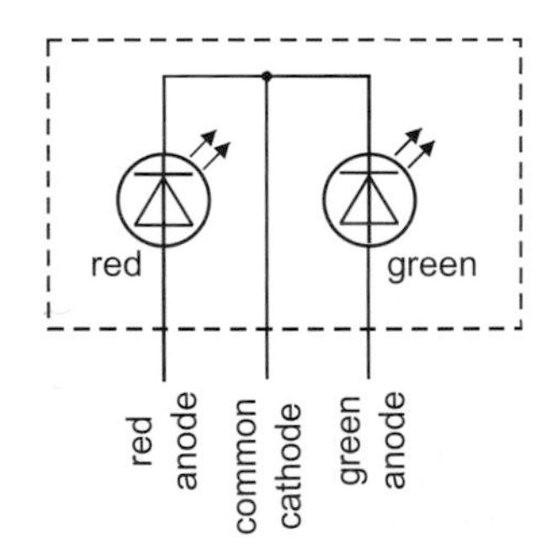

FIGURE 21.6

In the type with three terminal wires, the LEDs have a common cathode. A positive voltage applied to either of the other two wires is used to light the corresponding LED.

In the two-terminal type (see Figure 21.7) the LEDs are connected anode to cathode. Which LED lights depends on which terminal is made positive.

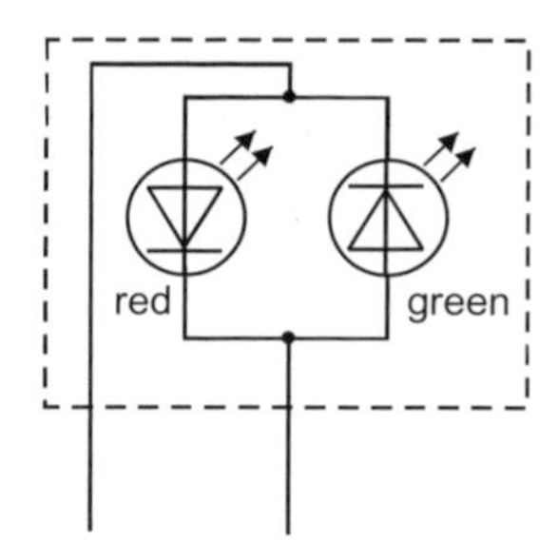

FIGURE 21.7

By switching the supply, we can light red and green alternately. If this is done at high speed, the LED appears to be emitting yellow light.

POWER DIODES

The diode illustrated in the photo on p. 67 is a signal diode. It is able to pass a maximum current of 100 mA. A rectifying diode for a power supply unit generally needs to pass more current than that. Special power rectifying diodes are made that can pass much larger currents.

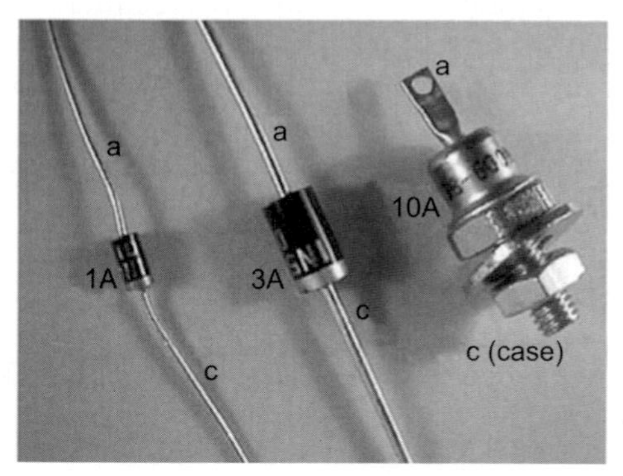

FIGURE 21.8

Below is a **bridge rectifier**. This consists of four power diodes in one package. They are already connected to make a full-wave rectifying bridge. It has four terminal wires, two for the AC input and two for the DC output.

FIGURE 21.9

The relationship between the voltage across a diode and the current through it is investigated below. First we test the diode when it is connected with forward bias.

Things to do

Set up this circuit on a breadboard:

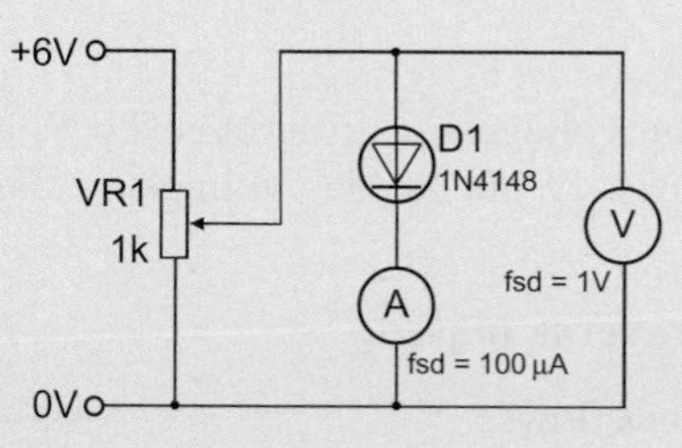

FIGURE 21.10

Check that the diode is connected the right way round, with its cathode wire to the ammeter.

1 Draw a table of three columns for the results. Head the columns 'Voltage, *V*', 'Current, *I*' and 'Ratio *V/I*'.
2 Connect the power supply and turn VR1 so that the voltmeter reads zero. Record the voltage and the current in the table.
3 Turn the knob of VR1 slowly until the voltage across the diode is 0.1 V. Record this value. Read and record the current.
4 Repeat step (3) for increasing voltages: 0.2, 0.3, 0.4, 0.5, 0.6, 0.7, 0.9, 1.0, 1.5 and 2 V. You may need to change the ammeter or the range of the multimeter to an fsd of 1 mA for later readings. Note: 'fsd' means 'full scale deflection', the maximum reading.
5 Calculate the ratio V/I for each pair of readings. Does this remain constant for all pairs? Compare this result with the result from p. 41.
6 Plot a graph of your readings of voltage and current. In your own words, describe how the current changes as voltage is increased.
7 Repeat steps 1 to 6 with the diode connected the other way round, so that it is reverse biased. Range the voltage up from 0 V to 10 V in steps of 1 V. Record your results.
8 How much current flows through a diode when it is reverse biased?

CURRENT THROUGH A DIODE

The investigation above tells us several things about the current through a diode:

When forward biased:

- No current flows when the voltage is less than about 0.6 V.

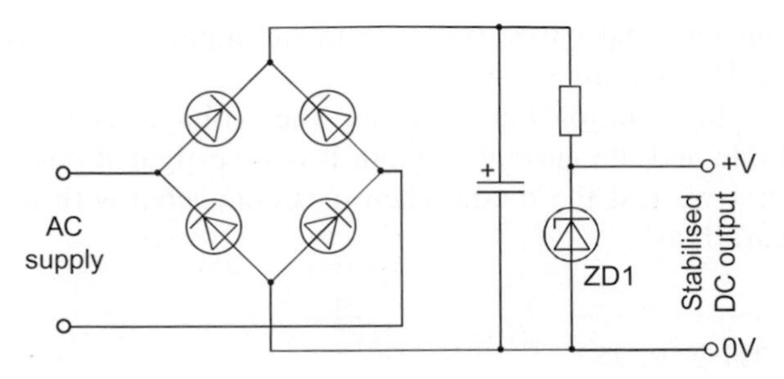

FIGURE 21.11

- When the voltage is a little over 0.6 V, a small current flows. With higher voltage, a large current flows.

When reverse biased:

- No current flows.

No diode is perfect, so when we said (above) that there is no current through the diode, there is always a small leakage current. This is only a few nanoamps, which is too small to measure easily.

If a Zener diode is reverse biased with a small voltage, it behaves like an ordinary diode. It does not conduct.

If a Zener diode is reverse biased with a voltage greater than a certain amount, called the **Zener voltage**, it conducts easily.

The Zener voltage of a Zener diode is fixed when the diode is manufactured. It may range from about 2.7 V to 20 V, with a tolerance of ±5%.

Zener diodes are used to regulate the output of power supply circuits. The circuit shown in Figure 21.11 has a rectifier with smoothed DC output (p. 70), followed by a Zener diode voltage stabiliser. The Zener diode is chosen with a Zener voltage equal to the required output voltage.

The output from the rectifier is several volts greater. Part of that voltage is dropped across the resistor. Its resistance is such that, when the load is drawing its maximum current, about 5 mA flows through the Zener to the 0 V line. The output voltage is equal to the Zener voltage. If the load is taking less than its maximum, or no current at all, the surplus current flows away through the Zener diode to ground. The Zener diode is still operating and the output voltage is still equal to the Zener voltage.

QUESTIONS ON DIODES

1 What are the properties of a diode?
2 The cathode of a diode is made 4 V positive of its anode. What is the name for this type of bias? Does current flow though the diodes?
3 Draw a circuit diagram of a half-wave rectifier. Describe how it works. Draw a sketch of 3 cycles of its output waveform.
4 Answer question 3 for a full-wave rectifier.
5 How do we smooth the output of a rectifier?
6 What are the properties of an LED? Describe some of the ways in which LEDs are used.
7 With the help of a diagram, describe one type of bicolour LED and how it works.
8 Describe how the current through a forward biassed diode varies as the voltage is increased from 0 V to 3 V.
9 State the properties of a Zener diode. How is a Zener diode used as a voltage stabiliser?

DESIGN TIME

Design time pages are scattered throughout the book. They provide collections of simple circuits, tips, problems, data and other things of interest to people designing circuits.

You are not expected to memorise these circuits. They are here to give you something to think about. The thinking will help you understand electronics better. You may find one or two of the circuits help you with your exam project. Or you might just like trying them out on a breadboard to see what happens.

The circuits are not described in detail. You are expected to work from the circuit diagram and think things out for yourself.

An LDR controls an LED. What happens? How does it work?

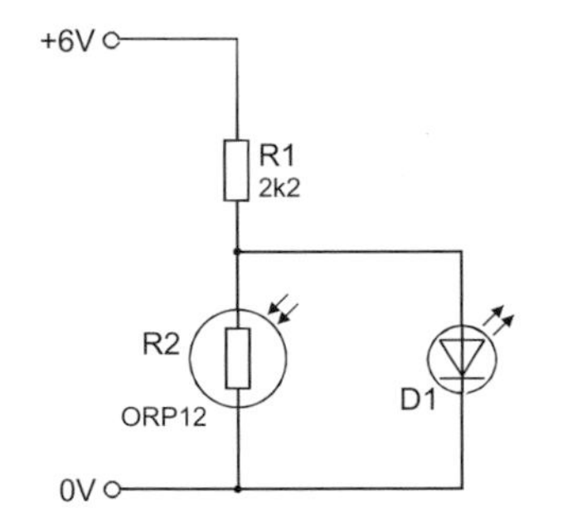

FIGURE 21.12

The LED in the circuit could be at the end of a long lead. It could be a **remote indicator**. This circuit could be improved (and will be later), but — for the moment — suggest some ways that this circuit could be used.

Try exchanging R1 and R2.

Shade the circuit, if necessary, so that the LED *just* goes out. Now change R1 for another resistor so that the LED comes on again. Try to find a resistor so that the LED is off in fairly dim light and comes on in very dim light.

Add a switch to the circuit so that you can switch it to operate in (1) bright light, or (2) dim light.

What is the purpose of this arrangement of switches?

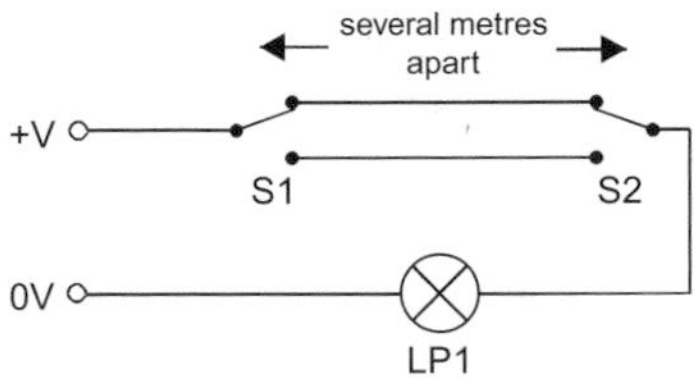

FIGURE 21.13

When a DPDT switch is wired like this, what does it do?

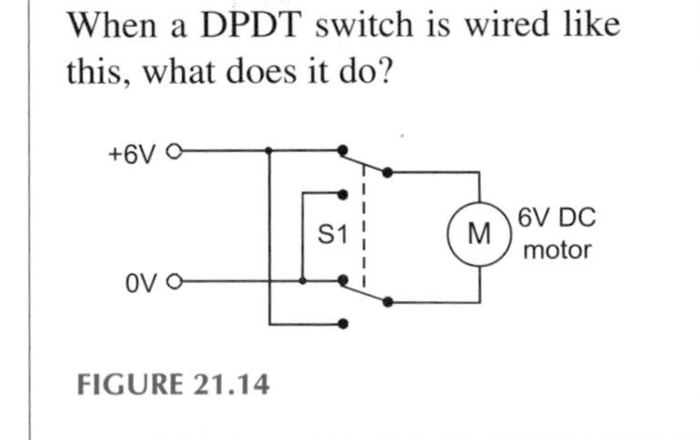

FIGURE 21.14

Transistors

There are several classes of transistor. The class that is described in this Topic is known as a **silicon npn transistor**. It is also known as a **bipolar junction transistor**, or **BJT**.

The BJT class of transistor described in this Topic was the first to be widely used. Another important class is the **MOSFET**, described in Topic 26. We shall not go into the reasons for these names, and you need not remember them, except for 'BJT' and 'MOSFET'.

All transistors have three terminal wires or connections.

Low-power transistors are enclosed in a plastic package or a metal case. The plastic case has a flat surface and the metal case has a tag on its rim. These are to help you to identify the terminal wires.

FIGURE 22.1

Seen from **below**, the wires are arranged like this in most (but not all) low-power transistors:

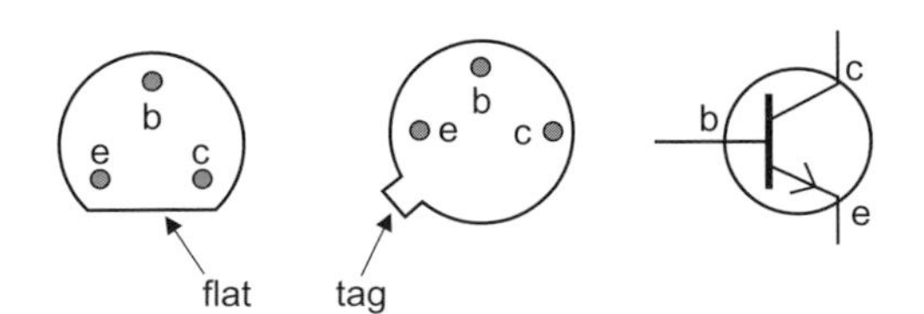

FIGURE 22.2

The symbol on the right is used to represent BJTs in circuit diagrams. The terminals are labelled with letters c, b and e, to identify the terminals as **collector**, **base** and **emitter**.

TRANSISTOR ACTION

To use a BJT, we connect it so that:

- its **emitter** is its most negative terminal.
- its **collector** is several volts positive of its emitter.
- its **base** is 0.7 V (or slightly more) positive of its emitter.

Under these conditions, we find that:

- a small base current flows **into the base**.
- a much larger current flows **into the collector**.
- the base and collector currents flow **out of the emitter**.

This diagram illustrates the way the currents flow:

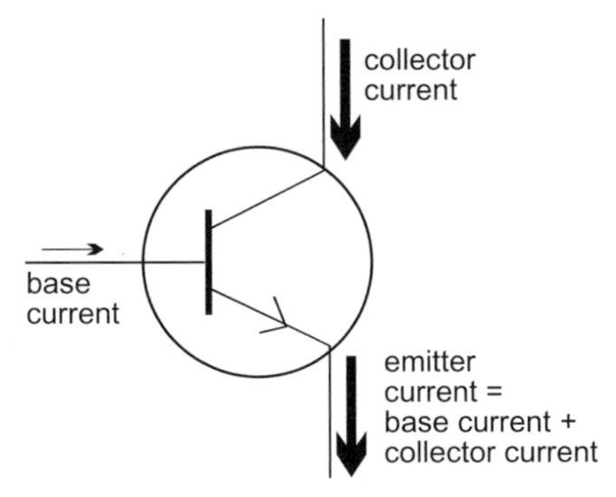

FIGURE 22.3

The base current is drawn with a thinner arrow because it is a much smaller current than the collector or emitter currents.

Things to do

This is a demonstration of the relative sizes of the base and collector currents. We demonstrate the sizes of the current by passing them through filament lamps. The bigger the current, the brighter the lamp.

(continued)

Things to do (continued)

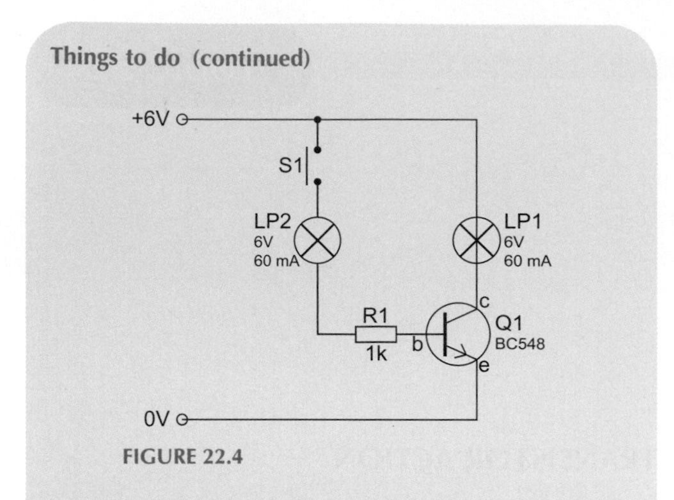

FIGURE 22.4

S1 is a push-to-make push-button. It has two short wires soldered to its terminals. Their other ends are stripped so that they can be pushed into sockets in the breadboard. LP1 and LP2 are two identical lamps, in sockets with connecting wires attached. Q1 is a transistor, type BC548; most other types can be used. A BC548 has the arrangement of terminal wires as in the diagram opposite.

1 Check that the three terminal wires of Q1 are inserted into the correct sockets.
2 Connect the power supply. Does LP1 light?
3 Press S1. Does current flow through LP1? Does current flow through LP2?
4 Keep S1 pressed down and unscrew LP2 from its socket. Watch what happens to LP1. Was LP2 carrying a current at step (3)?
5 If so, what can you say about this current?

TRANSISTOR SWITCH

The investigation above demonstrates one of the two important uses of transistors. A very small base current switches on a much larger collector current. We call this a **transistor switch**.

An an example, we can use the small current through an LDR sensor to switch a relatively large current through a filament lamp.

Here is the circuit:

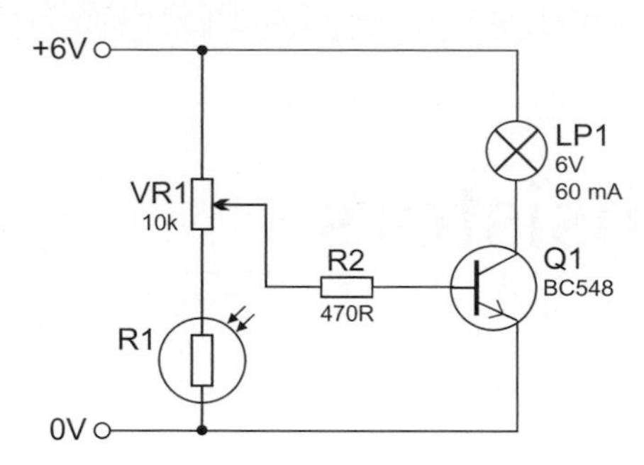

FIGURE 22.5

The LDR sensor consists of a voltage divider made up from VR1 and the LDR. A variable resistor is used so that the switching level can be set to different light levels. The transistor switch consists of R1 and Q1. R1 limits the amount of current being drawn from the potential divider. The collector current of Q1 flows through lamp LP1, and is about 60 mA.

With the LDR in normal room lighting, we adjust VR1 so that the light just goes out. When the LDR is shaded, its resistance increases. This increases the voltage across it.

An increase of voltage across the LDR raises the voltage at the wiper of VR1. More current flows to the base of Q1. More current flows through LP1 and into the collector of Q1. The lamp is switched on.

Things to do

Check that the switching circuit above works as described.

CIRCUITS AND SYSTEMS

The switching circuit is a three-stage system:

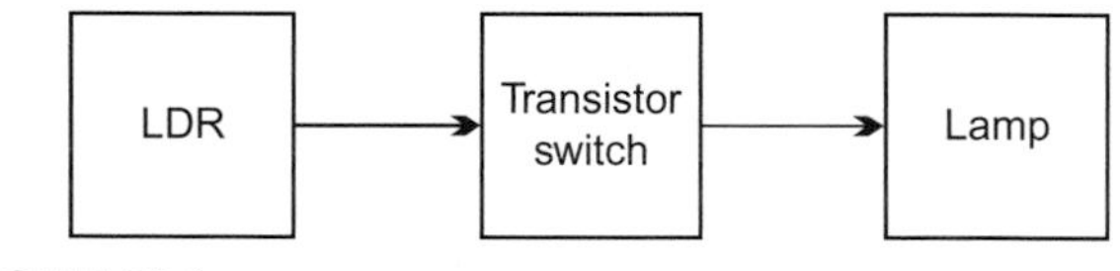

FIGURE 22.6

We will build several more systems that have this pattern.

DESIGN TIME

A voltage divider circuit is the basis of this simple electronic thermometer. If you want to measure temperatures around room temperature (25°C), R2 should have about the same resistance as the thermistor R1 has at that temperature.

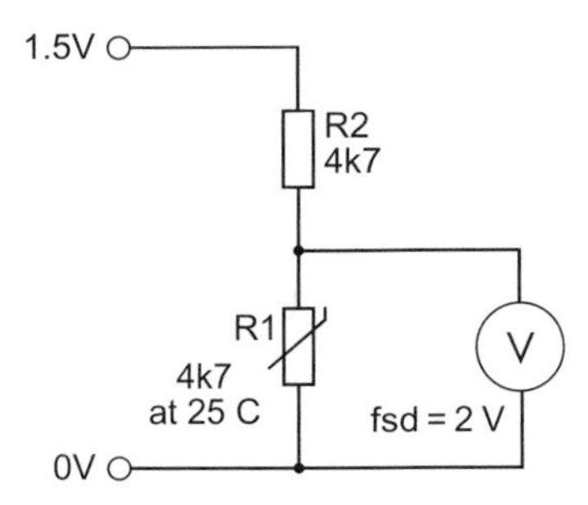

1 Stick an adhesive label on the outside of the meter to cover its scale. Use a removable label that can be peeled off later.

2 The first step is to **calibrate** the electronic thermometer. Put the thermistor and an ordinary room thermometer in various places that are at different temperatures.

3 Leave them for a few minutes in each place. Then draw a pencil line on the label, level with the needle of the meter. Mark the line with the temperature shown on the room thermometer.

4 Using the pencilled marks as a guide, draw a scale in ink to show temperatures in steps of, say, 5 degrees. Rub out the pencilled marks and figures.

5 Put the thermistor in several other places and read the temperature directly from your scale on the meter.

The range of the electronic thermometer depends on the value of R2. Design a circuit that includes a switch, so that there are two temperature ranges. Calibrate each range separately.

Add an indicator LED to this circuit, to show when the power is switched on.

It is a good idea to use a low supply voltage for these thermistor circuits. Why?

Design and build a **latching relay** circuit. The relay switches on a small siren when a microswitch is closed by an intruder. But the siren continues to sound when the switch is opened again. It sounds until a 'cancel' button is pressed.

Hints: The relay holds itself on. You need more than SPST contacts.

All the windows and doors of a house have a magnetic reed switch mounted on them. The switches are closed when the windows and doors are closed. Design a circuit that lights an LED only when *all* of the windows and doors are closed.

EVEN IF you are not following a Design and Technology course, these Design Time projects and problems will help you understand electronic theory.

IN THE LAB: BUILDING ON STRIPBOARD

Stripboard is a handy way of assembling an electronic circuit. It is particularly useful when you intend to build the circuit only once.

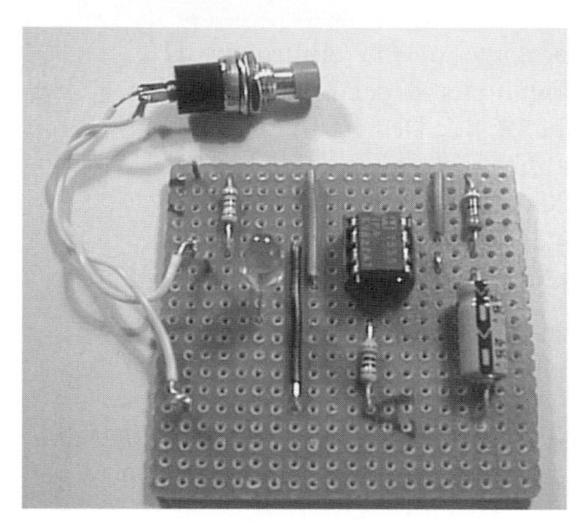

FIGURE 22.7

The components are mounted on the plain side of the board. The terminal wires and pins of the components are passed through the holes in the board.

The wires and pins are soldered to the copper strips on the rear side of the board.

FIGURE 22.8

In this way we build up a circuit with the copper strips as the connecting conductors. We may use wire links to connect one copper strip to another.

Sometimes a copper strip is cut across to separate it into two or more parts.

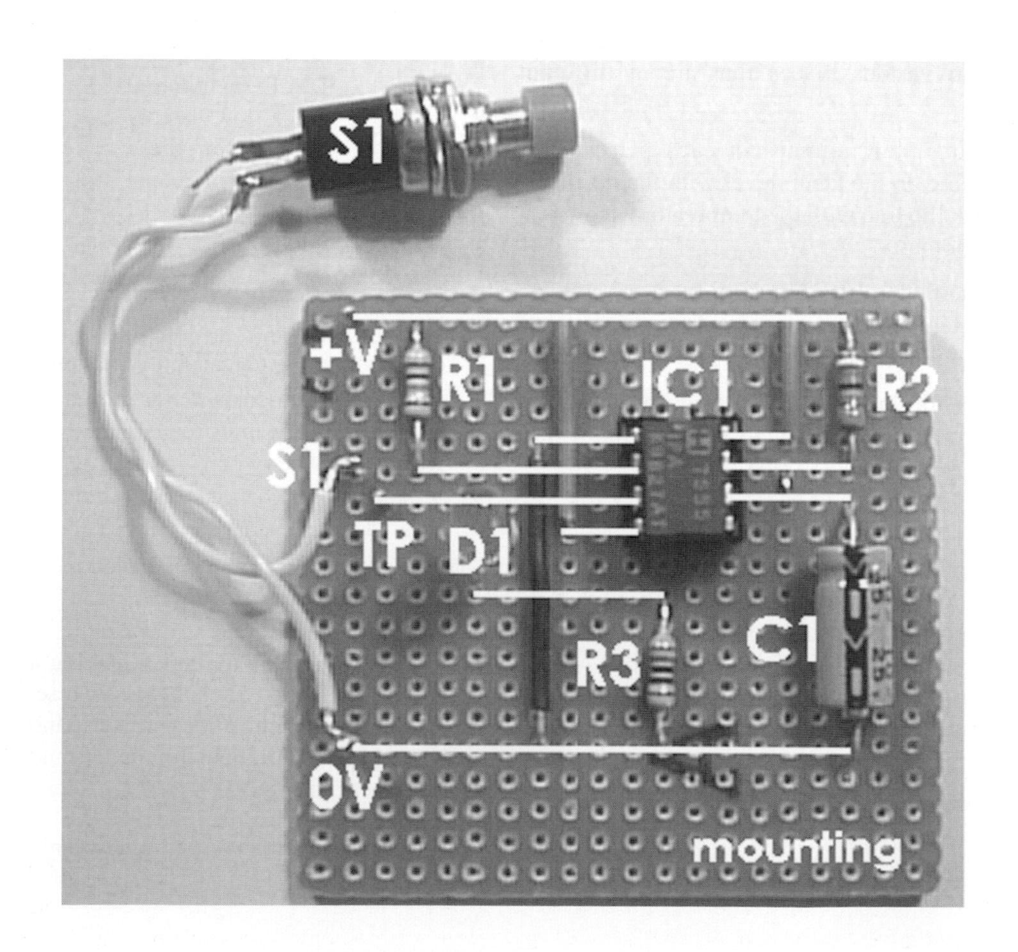

FIGURE 22.9

Stripboard has the important advantage that it is easy to make alterations to the circuit after it has been built. This makes stripboard very useful for building **prototype** circuits. These are the first versions of circuits that have been designed, but not tested in action. After testing, they may need modifying to improve their performance. For example, we may decide to change some of the component values, or to correct the logic. This is almost impossible to do when the circuit is built on a printed circuit board (PCB).

TOOLS

- **Junior hacksaw**: for cutting the stripboard to size.
- **Medium file**: for smoothing the cut edges of the board and (optionally) rounding the corners.
- **Magnifier**: preferably 8× or 10×; essential for examining soldering work at all stages.

FIGURE 22.10

- **Spot face cutter**: for cutting the strips where a break is needed.
- **Soldering iron**: Low-power (about 15 W), with a fine bit (not more than 2 mm diameter).
- **Wire stripper**: for removing insulation from the ends of connecting wires.
- **Wire cutter**: for cutting connecting wires and for trimming off the projecting ends after component and connecting wires have been soldered into the board.
- **Heat shunt**: to absorb heat and prevent it reaching delicate components such as transistors.

MATERIALS

- **Stripboard**: is sold in standard sizes. It is often more convenient (and cheaper) to buy a large board and cut it into pieces of suitable shape and size.
- **Solder**: Use cored lead-free solder, preferably 22 swg (0.7 mm diameter).
- **Wire**: For on-board connections: single-stranded (1/0.6) with PVC insulation. It is helpful to have several different colours, such as red, blue, black and yellow. For off-board connections to the control panel, other boards, and power supply: multi-stranded wire, with PVC insulation in different colours. For normal use, 10-stranded wire is best (10/0.12). For special purposes you may need: heavy-duty wire, sheathed wire, computer cable, telephone cable, ribbon-cable or mains cable.

LAYING OUT THE BOARD

The first stage is a schematic or circuit diagram of the circuit you intend to build. You may design this yourself or use a design from a book or magazine. The example we are following in this Topic is the 7555 monostable circuit. You may decide to check it first by building it on a breadboard, as in the photo. This gives you a chance to check component values before building a permanent version of the circuit.

Plan the layout of the stripboard on squared paper.

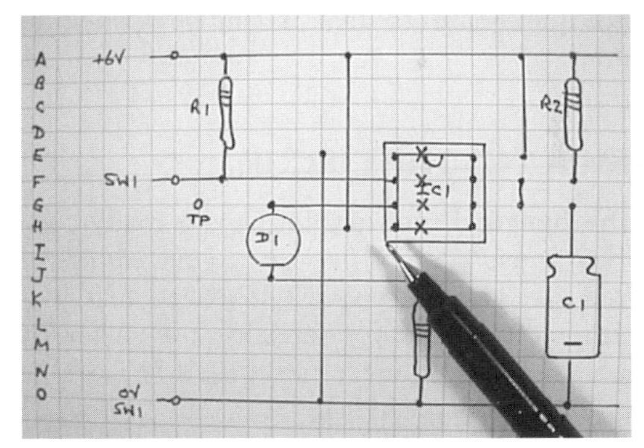

FIGURE 22.11

The layout shows:

- **Copper strips**: These are the horizontal lines of the squared paper. The corners of the squares are where the holes are located. To make the layout clearer and to avoid mistakes in the layout, we have inked over the supply line, the input line from S1, and the line from D1 to R3, a connection which might not be clear otherwise.
- **Components**: Allow for the size of the components that you will be using. If in doubt, place the component on a spare scrap of board and count the number

of holes required. Normally, a 0.25 W resistor lying flat on the board needs a minimum of four strips *between* the strips its wires are soldered to. Resistors may also be stood on end and can then be soldered to adjacent strips.

- **Terminal pins**: Labelled to indicate whether they are power connections or connections to off-board components. In some designs you may use an edge-connector or some other type of connection.
- **Test points**: It makes testing simpler if certain points in the circuit have a terminal pin to which test equipment can be connected. An example is the output of the timer IC at pin 3. The test point is marked 'TP' in the diagram.
- **Wire links**: As far as possible, run these at right angles to the strips.
- **Cut strips**: These are each marked with a bold 'X', preferably with a pen of different colour. In this circuit, we have had to cut the four strips that run beneath the IC. Apart from these, we have been able to arrange the layout so that no other strips need to be cut.

When the diagram is complete, you will know the minimum size of board required. You may need to increase the size of the board slightly to allow space for drilling holes for bolts or plastic stand-offs. Alternatively, you may leave a blank area along one edge of the board so that it can be slotted into a support.

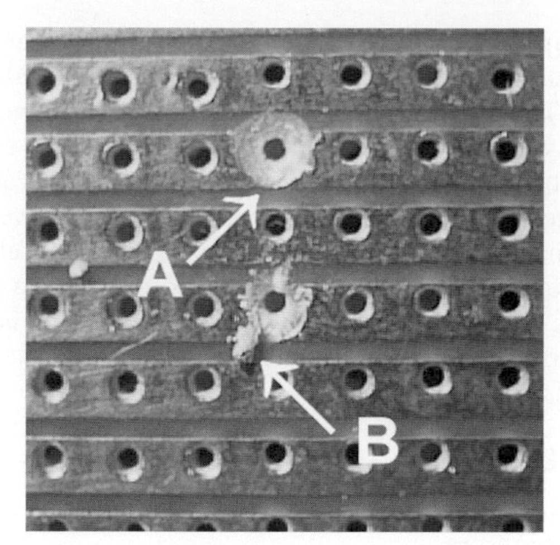

FIGURE 22.12

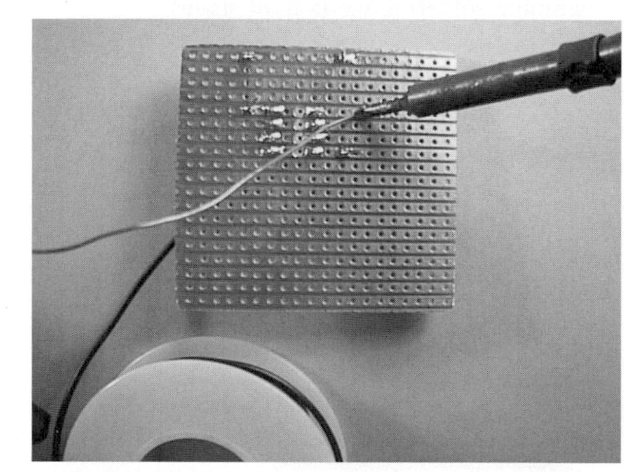
FIGURE 22.13

ASSEMBLING THE CIRCUIT

Cut the board: If you do not have a ready-cut standard board, use a junior hacksaw to cut a board to the required size. Before you cut it, check that it really will fit inside the box you intend to use. After cutting the board, smooth its edges with the file. Use the file to round the corners slightly.

Cut the strips: Use a spot face cutter for this. When you have finished, use a magnifier to examine the cuts. Look for incomplete cuts, where a 'wire' of copper still connects the strip across the intended cut (A, Figure 22.12). Also look for and remove any flakes of copper that might bridge the gap between adjacent strips (B, Figure 22.12).

Solder wire links: See p. 83 for soldering procedures. It is usually best to start by soldering one stripped end of the wire into its hole (see Figure 22.13).

Then cut the wire to length, strip its other end, bend it, and insert it in the other hole, pulling it tight. Solder the other end into its hole. Trim both ends.

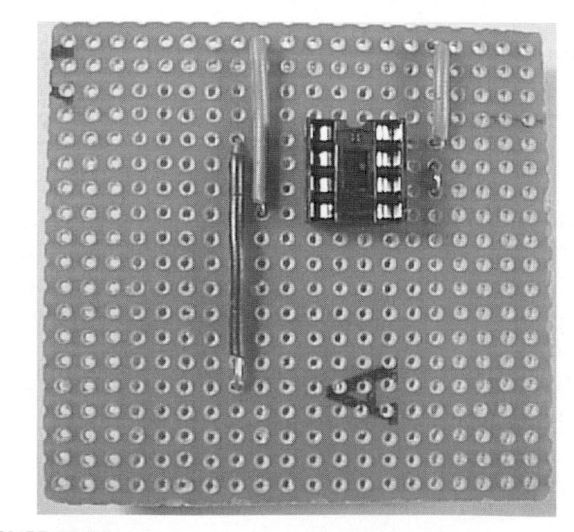
FIGURE 22.14

If there are IC sockets in the layout they can be soldered in place now. They act as 'landmarks' and help you to avoid soldering components in the wrong places (Figure 22.15).

When soldering IC sockets, begin by soldering two diagonally opposite pins. Then check that the socket is lying flat on the board. If not, remelt the solder on one or both pins and push the socket flat. Solder the remaining pins.

Solder resistors: Check values carefully and that their wires go to the correct holes.

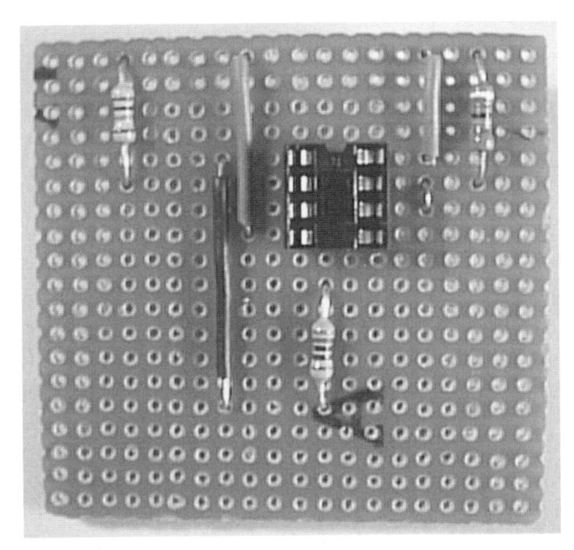

FIGURE 22.15

Solder other small components: These include terminal pins, small capacitors, preset pots, diodes, transistors, and sockets (if you have not assembled these earlier). Check that polarised components such as diodes and tantalum capacitors are inserted the right way round.

Solder larger components: These include electrolytic capacitors, which too must be the right way round. Solder leads to off-board components.

Last thing: Remove ICs from their packaging and insert them in their correct sockets (right way round).

VISUAL CHECK

Using a magnifier, look *again* for badly cut strips, badly made solder joints, and blobs or fine threads of solder causing short-circuits between adjacent strips. Check *again* that the components and links are in the correct holes. Check *again* that diodes, polarised capacitors, transistors and ICs are the right way round.

Refer to page 120 for testing procedures.

MODULAR APPROACH

The order of assembly outlined above is best for most projects, especially small ones. However, there are some projects where a different order is preferable. If a circuit has several stages it may be better to build it stage by stage. When each stage is built, test it before adding the next stage on to it.

Another way, which is a combination of both techniques, is to assemble the whole circuit as described above but to leave out the connecting links between each stage. Then test each stage separately before adding the links to join it to the other stages.

SOLDERING

Cleaning: Make sure that all surfaces are free from grease. Even handling the board with bare fingers can make it too greasy. If the copper strips or the wires are at all tarnished, rub them with fine emery paper until they gleam.

Heating: Let the soldering iron warm up before you start work. Touch the iron to the end of the solder coil. A *small* amount of the lead-free solder melts and coats the tip thinly. For best soldering, the combination is 'hot iron plus quick action'. Pass the wire link or component lead through the hole in the copper strip. Touch the tip of the iron against both the wire and the strip at the same time. After a few moments, feed a little solder into the angle between the strip and the wire (Note: on to the wire and strip, NOT on to the iron). With practice, you will learn to feed in enough but not too much. The solder must flow into the angle and wet all surfaces freely.

Finishing: Remove the iron. Do not move the board or wire until the solder has set (about 3 seconds, if you have not over-heated the joint). Use pliers or a wire cutter to cut off the excess wire protruding from the joint.

FIGURE 22.16

Inspecting: Use a hand lens to examine the joint. If there is not enough solder, or if the joint is 'dry' (Figure 22.16, arrowed), repeat the soldering. A 'dry' joint is often the result of not having the iron hot enough, or trying to solder on to a greasy surface.

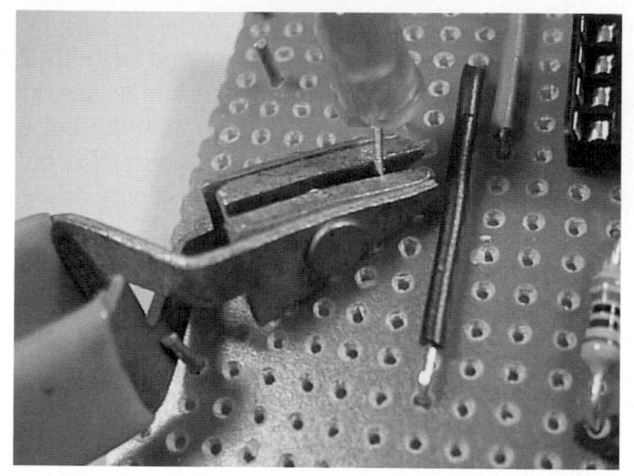

FIGURE 22.17

Heat shunt: When soldering heat-sensitive components such as diodes and transistors, clip a heat shunt to the wire leads of the device before soldering (above). The heat from soldering is diverted into the heat shunt instead of travelling up the wires to the device itself. Even with a heat shunt, still try to solder the joint as quickly as you can to avoid over-heating.

WARNING! A hot soldering iron does not look hot, but is hot enough to burn your skin, or your clothes, or the surface of the workbench. It is also hot enough to melt the plastic insulation on wires. Handle the iron with great care!

DESOLDERING

If a component or wire is soldered in the wrong place, it is usually easy to remove it. Simply re-heat the joint with the iron, and then pull the wire out. A pair of forceps (tweezers) helps to avoid burnt fingers.

This operation may leave too much solder on the strips. Use **solder braid** to remove excess solder. This consists of fine copper wire braided into a ribbon. Touch the end of the braid on to the soldered area. Melt a little solder on the tip of a *hot* iron to give good thermal contact.

FIGURE 22.18

Touch the iron against the braid. The solder beneath the braid melts. The braid acts as a wick and soaks up the molten solder. Remove the braid and, when the solder in it has solidified, clip off the used end of the braid.

SURFACE MOUNT DEVICES

Advances in electronic technology and the trend toward portability mean that designers are cramming more and more components into smaller and smaller space. A mobile phone is a good example of this. But smaller components have advantages in larger equipment too, such as desk-top computers.

One way of miniaturising circuits is to use integrated circuits (p. 143). Another way is to build the circuit with **surface mount devices**.

The smallest SMDs are the resistors, measuring only 1.6 mm long by 0.8 mm wide. They have metal contacts at their ends for soldering directly to the circuit board. Other devices such as transistors are slightly larger. Shown below are two SMT transistors. The labelled one is upside down, to show the 'legs' by which it is soldered to the circuit board.

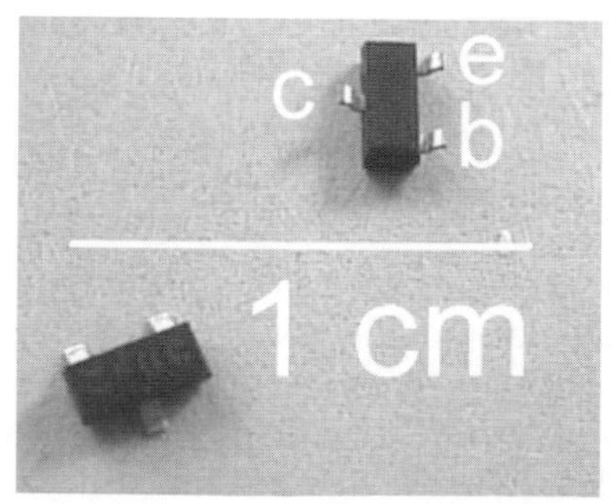

FIGURE 22.19

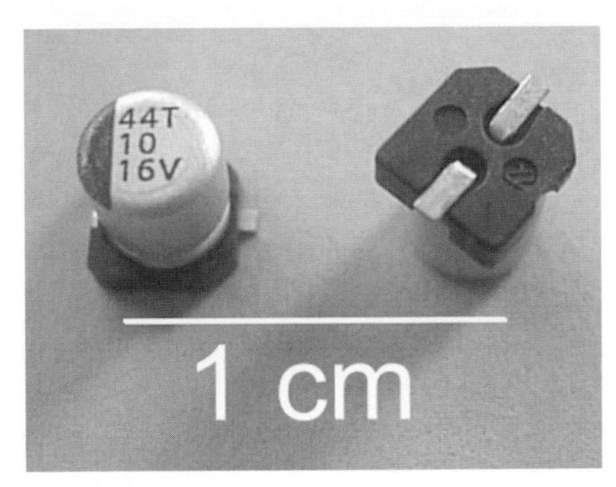

FIGURE 22.20

Above are two 10 μF electrolytic capacitors. Many integrated circuits are available as SMDs (see photo of SMD microcontroller.

SMDs have several advantages:

- small, and there can be tracks and components on *both* sides of the board (both features save board space).
- easily assembled on the board, no holes to be drilled for leads (both features reduce assembly costs).
- shorter tracks reduce the inter-track capacitance, and the lack of wire leads reduces self-inductance (both features improve performance in radio-frequency and high-speed digital circuits).
- shorter tracks reduce propagation delays (improves performance of digital circuits).

Transistor Action

PASSIVE AND ACTIVE DEVICES

Electronic components are of two kinds — **passive** and **active**. Passive devices are not able to generate an increase of power. Examples include resistors, capacitors and inductors. Resistors are able to convert electrical power to heat. Inductors are able to convert electrical energy into magnetic force. But neither of these devices is able to increase the power in the circuit. They are passive. In contrast, a transistor has a low-power input (small current) and converts this to a high-power output (large current). It is an active component. The energy for this activity comes from the electrical supply to the circuit.

CURRENT AND VOLTAGE CHANGES

The investigation below looks in more detail at the changes of current and voltage in a transistor switch circuit.

Things to do

This circuit has two meters, a microammeter to measure the base current and a milliammeter to measure the collector current. You will also need a digital voltmeter (fsd = 10 V) , but this is not to be connected into the circuit.

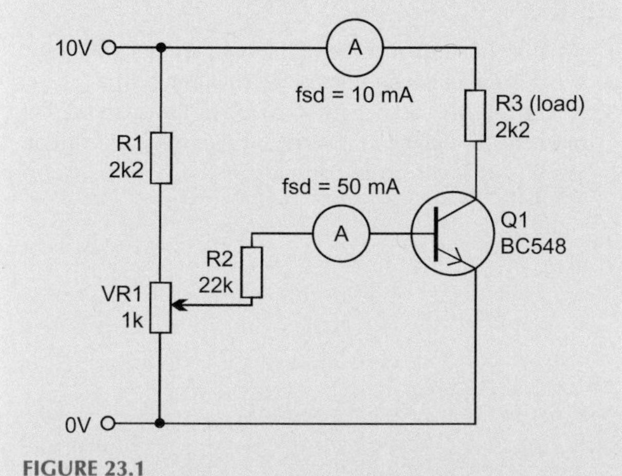

FIGURE 23.1

The diagram specifies a BC548 BJT, but you can try it with other types, such as BC337, 2N2222A, or 2N3904.

1. Adjust VR1 so that there is zero base current (i_b). Read the collector current (i_c). Record your results in a table.
2. Without altering the setting of VR1, use the voltmeter to measure v_{be}, the voltage difference between the base and emitter. Also measure v_{ce}, the voltage difference between collector and emitter.
3. Repeat steps (1) and (2) with i_b equal to 5 μA to 50 μA (if possible) in steps of 5 μA.
4. Plot a graph of collector current against base current. What does this tell you about the relationship between the currents?
5. Plot a graph of base voltage against base current. What does this show as base current increases?
6. Plot a graph of collector voltage against base current. What does this tell us about the voltage across the load as collector current increases?

TRANSISTOR ACTION

The results you obtain from the investigation may vary slightly depending on the type of transistor tested. With a typical transistor, the graph of collector current against base current looks like this.

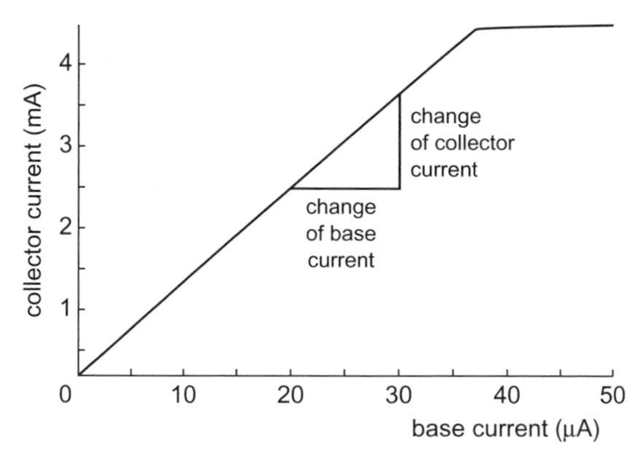

FIGURE 23.2

There is a straight-line (linear) relationship between the base and collector currents. In other words:

Collector current is directly proportional to base current.

We can find out something else of interest from this curve. We mark off a section of the curve and

Electronics: A First Course.

measure the *change* in base current. The section marked starts at 20 μA and runs to 30 μA. This is a change of 10 μA. Over the same section of the curve, the collector current changes from 2.5 mA to about 3.5 mA, a change of 1 mA (= 1000 μA).

Putting this all into microamps, we can say that a change of 10 μA in base current results in a change of collector current of 1000 μA. The change of collector current is 100 times bigger than the change of base current. Putting this in other words we can say that:

The current gain of the transistor is 100.

This current gain is usually called the **small signal current gain**, and it has the symbol h_{fe}.

The graph of base-emitter voltage against base current is like this:

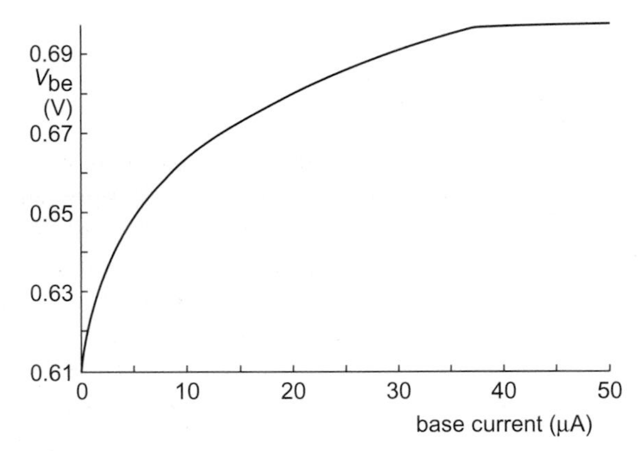

FIGURE 23.3

Noting the scale on the left, we see that the voltage between the base and emitter starts off a little less than 0.7 V and finishes up at 0.7 V. As an approximation we may say that:

The base-emitter voltage is close to 0.7 V.

In fact, it is equal to one diode drop, because there is the equivalent of a forward-based diode between the base and emitter terminals.

The collector-emitter voltage changes in this way:

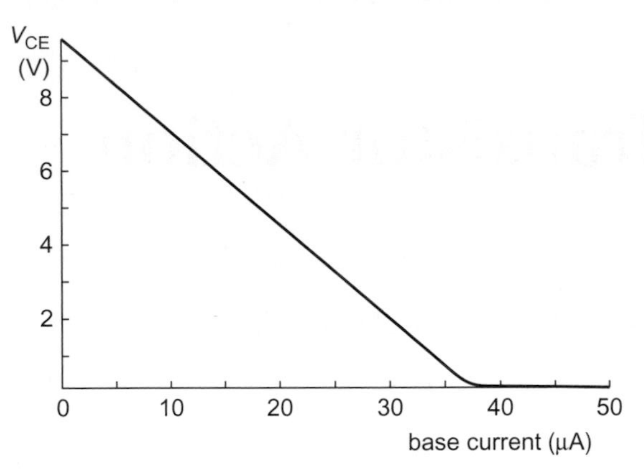

FIGURE 23.4

As base current increases, so does collector current. As the current through R3 increases, the voltage across it increases too (Ohm's Law). The voltage at one end of R3 is fixed at 10 V, the supply. The voltage at its other end (the collector end) must fall. The graph shows the falling voltage at the collector. It falls steadily until it is only slightly greater than zero. At this point it can not fall any further because the collector is only just positive of the emitter and the transistor would not work if it went down further. We say that the transistor is **saturated**. Or we can say that it has **bottomed out**.

In this circuit, the transistor saturates when the base current is about 37 μA. From the graph of collector current against base current, we can see that, when the base current reaches this value, the collector current levels off. It no longer increases in proportion to the base current.

Self Test

1 What is the normal value of the base-emitter voltage?
2 What is the collector voltage of a saturated BJT?
3 In a given BJT, a change of 20 μA in the base current produces a change of 2.4 mA in its collector current. What is its small signal current gain?

Transistor Switches

There is a basic transistor switch on p. 78. Now look more closely at these switching circuits and how to design them.

A transistor switch makes use of the BJT's most important property — **gain**. There is more than one way of defining gain, but here we mean the small signal current gain, h_{fe}, described on p. 88. Gain has no units. It is just a number, because it is a current divided by a current. The gain of a typical BJT is 100.

This circuit is an interesting but simple way of demonstrating transistor gain:

Things to do

The parallel lines A and B in the diagram represent two metal contacts spaced with about 1 mm between them. They could be two drawing pins pushed into a small scrap of wood, cork or plastic. Or they could be a pair of stripped copper wires, about 3 cm long, pushed into the sockets of a breadboard, and lying flat on the board. Select the sockets so that the wires have NO contact with each other, either electrical or physical.

Currents are small in this circuit, so there is no need for a resistor in series with the LED, or for one in the connection to the base of Q1.

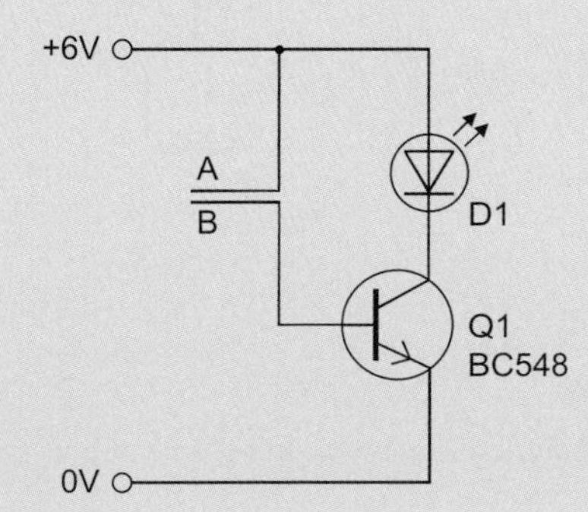

FIGURE 24.1

1. Check that there is no connection between A and B.
2. Connect the power supply. The LED should not light.
3. Press your finger on A and B to bridge the gap between them (but NOT to force them into contact with each other!)
4. The LED should light. You may not see it easily if you are close to a window. Try moistening your finger-tip before touching the contacts.
5. Estimate how much current is passing through the surface of your finger-tip.

POWER TRANSISTORS

Low power BJTs such as the BC548 are suitable for switching on LEDs and small filament lamps. They are rated to carry a collector current of up to 100 mA. Many other devices such as DC motors and bright lamps take more current than this. To switch these, we need medium-power or high-power transistors.

The high power BJT shown below is capable of passing up to 10 A. Like any other transistor, it has three terminals.

One of the problems with high power is that part of the power used appears as heat. With a current of several amps, the heat may be so great that the BJT is damaged.

To avoid this, we bolt a **heat sink** to the tag at the top of the transistor. The heat sink carries the heat away to the surroundings.

A heat sink is made of metal (usually aluminium) to conduct heat away. Most of them have fins, to allow convection currents in the air to carry the heat away. Also, they are painted matt black so as to radiate heat efficiently. A special heat-conducting paste is applied to the surface in contact with the tag.

FIGURE 24.2

However, a transistor usually does not need a heat sink

if it is switched fully off (no current) or fully on (saturated). When it is saturated, its resistance to the collector current is very low. A low resistance converts little power to heat.

Memo

$P = I^2R$. Because R is small, P is small.

DESIGNING A SWITCH

We need a circuit that switches on an LED when the light level falls. It could be the first stage in a security system for detecting intruders. Its system diagram has the typical three stages:

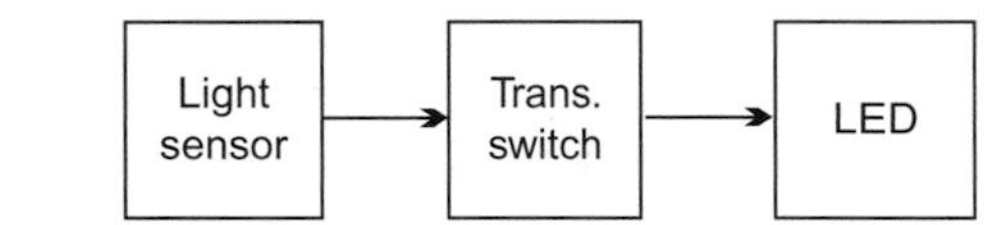

FIGURE 24.3

The light sensor can be a voltage divider, made from an LDR and a resistor. The output current from the sensor goes to a transistor switch, consisting of a transistor with a base resistor. This switches an LED with its series resistor. The complete circuit is:

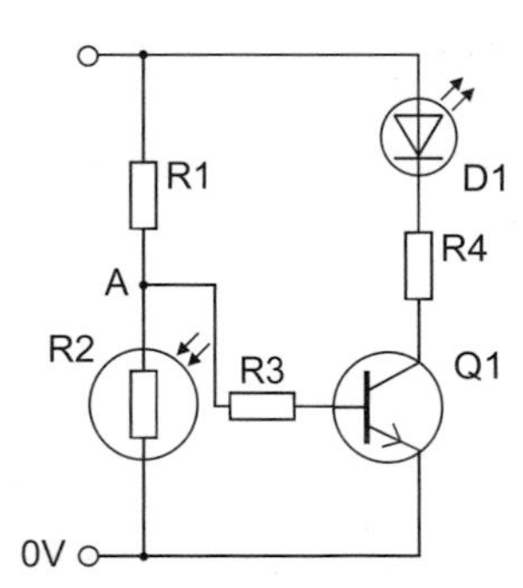

FIGURE 24.4

R1 and the LDR (R2) are arranged so that the voltage at point A rises as the LDR receives less light.

The power supply will be a **6 V** plug-in PSU because the circuit will be run night and day. The sensor will be the popular and easily available **ORP12**.

A typical LED takes **20 mA** when lit. A low-power transistor such as a **BC548** can switch up to 100 mA, so we will settle on this type for Q1. When Q1 is saturated, there will be almost 6 V across D1 and R4. The forward voltage drop is about 2 V, so we need to drop 4 V across R4. Ohm's Law tells us that R4 must be 4 V divided by 20 mA, which gives **200 Ω**.

Using a multimeter, we may find that the resistance of R2 in low light (the LED-ON level) is 1.3 kΩ. To switch on Q1, we need a voltage at point A that is over 1 V. A few calculations with some of the E24 values show that if R1 is **3.9 kΩ**, then the voltage at A is **1.5 V**. This gives us a margin to allow for resistor tolerance.

When Q1 is on, the voltage at A is 1.5 V (as just calculated) and that at the base is 0.7 V (p. 88). The voltage drop across R3 is 0.8 V. If the gain of Q1 is 100 (p. 88), the base current must be (20 mA)/100, or **200 μA**. So the resistance of R3 must be 0.8 V divided by 200 μA, which gives 4 kΩ. The nearest value is **3.9 kΩ**.

Here is the final design, including the critical voltages and currents:

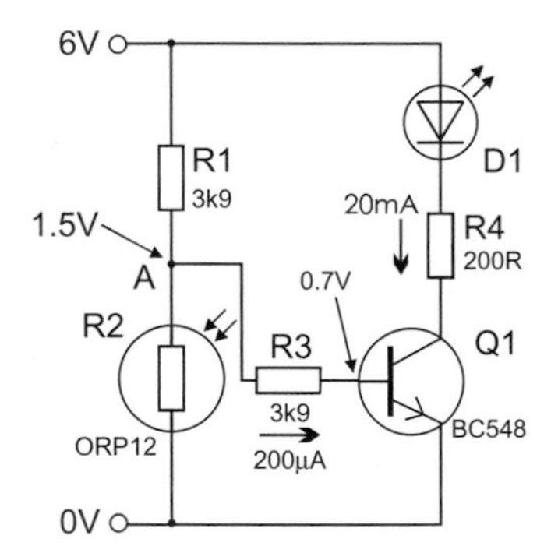

FIGURE 24.5

Things to do

Check the calculations by building and testing the circuit.

Thyristors

Thyristors belong to the group of components known as **silicon controlled switches** or **silicon controlled rectifiers** (shortened to **SCS** or **SCR**).

Their name suggests that their main use is for turning a current on and off. The name 'rectifier' indicates that, as in a diode, the current can flow only in one direction.

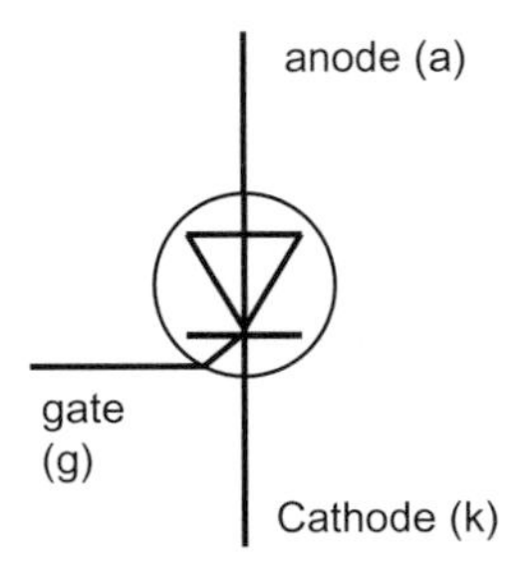

FIGURE 25.1

The symbol for a thyristor indicates its diode-like property, current flowing from the **anode** to the **cathode**. But a thyristor has a third terminal, the **gate**. The gate controls the flow of current through the thyristor.

Most often a thyristor is used for switching a large current. A typical example is the C106D thyristor (right), which can switch currents of up to 4 A. This is why the thyristor has a tag for bolting to a heat sink.

Although the anode to cathode current can be high, only a small current of 15 μA is required to trigger the thyristor into conduction.

FIGURE 25.2

Things to do

Build this circuit on a breadboard and use it to discover how a thyristor operates. The switches are not the same type.

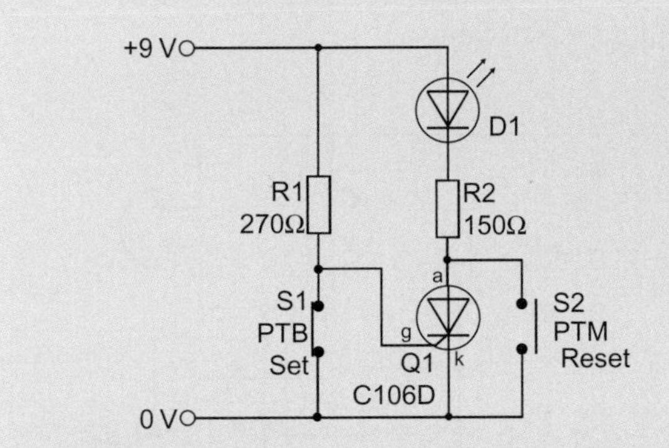

FIGURE 25.3

Note what happens to the LED (D1) at each stage:

1 Switch on the power.
2 Press and hold S1.
3 Release S1.
4 Press and hold S2.
5 Release S2.
6 Press S1 for an instant.
7 Pull one wire of R2 from its socket.
8 Replace the wire of R2 in its socket.
9 Press R1 for an instant.
10 Switch off the power.
11 Switch on the power.
12 Press S1 and S2 alternately.

What do you need to do to turn D1 on? State three ways of turning D1 off.

The circuit above demonstrates that the thyristor does not conduct until it has been triggered by a positive pulse applied to its gate. From then on, it conducts indefinitely. It ceases to conduct when the anode and cathode are connected, when the flow of current through the thyristor is interrupted, or when the power supply is turned off. It can also be shown that conduction ceases if the flow of current is reduced below a value known as the **holding current**.

BISTABLE CIRCUIT

As well as demonstrating the facts about a thyristor, the circuit opposite is an example of a type of circuit known as a **set-reset bistable circuit**. An S-R bistable is a circuit that can exist indefinitely in one of two states, **set** and **reset**. At any time it can be changed from one state to the other by sending a short pulse to one of its terminals. The pulses could come from a logic circuit or computer but, in the demonstration, they are produced by pressing S1 (set, high pulse to the gate) or S2 (reset, low pulse to the anode).

SWITCHING DC

The circuit below may be adapted to switch on a siren and an alarm lamp.

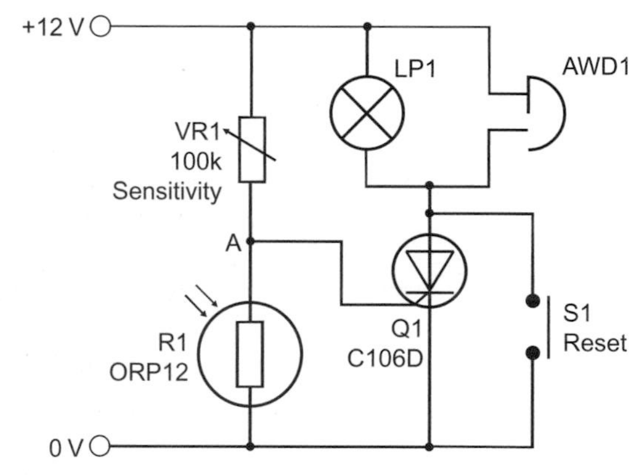

FIGURE 25.4

The relatively large current through the lamp and the electric bell (or other audible warning device) is switched on by the very small current from the VR1/R1 voltage divider.

The LDR (R1) has a beam of light shining on it. This means that the resistance of R1 is very low and the voltage at point A is too low to trigger the thyristor. When an intruder breaks the beam, even for an instant, the resistance of R1 increases. The voltage at A rises sharply, and the thyristor is triggered. Current flows through the thyristor the lamp and the bell.

Once the beam has been broken, restoring the beam has no effect. The alarm continues to sound and the lamp remains on. The only way to turn them off is to press the reset button (S1) or disconnect the power supply.

SWITCHING AC

Most thyristor applications are concerned with AC circuits. The circuit shown in Figure 25.5 is a lamp-dimming circuit which controls the power supplied to the filament lamp LP1.

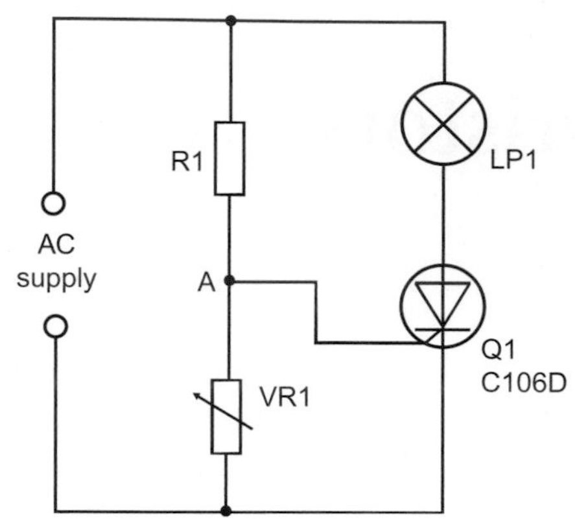

FIGURE 25.5

R1 and VR1 make up a voltage divider. As the input voltage alternates, the voltage at A rises a few volts above zero and falls a few volts below. It rises and falls by only a few volts, because R1 is much bigger than VR1. We can adjust VR1 to vary the size of the rise and fall.

Rises and falls are greatest if VR1 is set to its maximum value. The values of R1 and VR1 are such that, very early in the positive half-cycle, the voltage at A becomes large enough to trigger the thyristor. The lamp is turned on very early in the half-cycle. Because of the thyristor action, the thyristor stays on for the remainder of the half-cycle. So the lamp is on for almost the whole of the half-cycle. Maximum power is delivered to the lamp and it glows brightly.

Remember that this is a half-wave rectifier so, as explained on p. 69, half the available power is wasted during the negative half cycle.

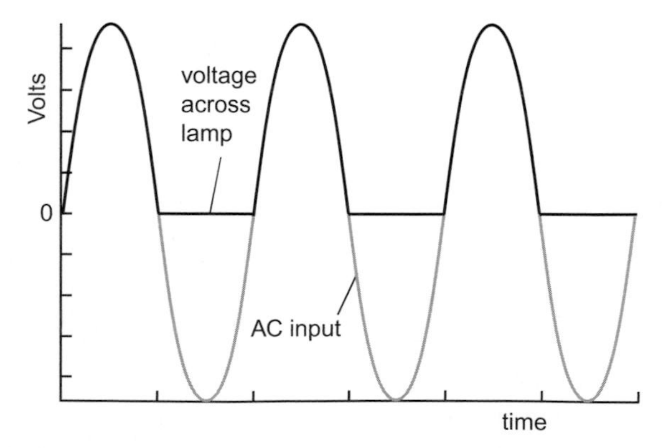

FIGURE 25.6

The graphs above show the changes of voltage at the input (grey) and across the lamp (black). The graphs are the same as those on p. 69, except that they do not show the effect of forward voltage drop across the diode. There is still a 2 V drop across the thyristor but this does not show in the graphs above because the input voltage is several hundred volts.

If VR1 is set to a lower value, the rises and falls of voltage at A are smaller. This means that it takes longer for the voltage to rise to the level that switches on the thyristor.

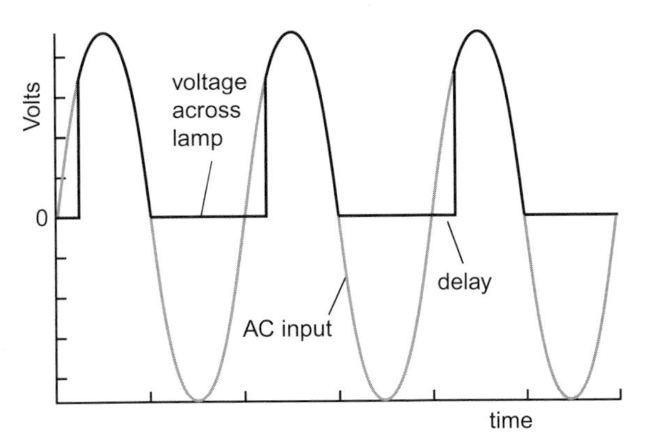

FIGURE 25.7

The effect of this is that there is a delay between the beginning of the positive half-cycle and the triggering of the thyristor. This is seen in the graphs above.

Switching on the lamp is delayed by a fraction of a second. Once the lamp is on, it stays on for the remainder of the positive half cycle. Power is delivered to the lamp for a shorter period in each cycle and therefore the lamp is partly dimmed.

FULL-WAVE CONTROLLED RECTIFIER

The half-wave rectifier wastes 50% of the power. A full-wave rectifier can be made by connecting two thyristors in parallel, so that one conducts during the positive half-cycle and the other conducts on the negative half cycle. Each thyristor has a voltage divider to trigger it.

Another full-wave rectifier is shown below.

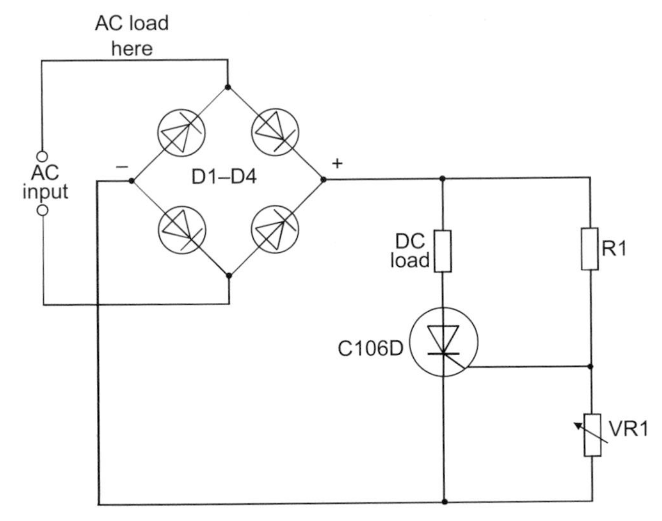

FIGURE 25.8

The AC supply is rectified by a full-wave diode rectifier, D1–D4. This gives pulsed DC in which *all* half-cycles are positive. The diode is switched on during both half-cycles so the DC load (represented by a resistor) can receive up to full power.

Alternatively, the load can be placed on the input side of the rectifier, where marked in the diagram. Here the current is alternating. The current still has to flow through the thyristor to complete its path through the circuit. It is turned on and off by the thyristor during both positive and negative half-cycles.

This circuit is used in mains-powered circuits for controlling brightness of lamps and the speed of DC or AC motors.

Operating Voltage

You could study these rectifier circuits by connecting them on a breadboard. But, **use a low-voltage AC supply such as 12 V AC.**

NEVER power your circuits from the mains.

Field Effect Transistors

These are known as **FETs**, for short. They are active devices (p. 87) and there are several kinds of them, including junction FETs (JFETs) that were widely used in the past. Nowadays an enormous range of n-channel metal oxide silicon FETs (n-channel MOSFETs) is available and, with BJTs, are now the most commonly used types. We deal only with MOSFETs in this book and will simply call them FETs.

There are low-power, medium-power and high-power FETs, enclosed in the same kinds of package as BJTs. The photo shows a typical high-power FET. The case is the that of the high-power BJT seen on p. 89, but this photo shows the other side of the transistor.

FIGURE 26.1

We can see that the metal tag extends down the body of the transistor. This gives a large area for contact with a heat sink. It also means that the tag is in close contact with the actual transistor in the body of the device.

The photo also shows that the FET has three terminals. These are named **source, drain** and **gate**. They roughly correspond with the emitter, collector and base of a BJT, but there are important differences. The most important practical difference is that virtually no current flows into the gate of an FET.

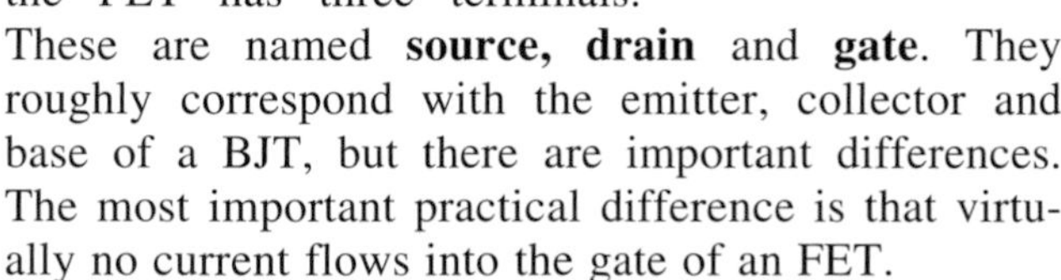

In normal use, an FET is connected in a similar way to a BJT. The source is the most negative terminal and the drain is the most positive. When a positive voltage is applied to the gate, a current, the **drain current**, flows in at the drain and out at the source. The next discussion looks more closely at this.

TRANSISTOR ACTION

An FET can be used as a transistor switch, as in this circuit:

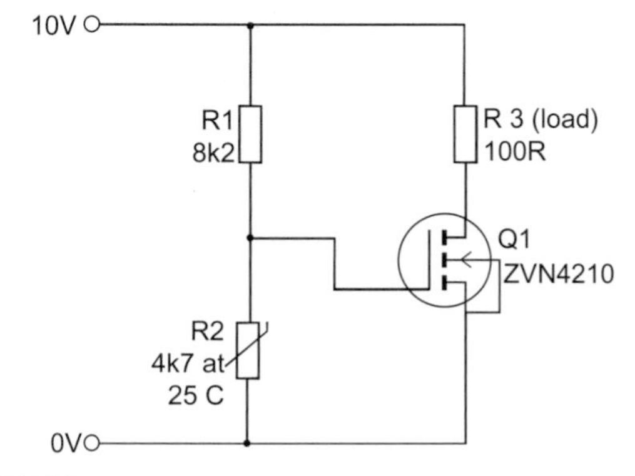

FIGURE 26.2

The transistor switches on the load when the temperature falls below a given level. The transistor is switched on by the *voltage* output of the voltage divider. As temperature falls, the resistance of R2 increases. When the voltage across R2 exceeds the **threshold voltage** of the FET, the transistor begins to turn on. The threshold voltage varies with different types of FET, but is often between 2 V and 4 V.

Once the threshold voltage is exceeded, further increase of voltage may rapidly saturate the transistor. Here is a graph of the drain current (I_D) through a 100 Ω load against the gate-source voltage (V_{GS}):

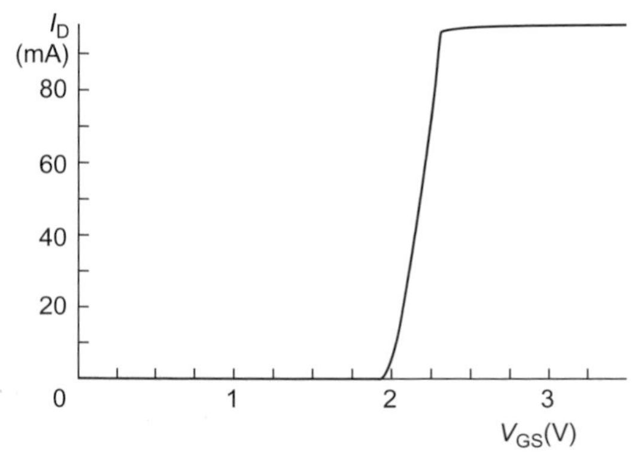

FIGURE 26.3

As V_{CS} increases above the threshold (2 V), the current increases rapidly. The FET saturates at about 2.3 V.

The FET saturates because the voltage across the load resistor brings the drain terminal almost down to zero. With a smaller load, the voltage drop is smaller and we can see the shape of the voltage-current curve more easily:

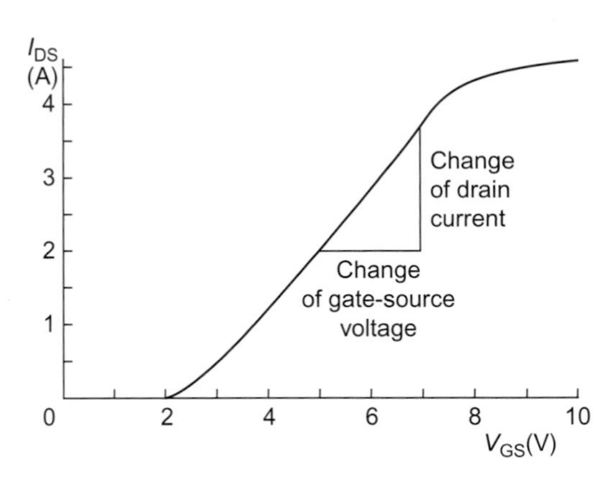

FIGURE 26.4

Compare this graph with the graph for base current and collector current of a BJT (p. 87). The line is reasonably straight in the middle, so that current is proportional to voltage. However, it is not straight at the ends, which means that current is only approximately proportional to voltage if the signal voltage varies over a wide range.

What might be called the 'gain' of the FET is the change of drain *current* for a given change in gate-source *voltage*. In the central region of the graph above, the current increases by about 1.7 A when the voltage increases by 2 V. We can say that the 'gain' is:

$$1.7/2 = 0.85 \text{ amps/volts}$$

We have already said (p. 41) that volts/amps is called resistance. The inverse, amps/volts, is known as **conductance**. In the case of an FET, where the voltage is the *input* and the current is the *output*, we call this the **transconductance**, g_m, of the FET:

$$g_m = \frac{\text{change of drain current}}{\text{change of gate} - \text{source voltage}}$$

The unit of conductance and transconductance is the **siemens**, symbol **S**. The FET illustrated by the graph above has a transconductance of 0.85 S, or 850 millisiemens.

Self Test

1. The collector current of a BJT increases by 35 mA when its base current is increased by 140 μA. What is its gain?
2. The drain current of an FET increases by 2 A when its gate-source voltage is increased by 1.6 V. What is its transconductance?

BJTs AND FETs

BJTs and FETs have similar actions, but distinctive properties:

- **Conversions**: BJTs convert current to current. FETs convert voltage to current.
- **Input current**: A BJT needs an input current. An FET requires no input current.
- **Input/output**: The input/output relationship of a BJT is linear (a straight-line graph), but that of an FET is not linear for large signals. This may lead to distortion of large signals by an FET.
- **Speed**: FETs switch faster than BJTs, though both types are fast enough for most applications.
- **Input voltage**: An FET switches on when the gate-source voltage exceeds the threshold voltage. The gate voltage can be any voltage between the threshold and the supply voltage when the FET is on. The base-emitter voltage of a BJT remains close to 0.7 V when it is on, whatever the size of the input current.
- **Input resistor**: An FET does not require a resistor in its gate circuit. This simplifies circuit-building.
- **Output resistance**: Most FETs have very low resistance when switched on, usually less than 1 Ω. This makes them good transistor switches.

Things to do

Design, build on a breadboard, and test a circuit that uses an FET to switch an LED on when light is high, and off when light is low.

SWITCHING A RELAY

Sometimes we want to use a transistor switch to control a high-power device. It may require larger current or higher voltage than a power transistor can handle. Or we may want to switch an alternating current, which is something that transistor switches can not do. This is when we use a relay (p. 108).

In this example, the system has four stages. The sensor is a thermistor. This controls an FET switch, which switches a relay coil. The relay contacts control a mains-powered heater.

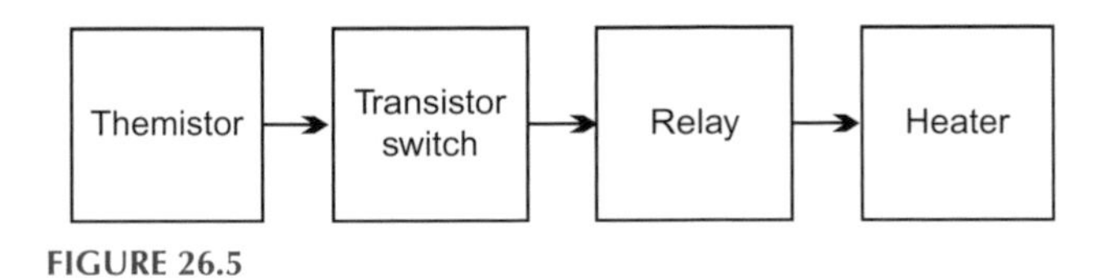

FIGURE 26.5

As in many switching circuits, the first stage is a voltage divider. There is a variable resistor to set the switch-on temperature. There is no resistor between the divider and the gate of the FET.

When the temperature falls, the resistance of R2 increases. The voltage at the wiper of VR1 increases. Q1 is switched on when this voltage exceeds the threshold voltage. Current flows through Q1 and the relay coil (RL). D1 is a protective diode.

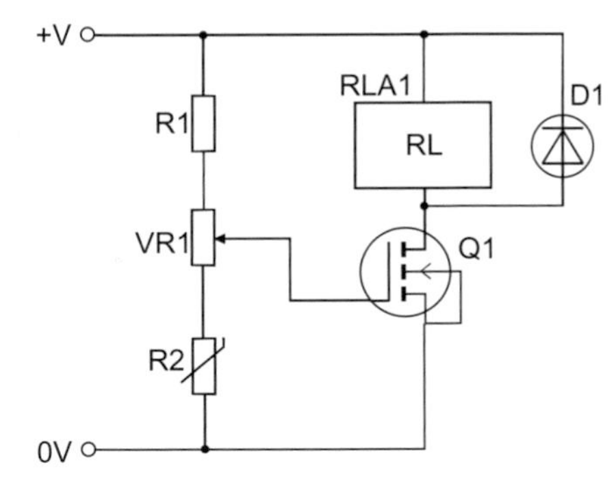

FIGURE 26.6

The relay switches a separate mains circuit, using its normally-open contacts. When current flows through the relay coil, the normally-open contacts close. The heater circuit is completed and the heater comes on. It stays on until the temperature reaches the set level.

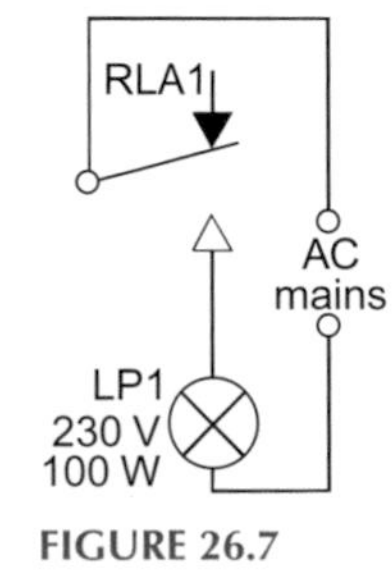

FIGURE 26.7

PROTECTIVE DIODE

A protective diode is essential when switching a load that has high inductance. Inductive loads include relay coils, electric bells and electromagnets. Figure 26.8 illustrates the effect of self-induction (p. 63) in the coil.

(a) When the transistor (BJT or FET) is on, current is passing through the coil. Possibly the current varies slightly and, if so, the magnetic field varies slightly. But there are no dramatic changes.

(b) When the transistor switches off, the current ceases *instantly* and the field collapses *rapidly*. The effect of induction depends on the *rate of change* of the field. The faster the change, the greater the effect.

Total switch-off is a very fast change. A very large electric force is induced, to try to keep the magnetic field as it was. This force may be equal to several hundred volts, even though the original voltage across the load was only, say, 10 V. A current of several amps surges through the transistor and burns it out.

(c) The solution is to wire a diode as shown, to conduct the excess current safely away.

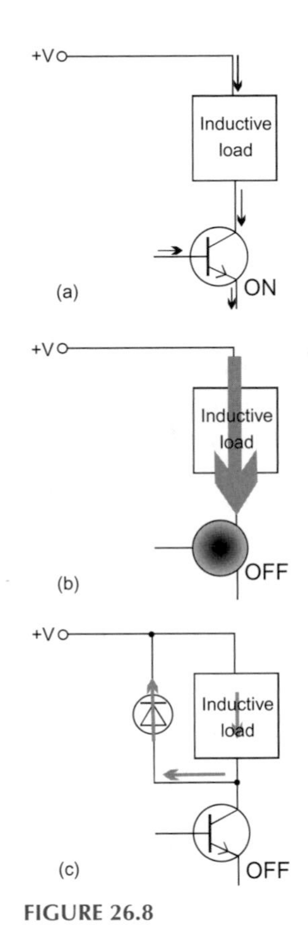

FIGURE 26.8

QUESTIONS ON TRANSISTORS AND THYRISTORS

1 Name the three terminals of a BJT. Out of which of these does the current flow when the BJT is connected into a circuit?
2 Describe how you would demonstrate transistor action and what results you would obtain.
3 Which is the biggest, collector current, emitter current, or base current?
4 Describe how you would measure the small signal current gain of a BJT.
5 When the base current of a BJT is 40 μA, its collector current is 35 mA. When the base current is 65 μA,

the collector current is 38 mA. Calculate its small signal current gain.

6 What is the typical small signal current gain of a BJT? What is its base-emitter voltage when it is switched on?
7 What is meant when we say that a BJT is saturated?
8 Design a transistor switch to be used with an ORP12 as sensor, to switch a 12 V 6 W lamp when the light level increases above a set level.
9 What are advantages of FETs over BJTs when used as switches?
10 What do we mean by the threshold voltage of an FET?
11 What is the difference between gain and transconductance?
12 An FET has a transconductance of 3 S. Its gate voltage is above the threshold and increases by 1.5 mV. What change occurs in the drain current?
13 Draw the symbol for an FET and name the terminals.
14 Name the three terminals of a thyristor. Through which terminal does current flow out when the thyristor is triggered?
15 Once a thyristor is conducting, what can we do to stop it?
16 Draw a circuit diagram of a thyristor half-wave rectifier and explain how it works.

MULTIPLE CHOICE QUESTIONS

1 The symbol for small signal current gain is:
A h_{FE}
B g_m
C h_{fe}
D I_D

2 Over its operating range, the collector current of a transistor is directly proportional to:
A the change in base current.
B the collector voltage.
C the base current.
D the emitter voltage.

3 The normal voltage difference between the base and the emitter is:
A 0 V.
B about 0.7 V.
C equal to the collector voltage.
D variable.

4 Figure 26.9 shows an experimental circuit.

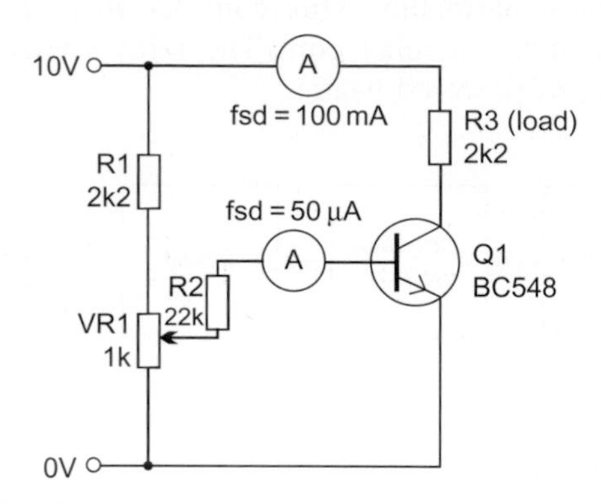

FIGURE 26.9

Plotting v_{CE} (the voltage between collector and emitter) against the base current gives:

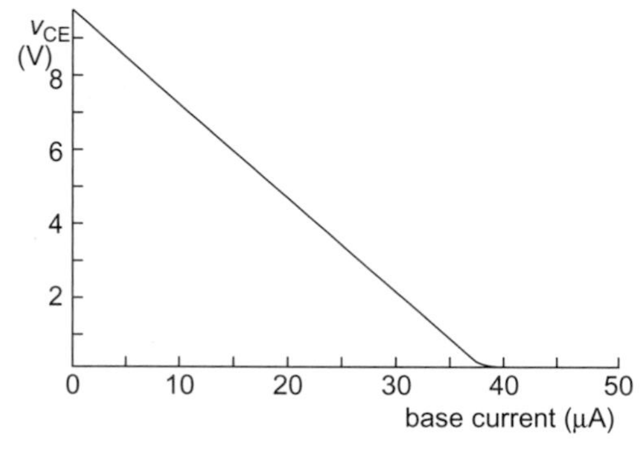

FIGURE 26.10

The graph shows that as base current increases to 37 μA:
A the voltage across R3 decreases.
B v_{CE} is constant.
C the transistor is saturated.
D the voltage across R3 increases.

5 The current through an FET is:
A constant.
B much greater than the base current.
C greater than that through a BJT.
D proportional to the gate voltage.

6 A thyristor is triggered by a:
A negative pulse applied to the gate.
B negative pulse applied to the anode.
C positive pulse applied to the cathode.
D positive pulse applied to the gate.

7 A thyristor is NOT turned off by:
A short-circuiting anode to cathode.
B a negative pulse to the gate.

C turning off the power supply.
D interrupting the current through it.

8 In a thyristor half-wave rectifier the lamp can be switched on at any time:
A in the first half of the positive half-cycle.
B after the peak of the positive half-cycle.
C in the cycle.
D in the second half of the positive half-cycle.

9 An FET is used to convert:
A a small current to a larger current.
B a voltage to a current.
C a small current to a voltage.
D a small voltage to a larger voltage.

10 When an FET is conducting current the gate voltage may be:
A 0 V.
B greater than the threshold voltage but less than the supply.
C close to 0.7 V.
D between 0 V and the threshold.

Part 3

Electronic Systems

The Structure of a System

An electronic kitchen scale is a simple example of a system.

FIGURE 27.1

It consists of:

- a force sensor,
- a microcontroller for taking a signal from the sensor and producing signals to send to the display,
- a display.

Here is the system diagram:

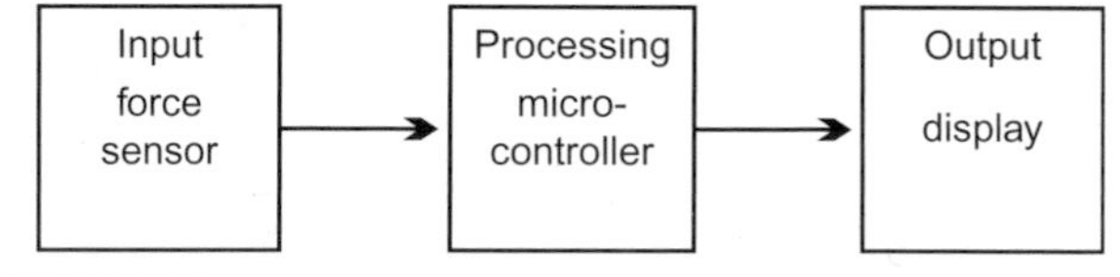

FIGURE 27.2

The system has three parts, just like the systems drawn on pages 78 and 90. In the diagram above we have named the parts as force sensor, microcontroller and display. We can understand the *system* without knowing anything about force sensors, microcontrollers or displays. We know what each part does, even though we do not know how it does it. This is described later in the book.

The diagram above also names the parts of the system in more general terms: input, processing and output. We can apply these names to many different systems. On p. 78, for example, the LDR provides the input, the transistor switch does the processing, and the lamp indicates the output.

Remember that the arrows in the system diagram do not represent wires carrying currents. They represent a **flow of information** through the system. At each stage this information is represented by forces, voltages or currents until, in the output stage, it is represented by numerals on the display.

The sliding doors of a shopping centre are opened and closed by a system that has a similar structure.

FIGURE 27.3

The system diagram is:

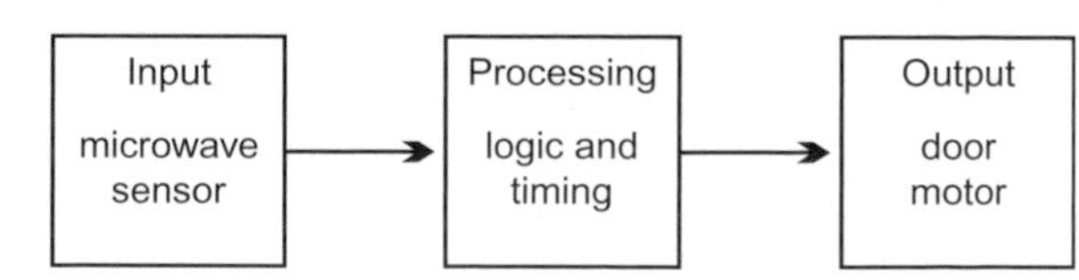

FIGURE 27.4

As before, there is a flow of information through the system. This time the information is 'A person is approaching the door'. The sensor detects the pattern of reflected microwaves that represents this information. The information is then converted into electrical signals. Finally the information is acted upon and the door motors are turned on.

A more complicated system is needed to run a supermarket checkout.

FIGURE 27.5

It has four inputs: the bar code scanner (1), a keyboard for use by the checkout assistant (2), an electronic scale for weighing vegetables (3), and an input from the in-store computer which holds the prices of the goods.

The checkout has four outputs: a display for the checkout assistant (4), a display for the customer, a printer to print the invoice, and an output to the in-store computer to update the stock list.

A small computer is at the centre of the system to process the information from the inputs and send information on to the outputs.

Things to do

Draw system diagrams of (a) a supermarket checkout, (b) an automatic teller machine, and (c) any other electronic system that you know about.

Figure 27.6 shows an Automatic Teller Machine (ATM). It has two inputs: a card reader (1) and a small keypad (2). It has three outputs: a screen (3), a receipt printer (4), and a banknote counter (5). It has a modem to connect it by telephone line to the bank. Processing is done by its own computer.

The ATM can give you cash drawn from a bank account almost anywhere in the world.

FIGURE 27.6

DESIGNING AND BUILDING SYSTEMS

This part of the book describes components and circuits that are useful in building systems. The topics are listed below, under the headings Input, Processing and Output.

Use this part as a source of ideas and practical help for the systems and projects you build. Learn about other systems that you do not have time to build.

INPUT

Sensors for temperature, light, force, sound, magnetic field, position, vibration, moisture.

Interfacing sensors Ways of connecting sensors to systems.

PROCESSING

Amplifying signals
Timing
Logic The basic logic gates
Logical systems
Logical sequences
Storing data
Microcontrollers
Programs

OUTPUT

Visual output Lamps, LEDs, displays. Data output. Transducers.

Audible output Bells, buzzers, sirens, loudspeakers, sounders.

Mechanical output Motors, solenoids.

Switches

Switches are used to control the flow of current in a circuit. Current flows when the switch contacts come together. We say that the switch is **closed**, or contact is **made**. Current can not pass through the switch when the contacts are apart. We say that the switch is **open**, or contact is **broken**.

There are many different types of switch, used for different purposes. Here we describe a few of the most useful types.

TOGGLE SWITCH

A toggle switch is a basic switch, operated by a toggle lever that can be pushed up or down. By convention, the down position is the 'on', or 'closed', or 'made' position. The toggle switch in the photo has its toggle lever up. Behind the lever is a threaded dolly with a large nut. This is for mounting the switch in a circular hole cut in a panel.

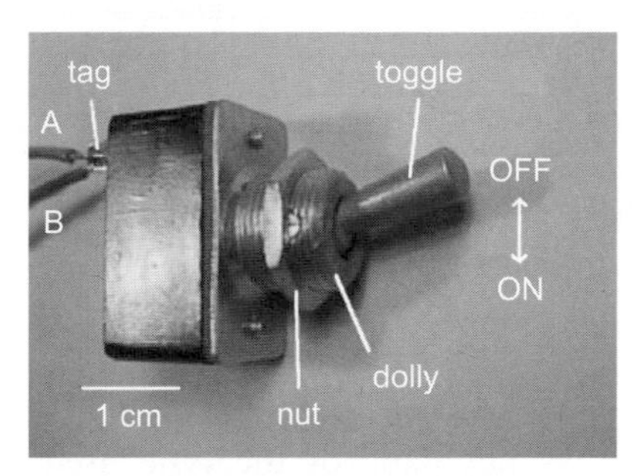

FIGURE 28.1

At the rear are two terminal tags to which the connecting wires are soldered. This switch is a heavy-duty type, rated to switch currents as large as 10 A and 250 V AC. Heavy-duty toggle switches are often used for switching the mains power supply to appliances and equipment. However, they can be used for switching smaller currents too.

FIGURE 28.2

The miniature toggle switch shown in Figure 28.2 is suitable for mounting on a control panel.

It is able to switch 1.5 A at 250 V AC.

Design Tip

When selecting a switch for a project, remember to check the contact rating. If you use the switch for too high a current or voltage, the contacts will become corroded and pitted by sparking. They may fail to make good contact, increasing the resistance of the switch. With much sparking, the contacts may melt and fuse together. Then the switch becomes permanently closed and is useless.

The larger toggle switch has two solder tags, showing that it has **single-pole, single throw** contacts (**SPST**). Its symbol shows how it works. It switches a single circuit and it is either open or closed.

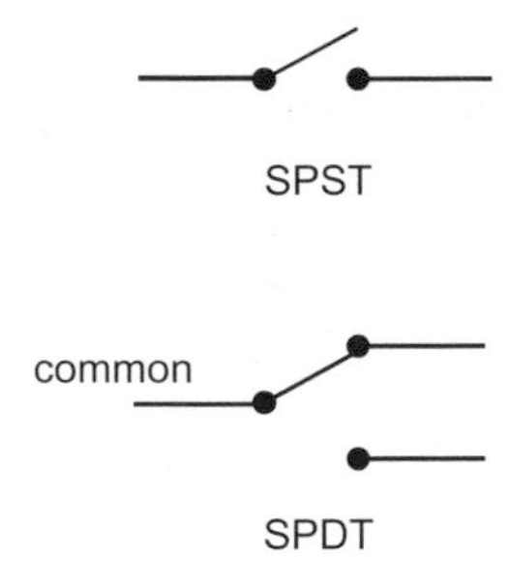

FIGURE 28.3

The miniature toggle switch has **single-pole, double-throw** contacts (**SPDT**). The centre tag is common and can be switched into contact with either of the other tags. Such contacts are called **changeover contacts**.

MICROSWITCH

The 'micro' part of its name does not mean that the switch itself is necessarily small. It means that the operating button moves only a small distance.

The switch is very sensitive. A light pressure on the lever causes the switch to click over from one position to the other. Most microswitches have SPDT contacts so that they can either switch something on or switch something off, perhaps both at the same time.

The contacts on a microswitch are sprung so that normally the common contact connects to what is

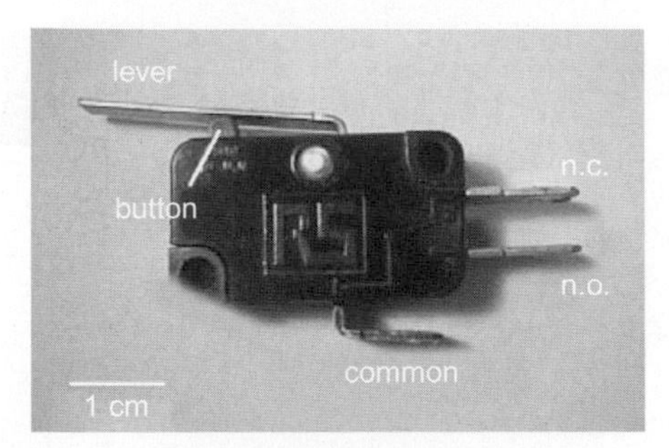

FIGURE 28.4

called the **normally closed (n.c.)** contact. The third contact is the **normally open (n.o.)** contact.

Microswitches are used where a switch has to be operated mechanically. For example, a microswitch is mounted inside a cupboard, so that the lever is held down when the door is closed. Its common and normally closed contacts are wired into a lamp circuit. When the door is closed, the contacts are open and the lamp is off. When the door is opened, the contacts close and the lamp is switched on.

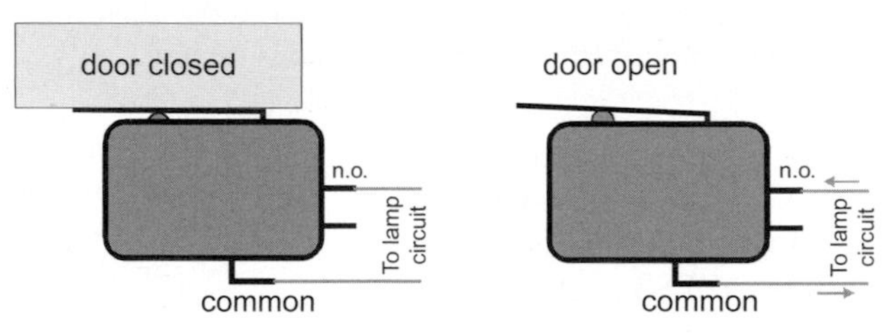

FIGURE 28.5

REED SWITCH

This consists of two springy metal contacts (the reeds) sealed into a capsule with their ends overlapping. Normally the contacts are open. In a magnetic field, the reeds become magnetised. Their North and South poles attract each other. The reeds come together and make contact. When the field is removed, their springiness moves the reeds apart.

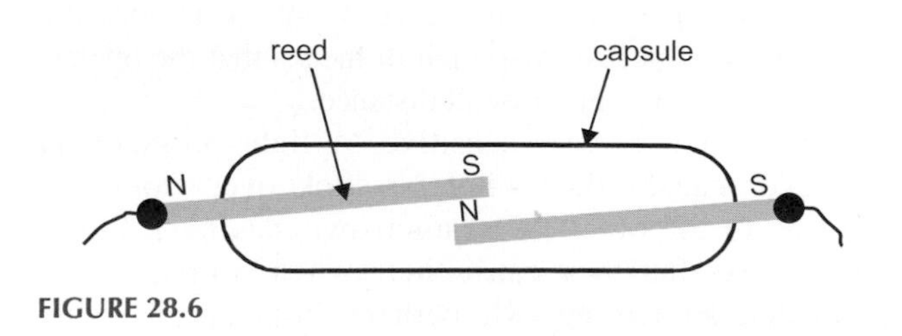

FIGURE 28.6

Larger reeds switches can switch mains currents of up to 2 A.

The magnetic field may be provided by a permanent magnet or by a coil. For example, in a security system, a reed switch is mounted on a door-frame. A magnet is mounted on the door. When the door is shut, the magnet is near to the switch. Contact is made. If the door is opened, the magnet is moved away from the switch. The contacts open, breaking the circuit and triggering an alarm. If the reed switch is operated by a coil wound round its capsule, the switch acts as a relay (see p. 108).

TILT SWITCH

The switch closes when it is placed in its normal vertical position. The switch opens when it is tilted. If a switch is attached to a movable part of a machine, it can be used to detect whether that part is in the correct position or not.

MEMBRANE SWITCH

When the button is pressed, a flexible, conductive plastic film below the button is pressed down. It bridges the gap between two metal contact pads, so that current can flow from one to the other.

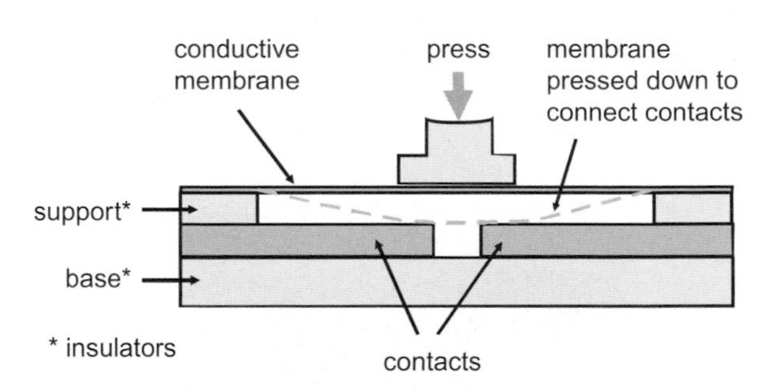

FIGURE 28.7

Membrane switches are available as single switches or as key-pads of the type used in pocket calculators and security systems.

Self Test

What type of switch would you use for:

(a) detecting if a protective grid on a lathe is in position.
(b) a power switch on a PSU.
(c) detecting if a mobile robot has fallen over.

PUSH SWITCHES

A push switch is operated by pressing a button. There are two types of action. Most switches are

FIGURE 28.8

push-to-make (or **PTM**) switches. Pressing the button pushes the contacts together and the switch closes. The other type are **push-to-break** (or **PTB**) switches. The contacts are normally closed but are forced apart when the button is pressed.

Either type of switch may be momentary or latching. A switch that is **momentary** acts only for as long as the button is pressed. When the button is released, the switch springs back to its normal state.

In a **latching** switch, the button stays down when pressed. The contacts remain closed or open, depending on the type of switch. You need to press the button *again* to return the button to its normal state.

Push buttons are used for a wide variety of control purposes, and may be used for power switching for lamps, radio sets and other appliances.

ROCKER SWITCHES

These are similar in action to toggle switches and are used for the same purposes. The difference is that they are operated by a rocker.

FIGURE 28.9

SLIDE SWITCHES

These have uses similar to those of toggle switches, but they are operated by a sliding knob.

The slide switch in the photo is an SPDT switch. The common connection is made to the central terminal.

FIGURE 28.10

DPDT SWITCHES

Toggle, rocker and slide switches are also made as double-pole double throw versions. This gives two separate switches within the same unit, but operated together.

DPDT switches can be used to switch on two circuits at the same time. They may also be wired into both the live and the neutral mains lines.

When the switch is off, the appliance is completely isolated from the mains. DPDT switches are often used with electric blankets for safety.

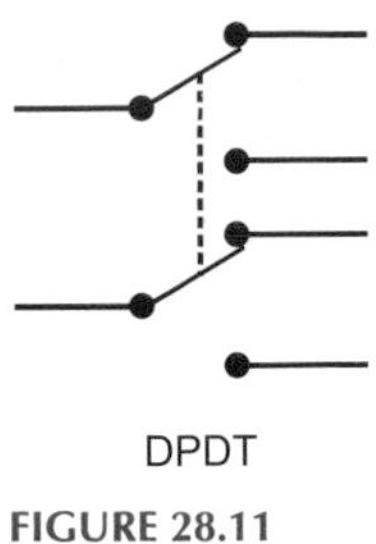

FIGURE 28.11

KEY SWITCHES

A key switch can be turned on and off only by using a key. Only the pair of keys sold with the lock will allow the switch to be operated. Key switches are used when security is important.

KEYBOARD SWITCHES

These are PTM, momentary switches with large square tops, often marked with letters, number or symbols. They are used for building keyboards for computers. Cheaper keyboards use **membrane switches** (see p. 106).

ROTARY SWITCHES

Rotary switches are used for switching one line to any one of several other lines. Often several such switches are combined in one unit.

They are used for functions such as selecting

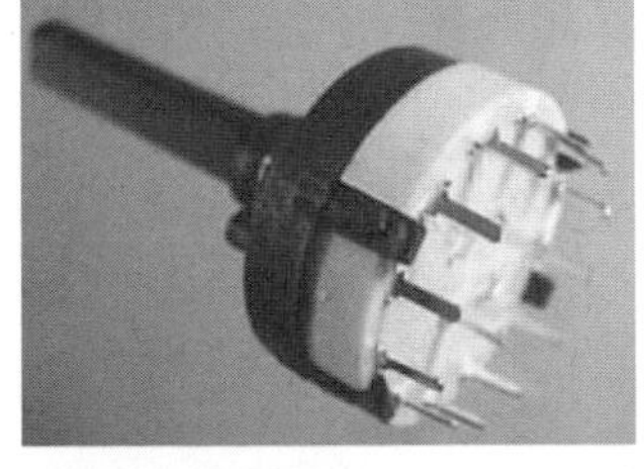
FIGURE 28.12

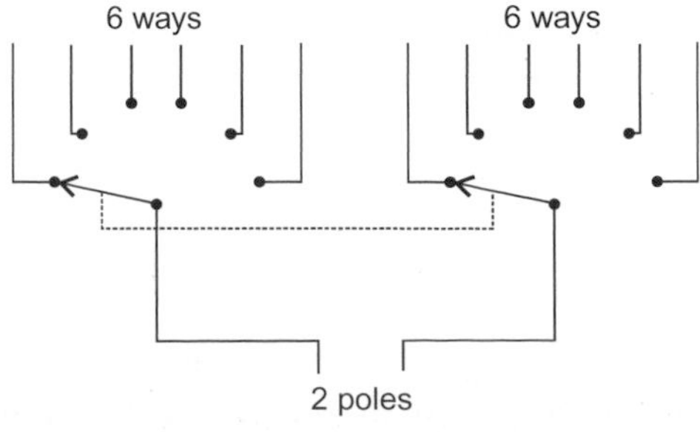

FIGURE 28.13

wavebands on a radio receiver, or selecting the measurement ranges of a multimeter.

The switch has one or more rotating contacts surrounded by a ring of usually 12 stationary contacts. Switches are produced with several different arrangements of contacts. These are 1-pole-12-way, 2-pole-6-way (see drawing), 3-pole-4-way and 4-pole-3-way.

RELAYS

A relay is a current-controlled switch. Compare with a reed switch, p. 106.

A relay has a low-voltage coil wound on a core. There is an iron armature that is attracted toward the core when current passes through the coil. This is attached to a sprung lever. The common contact moves across from the normally closed contact to the normally open contact.

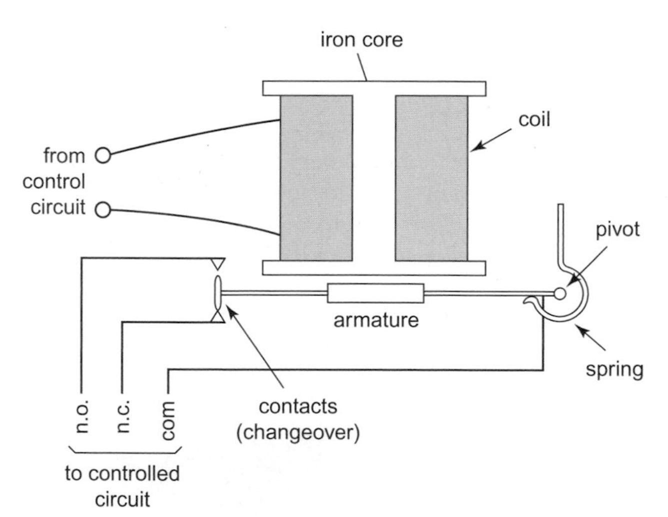

FIGURE 28.14

A typical relay of this type operates in about 10 ms.

Most modern relays are totally enclosed and sealed in, like the one shown top right. Most have SPDT contacts, but there are also DPDT versions. The larger ones will switch 10 A at 250 V AC. The maximum voltage for switching DC is always much less, often only a half of the maximum AC voltage. There are also miniature relays like the one below that are suitable for mounting on circuit boards.

The symbols shown in Figure 28.16 are used in circuit diagrams for the relay coil (left) and the changeover contacts (right). The solid arrow indicates the normally closed contact.

FIGURE 28.15

Build the latching relay circuit, p. 118.

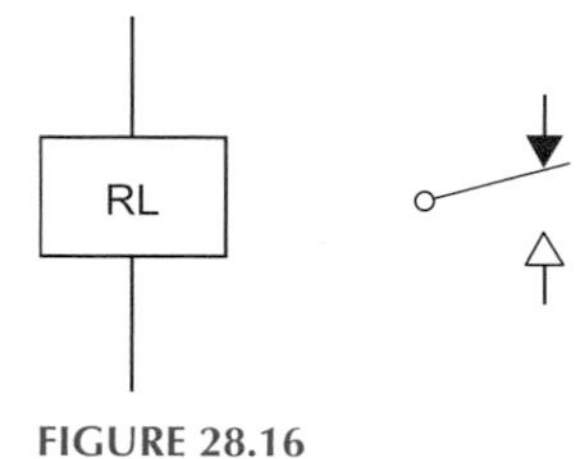

FIGURE 28.16

IN THE LAB: TESTING PROJECTS

The hints in this section are to help you test a project that you have built but which does not work. Very often the fault is either a:

- **Short circuit**: an electrical connection that should not be there, or an
- **Open circuit**: a break in a connection.

CHECKLIST

Run through the points in the list below:

- Is the power supply on?
- Is the voltage correct at the positive supply terminal? If it is lower than expected, you probably have a short-circuit somewhere. Switch off immediately.
- Is there a component missing? This includes a missing wire link. Check against the circuit diagram.
- Are all components soldered in correct holes? Applies particularly to projects on stripboard.
- Are components the right way round? Applies to diodes (including LEDs), electrolytic and tantalum bead capacitors, transistors, and ICs in sockets.
- Are all solder joints good? Check with a magnifier.
- Have IC pins become bent when being inserted in their sockets? Remove ICs and check.

CONTINUITY

Check that all points that should be directly connected are in fact connected. Use a multimeter set to its continuity range. A bleep is heard when the two probes are touched to points that are connected. Alternatively, use the LED continuity checker, shown in Figure 28.17.

Always disconnect the power supply when making continuity checks.

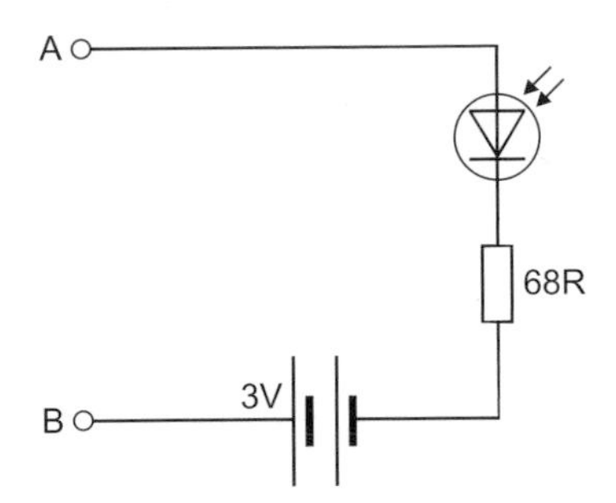

FIGURE 28.17

A and B are probes or crocodile clips on the ends of flexible leads about 25 cm long. The LED lights when there is continuity between A and B.

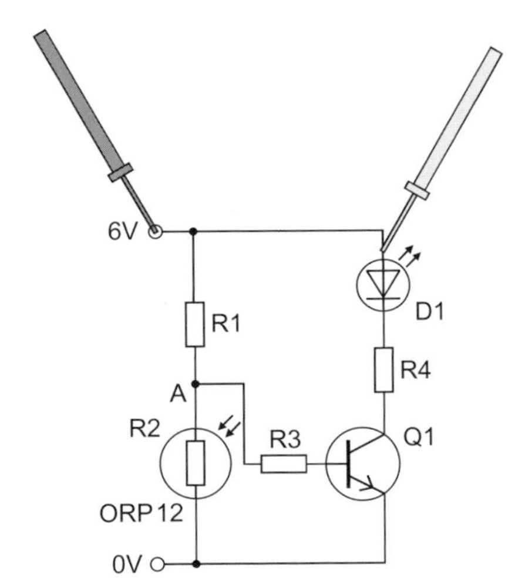

FIGURE 28.18

As an example, take the circuit above. The probes should be placed where shown as one of the tests for continuity in the positive supply line. Touch them against the positive supply input terminal and against the anode wire of the LED. Then check between the positive supply input terminal and the terminal wire of Rl. In another circuit there might be other direct connections to the positive supply line. These must all be checked for continuity.

In the circuit above, check the negative supply line by placing one probe against the 0V terminal and the other against the terminal wire of R2, and against the emitter terminal wire of Ql.

MOISTURE

Moisture sensors are generally home-made. A **water level sensor** has two probes of thick copper wire mounted close together.

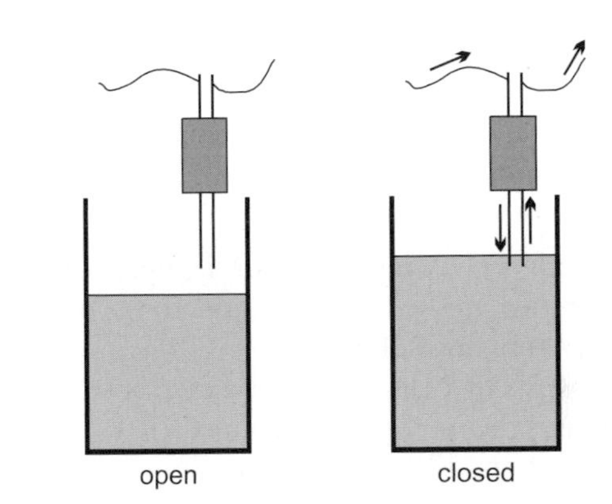

FIGURE 28.19

Normally the resistance between the wires is so high that it is like an open switch. When water partly covers the probes, conduction occurs and the resistance between the probes is low. Current flows, as through a closed switch.

Another type of moisture sensor can be made from a small rectangle of stripboard.

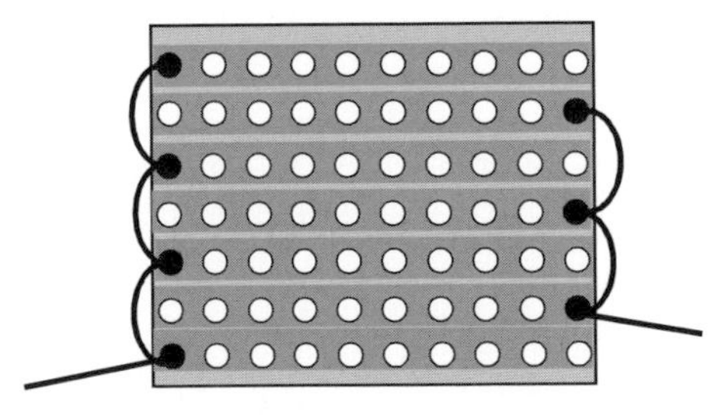

FIGURE 28.20

Connect alternate strips by soldered wires. This type of sensor is useful for detecting rain or water sprays.

POSITION SENSORS

Position sensors are available but are very expensive. A microswitch (p. 105) can be used to detect whether an object is in position or not. For example, it can

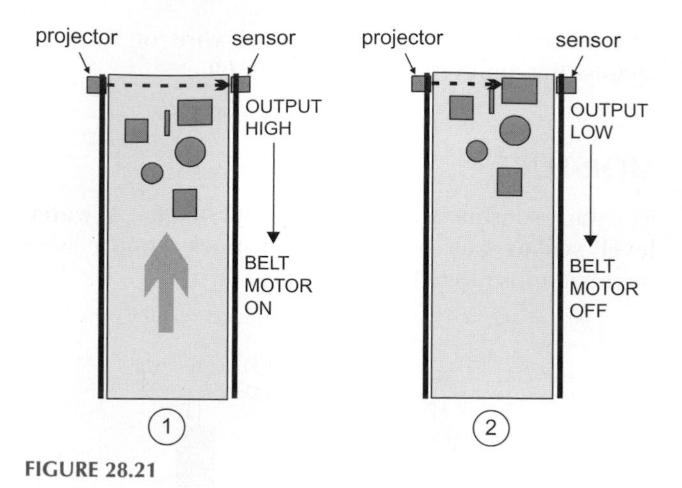

FIGURE 28.21

tell us whether a cupboard door is open or closed. Better, use *two* microswitches, one for 'door open', and another for 'door closed'.

If you want to determine the exact position of an object, it may be possible to connect it to the wiper of a slider pot (p. 47). As the object moves, the wiper moves. The resistance changes between the wiper and one end of the pot. In this way position is converted to current. The information can then be processed by a logic or amplifying circuit. A rotary pot can be used in a similar way to sense rotary position or angle.

Another way to sense position is to use a beam of light. Preferably use infrared. Direct the beam so that it is broken when the object reaches a certain position. The voltage output from an infrared photodiode sensor (p. 112) falls when the object is in position. In Figure 28.21, a beam is being used at a supermarket checkout. It moves the belt until the first item reaches the end near the operator.

SWITCHES AS SENSORS

Microswitches, magnetic reed switches and tilt switches are often used as sensors.

Things to do

Try to think of an unusual sensing application for a switch or a sensor. For instance, how could we use a photodiode to sense force? Or how could we use a microswitch to sense water level?

Set up a temporary version of your sensor design, and test it. What are its good points? What are its bad points?

Light Sensors

The two most commonly used light sensors are the light dependent resistor (LDR) and the photodiode.

LIGHT DEPENDENT RESISTOR

A light dependent resistor (or **LDR**) consists of a disc of semiconductor material with two electrodes on its surface.

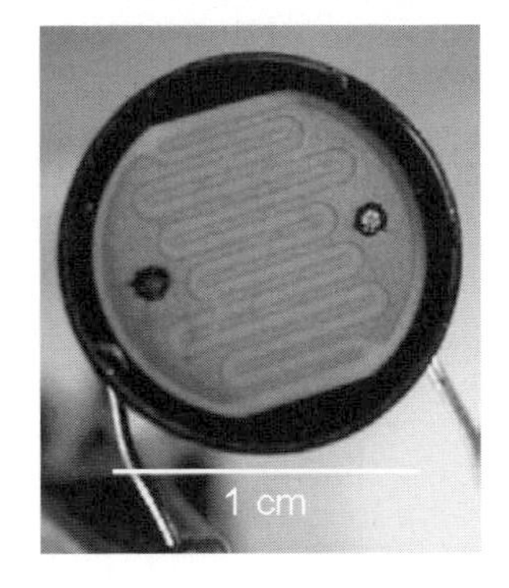

FIGURE 29.1

In the dark or in dim light, the material of the disc has a relatively small number of free electrons in it. There are few electrons to carry electric charge. This means that it is a poor conductor of electric current. Its resistance is high.

In the light, more electrons escape from the atoms of the semiconductor. There are more electrons to carry electric charge. It becomes a good conductor. Its resistance is low. The more light, the more electrons, the lower the resistance.

Things to do

This investigation is to measure the changes in the resistance of an LDR when we change the amount of light falling on it. It is better to use a battery for the power supply, so that the circuit can be carried from place to place.

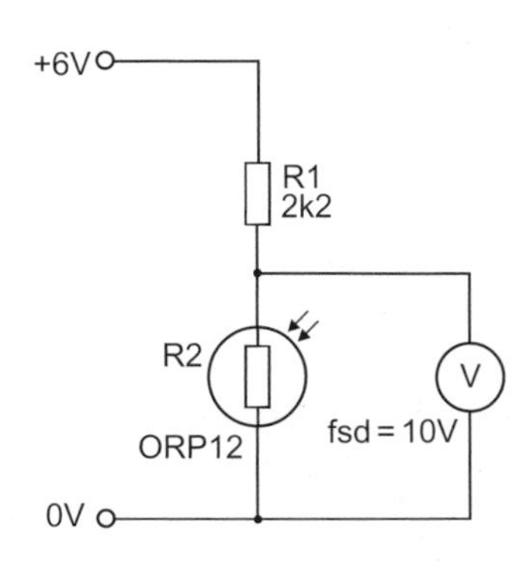

FIGURE 29.2

1 Set up the circuit on the workbench to begin with. Connect the power supply and measure the voltage. Record your result.
2 Cover the LDR with your hand. Repeat the reading.
3 Move the circuit to other places and measure the voltage when it is in each place. Suitable places are: by a large window, outdoors, under a bright bench-light, under the workbench, in a cupboard.
4 In what kind of place is the voltage reading greatest? In a bright place or in a dark place?
5 What is the effect of light on the resistance of the LDR?
6 What kind of network is made by R1 and the LDR?
7 What result would you get if you exchanged R1 and the LDR? Try it.
8 How could you alter the circuit to make it more sensitive to changes in light when it is in shaded places?

CIRCUITS AND SYSTEMS

The circuit described above could be used for measuring light intensity, or at least for comparing the light intensity in one situation to that in another. Figure 29.3 shows the system diagram of this circuit.

The circuit is a voltage divider (see p. 50). The output from the LDR is a voltage which varies

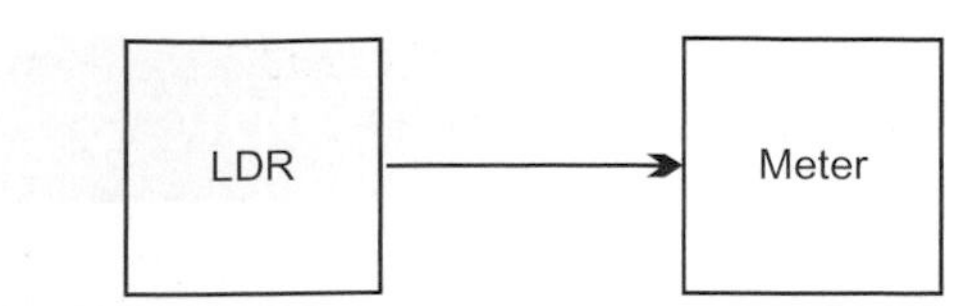

FIGURE 29.3

according to the amount of light falling on it. The amount of light may vary from complete darkness to brilliant sunshine. Consequently, the voltage across the sensor may vary from 0 V up to the supply voltage of the system. It may take any value in that range, so it is an analogue quantity. The LDR is an analogue device. Compare this with the action of switches which must be either on or off, and are therefore digital devices.

The output stage of this system is an electronic numeric display. The processing stage converts the input analogue voltage to its digital equivalent. This is then used to drive the display.

Although the LDR is an analogue device, it can also be used as a source of 1-bit digital signals. To obtain such signals, the voltage across the LDR is compared with a fixed voltage generated in the system. The system then determines whether the voltage across the sensor is greater than or less than the fixed voltage. This type of circuit has two possible states. Either the sensor output voltage is greater than the fixed voltage, or it is less than the fixed voltage. We will take the case of its being equal to the fixed voltage as having the same meaning as greater than the fixed voltage.

This gives us a two-state system:

- input greater than or equal to fixed voltage
- input less than the fixed voltage.

Analogue values generated by other input devices can be converted to binary values in a similar way.

FIGURE 29.4

FIGURE 29.5

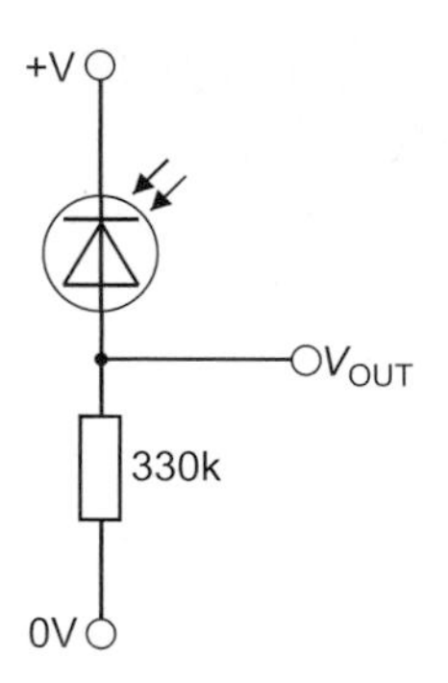

FIGURE 29.6

PHOTODIODE

A photodiode has the properties of an ordinary diode and is sensitive to light. The photodiode in the upper photograph is housed in a metal can. The diode is visible through the lens at the top of the can, as a square chip of silicon.

The photodiode in the lower photograph is housed in an opaque plastic case. The case is transparent to infrared light. This type of diode is useful in security systems to detect the breaking of an infra-red beam, which is invisible to an intruder.

A photodiode is connected so that it is reverse biased. Only a small leakage current of a few microamps passes through it. The current is proportional to the amount of light falling on the photodiode. The current passes through a resistor and a voltage develops across it.

The voltage across the resistor (V_{OUT}) is proportional to the amount of light falling on the photodiode.

A SYSTEM WITH PHOTODIODE INPUT

To demonstrate the action of a photodiode, we use the light sensor from the LEGO Mindstorms kit.

In the upper photo, the sensor is mounted on the *Demobot* robot pointing upward. Its purpose is to measure the amount of light falling on the robot from above. The lower photo is a closer view of the sensor showing it has both a photodiode and a light emitting diode. The LED is not used in this application.

FIGURE 29.7

FIGURE 29.8

The sensor produces an analogue voltage signal, the value of which is proportional to the amount of light falling on the sensor. The processing stage converts this to its digital equivalent. This process is repeated indefinitely, so that the voltage signal received by the processing stage is continually updated. The processor is programmed to respond *not* to the actual light intensity at any given time, but to the *rate of change* of intensity. If the processor detects a *sudden sharp decrease* in light intensity, it switches the motors on very fast for a short period. Slow decreases in light intensity have no effect. Neither do any increases in light intensity. Below is the system diagram:

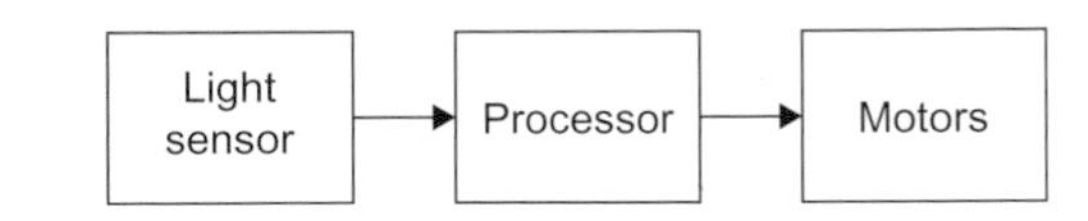

FIGURE 29.9

Here is the flow chart of this system:

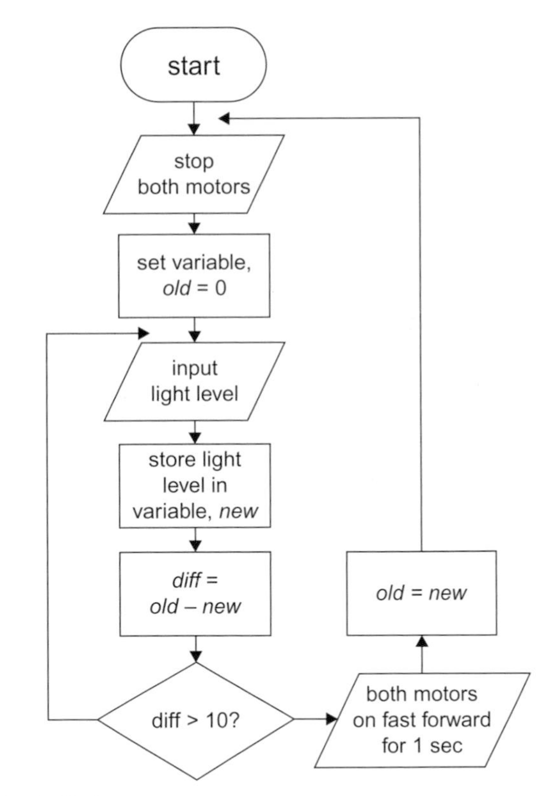

FIGURE 29.10

(This behaviour is very much like that of a house fly: if you try to swat a house fly, it nearly always manages to get away. Normally, the house fly does not respond to small and slow changes in the amount of light falling on it. But if you try to swat it, the rapid decrease in light as the swat comes down on the fly evokes a rapid response. It jumps into the air and flies away at high speed before the swat can strike it.) It is amusing to play with this robot. Try to pick it up by hand and watch it detecting the approach of your hand and scooting away.

There are movies of this robot on the companion website.

A SYSTEM WITH TWO INPUTS

This security system has two light-sensitive inputs. Its output stage is a floodlight which illuminates a garden.

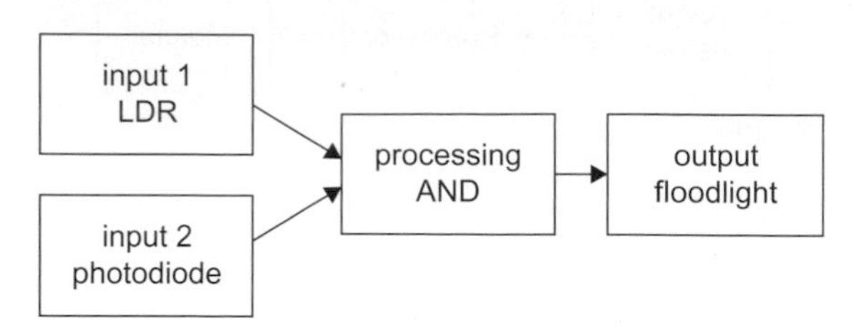

FIGURE 29.11

One input comes from a light dependent resistor and the other input is from a photodiode. The LDR is positioned so as to receive light from the sky. Its purpose is to determine whether it is daytime or nighttime. The photodiode is positioned to receive a beam of light from a nearby lamp. When an intruder breaks the beam the amount of light received by the photodiode is reduced.

The function of the system is to switch on the floodlight at night when an intruder is detected. The processing part of the system has to switch on the lights only when:

It is night AND An intruder is detected

This is a logical AND operation, so the inputs from the light sensitive devices must be converted to logical form and ANDed together to determine whether the floodlights should be switched on or not.

QUESTIONS ON SWITCHES AND LIGHT SENSORS

1 What is meant by the abbreviation 'DPST'?
2 What type of switch consists of two springy metal strips which make contact together in a magnetic field?
3 Describe a slide switch and explain two possible uses for this type of switch.
4 Name three possible applications for a microswitch.
5 A television receiver is controlled by a hand-held remote controller. Draw system diagrams to illustrate the action of the controller and the receiver.
6 This is a diagram of a system that gardeners sometimes place beside the paths of a garden. What is it and what does it do?
7 List three examples of systems that have light sensitive inputs. Draw system diagrams and explain what the system does.

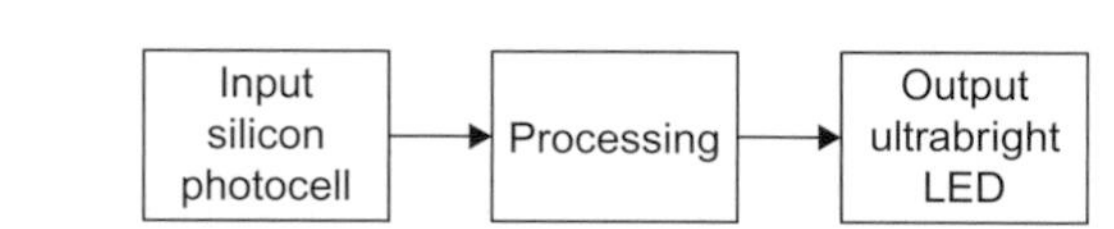

FIGURE 29.12

DESIGN TIME

Design time pages are scattered throughout the book. They provide collections of simple circuits, tips, problems, data and other things of interest to people designing circuits.

You are not expected to memorise these circuits. They are here to give you something to think about. The thinking will help you understand electronics better. You may find one or two of the circuits help you with your exam project. Or you might just like trying them out on a breadboard to see what happens.

The circuits are not described in detail. You are expected to work from the circuit diagram and think things out for yourself.

An LDR controls an LED. What happens? How does it work?

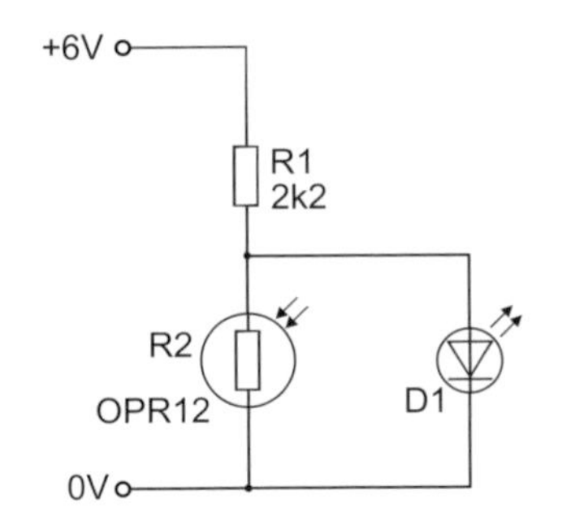

FIGURE 29.13

The LED in the circuit could be at the end of a long lead. It could be a **remote indicator**. This circuit could be improved (and will be later), but — for the moment — suggest some ways that this circuit could be used.

Try exchanging R1 and R2.

Shade the circuit, if necessary, so that the LED *just* goes out. Now change R1 for another resistor so that the LED comes on again. Try to find a resistor so that the LED is off in fairly dim light and comes on in very dim light.

Add a switch to the circuit so that you can switch it to operate in (1) bright light, or (2) dim light.

What is the purpose of this arrangement of switches?

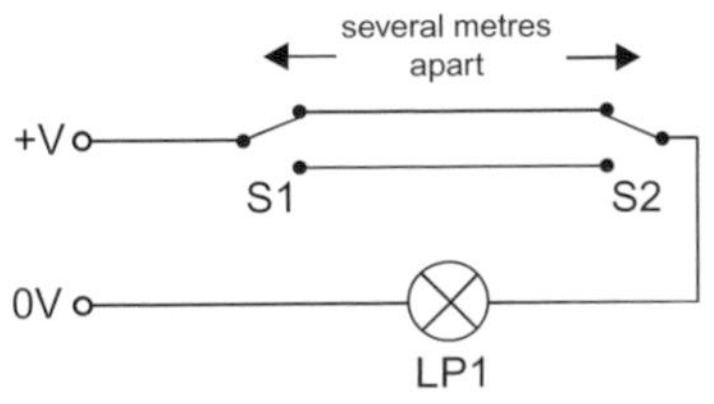

FIGURE 29.14

When a DPDT switch is wired like this, what does it do?

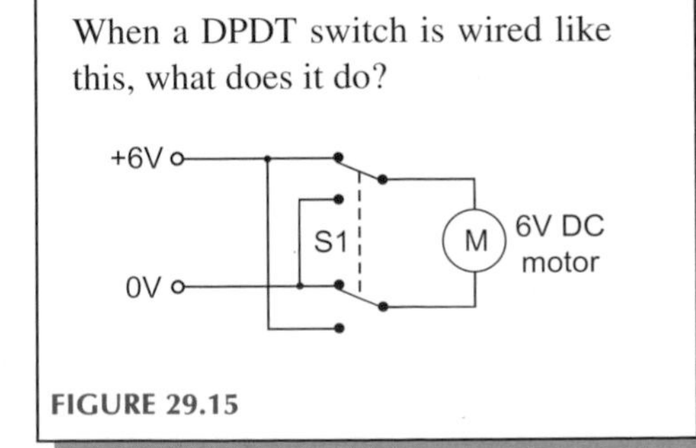

FIGURE 29.15

Temperature Sensors

There are several ways of sensing temperature, including:

- **Temperature-sensitive switches**. These are purely mechanical devices that open or close their contacts at a preset temperature. They are commonly used in circuits controlling room temperature, and the temperature of refrigerators and ovens.
- **Platinum resistance thermometer**. A high-precision instrument for measuring temperature, based on the change of resistance of a platinum wire.
- **Thermistor**. The most commonly used sensor in electronic systems.

THERMISTORS

A thermistor is made from a semiconductor material. It is shaped into a disc, a rod or a bead. Bead thermistors may be only a few millimetres in diameter. Some bead thermistors have the bead enclosed in a glass capsule.

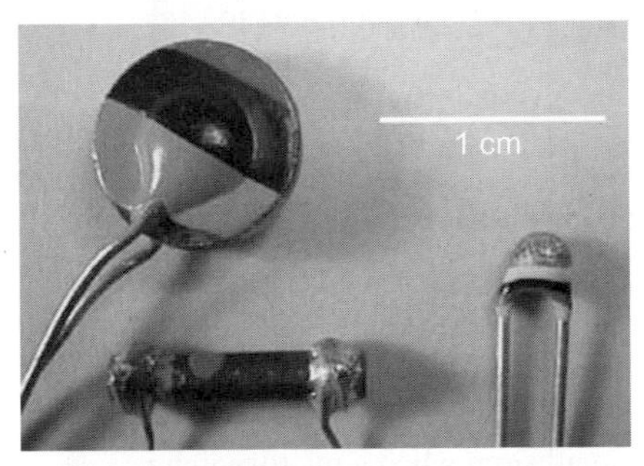

FIGURE 30.1

Because of their small size, bead thermistors respond very rapidly to changes of temperature. Thermistors have two terminal wires. The resistance of most thermistors decreases as temperature increases. These are **negative temperature coefficient** thermistors, or **ntc** thermistors. Thermistors with a positive temperature coefficient (ptc) are also made, but are less often used.

Thermistors are used in circuits that measure temperature or respond in other ways to temperature. They may also be used in circuits that could be put out of adjustment by changes of temperature. The thermistors automatically compensate for the temperature changes.

Things to do

This is a circuit for investigating the change of resistance of a thermistor with temperature. Preferably use a bead thermistor.

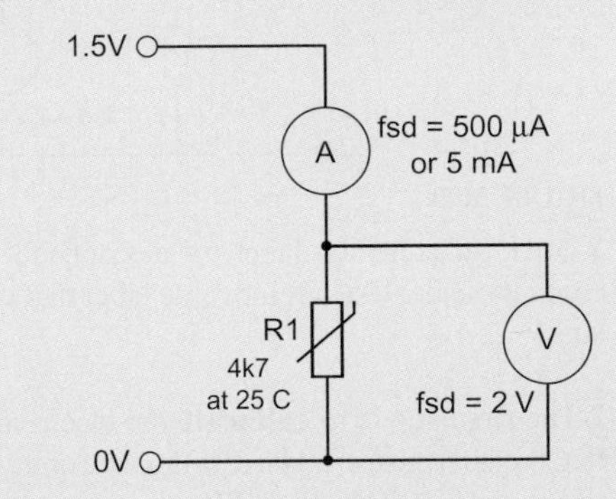

FIGURE 30.2

The thermistor is soldered to a pair of leads about 20 cm long, so that it can more easily be placed in various situations.

1. Put the thermistor in different places, with a thermometer beside it. Leave them there for 2 minutes. Then measure the temperature, current and voltage.
2. Record the results in a table that has a fourth column for the calculated resistance of the thermistor (V/I).
3. Plot a graph of resistance against temperature. Is the graph a straight line?

CIRCUITS AND SYSTEMS

The system diagram for the circuit on the left is:

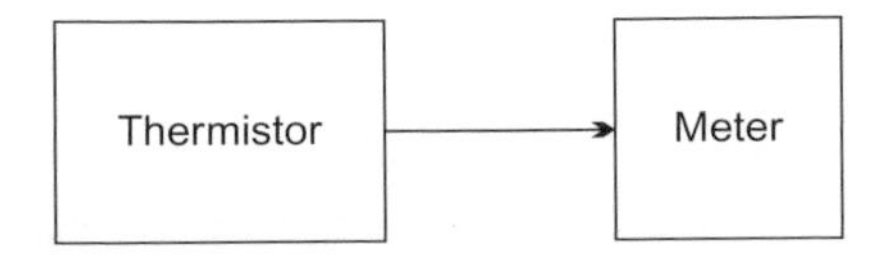

This has the same shape as the system diagram for the light meter (p. 112). Both diagrams can be drawn like this:

Sensor → Actuator

A **sensor** senses what is happening. Examples of sensors are switches, LDRs, and thermistors. An **actuator** makes something happen. Examples of actuators are motors, meters, lamps and LEDS.

DESIGN TIME

A voltage divider circuit is the basis of this simple electronic thermometer. If you want to measure temperatures around room temperature (25°C), R2 should have about the same resistance as the thermistor R1 has at that temperature.

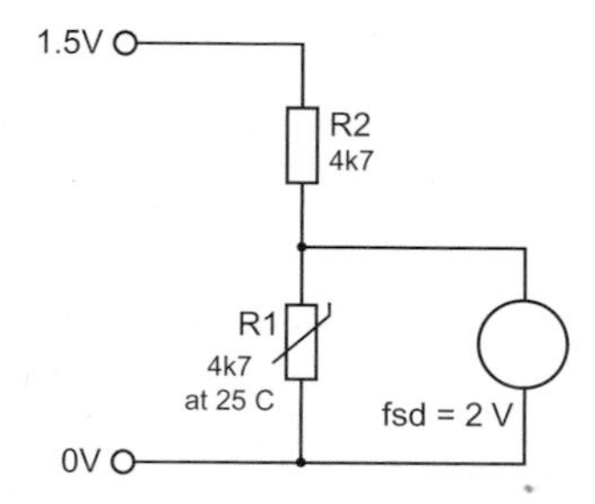

FIGURE 30.3

1 Stick an adhesive label on the outside of the meter to cover its scale. Use a removable label that can be peeled off later.

2 The first step is to **calibrate** the electronic thermometer. Put the thermistor and an ordinary room thermometer in various places that are at different temperatures.

3 Leave them for a few minutes in each place. Then draw a pencil line on the label, level with the needle of the meter. Mark the line with the temperature shown on the room thermometer.

4 Using the pencilled marks as a guide, draw a scale in ink to show temperatures in steps of, say, 5 degrees. Rub out the pencilled marks and figures.

5 Put the thermistor in several other places and read the temperature directly from your scale on the meter.

The range of the electronic thermometer depends on the value of R2. Design a circuit that includes a switch, so that there are two temperature ranges. Calibrate each range separately.

Add an indicator LED to this circuit, to show when the power is switched on.

It is a good idea to use a low supply voltage for these thermistor circuits. Why?

Design and build a **latching relay** circuit. The relay switches on a small siren when a microswitch is closed by an intruder. But the siren continues to sound when the switch is opened again. It sounds until a 'cancel' button is pressed. *Hints:* The relay holds itself on. You need more than SPST contacts.

All the windows and doors of a house have a magnetic reed switch mounted on them. The switches are closed when the windows and doors are closed. Design a circuit that lights an LED only when *all* of the windows and doors are closed.

EVEN IF you are not following a Design and Technology course, these Design Time projects and problems will help you understand electronic theory.

Topic 31

Sound Sensors

CAPACITOR MICROPHONE

Sound is detected with a microphone such as a capacitor microphone. For project work, it is often cheaper to use a 'microphone insert', as illustrated in the photo. This is the basic microphone without case and stand. It has two solder pads behind.

FIGURE 31.1

The sound quality from a capacitor microphone is very good, but the voltages it produces are small. Often there is a small amplifier built in.

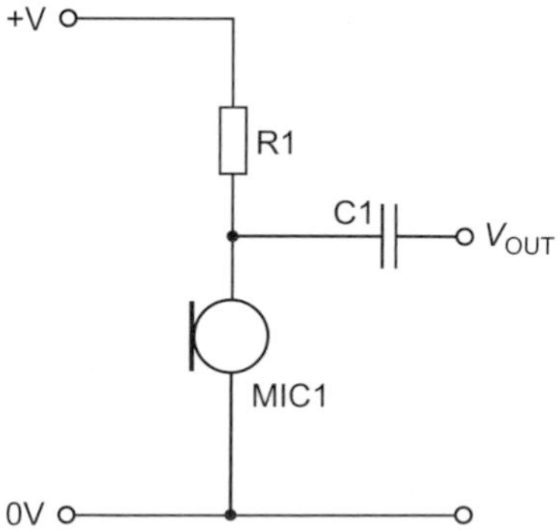

FIGURE 31.2

The complete unit is connected as shown in the circuit diagram. The resistor required depends on the supply voltage. Check the values on the data sheet for the microphone. Usually the signal from the microphone is passed through a capacitor to the next stage of amplification. This is because a direct connection might draw so much current that the amplitude of the voltage signal would be reduced.

CRYSTAL MICROPHONE

A crystal microphone produces a voltage output signal without the need for a power supply. Sound causes the crystal to vibrate and this generates the output signal, owing to the piezo-electric effect. The photo shows a crystal 'microphone insert'.

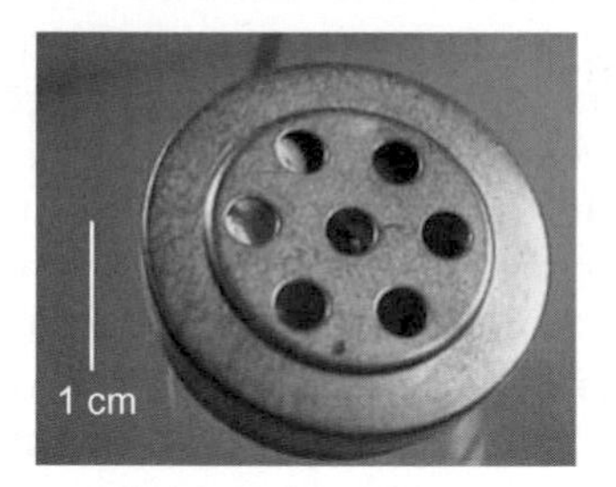

FIGURE 31.3

Often the microphone has a metal case intended to shield it from stray magnetic fields. This case is connected to one of the two terminals. To make use of the shielding effect, always connect that terminal to the 0 V supply line. The other terminal is connected to the amplifying stage by a capacitor. The reason for this is the same as for the capacitor microphone. Crystal microphones are cheap but the quality of their output is poor.

ULTRASONIC TRANSMITTER AND RECEIVER

Ultrasound is sound of high frequency, too high to be audible to the human ear. Typically, ultrasonic transmitters and receivers operate at 40 kHz.

An ultrasonic system comprises a transmitter and a receiver, both of which are specialised crystal microphones. The two microphones operate in opposite ways. The transmitter microphone receives an electrical signal at (usually) 40 kHz. This makes the crystal vibrate, generating a beam of ultrasonic waves in the air around. On arrival at the receiver microphone, the vibrating waves in the air cause the crystal to vibrate, generating an electrical signal which can then be further amplified and processed.

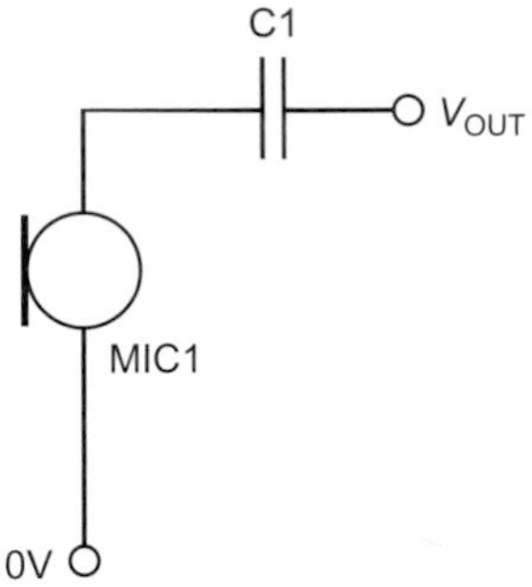

C1
V_{OUT}
MIC1
0V

FIGURE 31.4

Ultrasound is often used for measuring distances, and is sometimes referred to as Sonar. Bursts of ultrasound are emitted by the transmitter, reflected from surrounding objects, and detected by the receiver. The time interval between the emission of bursts of ultrasound and their reception is measured by special circuitry. Given the speed of sound in air, the distance from the transmitter/receiver to the reflecting surface can be calculated.

Ultrasonic sensors are used in security systems for detecting moving persons. They can also be used to detect when stationary objects are near. Another application is in equipment for measuring distances, such as electronic tape measures and range finders.

Figure 31.5 shows an ultrasonic sensor unit mounted on *Demobot*, the LEGO Mindstorms robot shown in Figure 29.7, page 113. Figure 31.6 is a closer view of the sensor showing the transmitting and receiving microphones.

One of the programs written for use with this sensor is an avoidance routine similar to the one used with a touch sensor. The robot moves forward a short distance, and then reads the value returned by the ultrasonic sensor. If this is less than a given distance (for example, 30 cm), it indicates that the robot is approaching an obstacle or wall. The robot then reverses a short distance, spins through an angle of about 40° to the right, and then proceeds in a forward

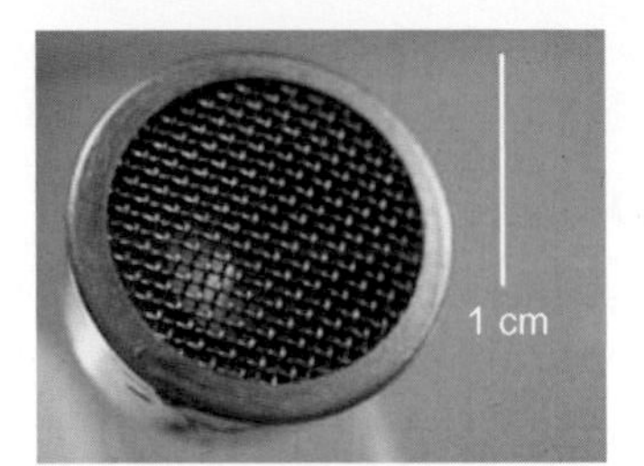

FIGURE 31.5

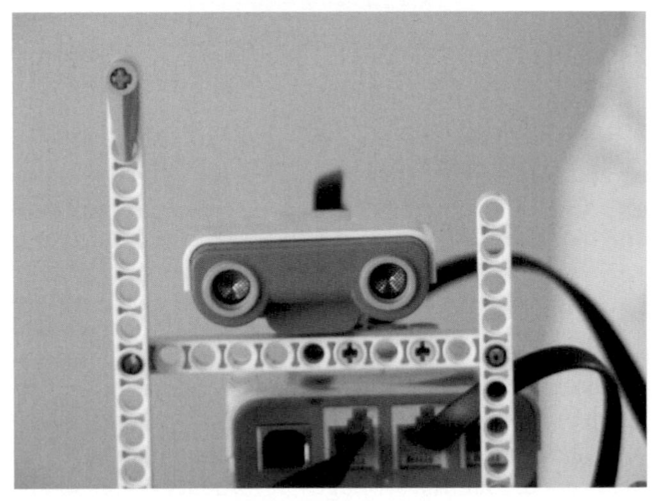

FIGURE 31.6

direction again. With this program operating the robot can roam around the room, avoiding the walls and furniture.

A movie of the robot performing this routine is available on the companion website.

Force Sensors

The device most often used for detecting mechanical force is the **strain gauge**. It consists of thin metal foil etched to form a number of very fine wires. This is embedded in a plastic film.

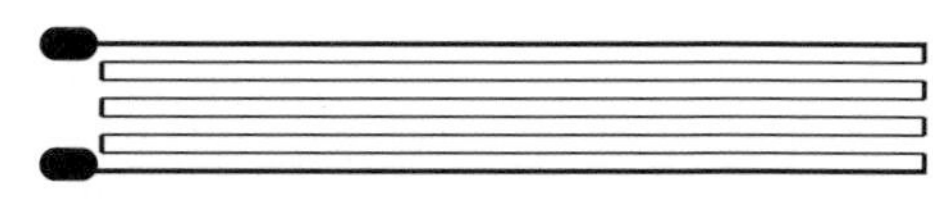

FIGURE 32.1

The gauge is cemented to the object that is to be stressed. When the object is stressed there is tension on the foil which stretches the wires. They become longer and thinner (strained) so their resistance increases. The change in resistance is very small, so a special circuit is used to measure it.

Stress and Strain

Stress: the force exerted on an object.
Strain: the change in the shape of the object due to stress.

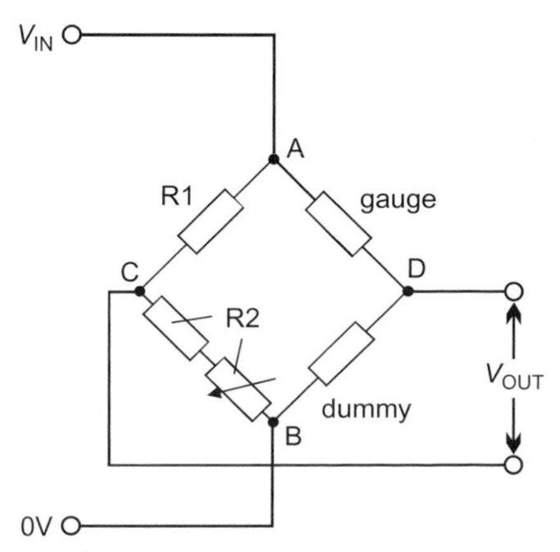

FIGURE 32.2

The circuit is a **Wheatstone Bridge**. One arm of the bridge is the gauge and another is an identical gauge, the dummy. This is not put under stress but is included in the bridge to compensate for changes in resistance of the gauge and its leads due to temperature. *R*2 is a fixed resistor and variable resistor in series.

Think of the bridge as two voltage dividers (ACB, ADB) side by side. The voltage at C is proportional to R1/R2. The voltage at D is proportional to the resistance of the gauge divided by the resistance of the dummy.

One way to read out the result is to adjust the variable resistor until the voltage at C equals the voltage at D. When this is done, the bridge is said to be balanced and V_{OUT} is zero. Then we calculate the resistance of the gauge using this equation:

$$\frac{R1}{R2} = \frac{R_{gauge}}{R_{dummy}}$$

R1 and R2 are known. The resistance of the dummy at a standard temperature is taken from a data sheet, so we can calculate the unknown resistance of the gauge under strain. The final step is to calculate the force from the change in resistance of the gauge. Usually the circuit is calibrated by applying known forces, measuring the changes in resistance, and plotting a graph to relate force to resistance.

A **load cell** consists of one or more strain gauges cemented to a metal bar or ring. The load cell is calibrated by the manufacturer. They are designed to measure tension, compression, or twisting forces. When the bar or ring is strained, the voltage at the gauge terminals is used to find the force. Special electronic devices automatically calculate and display the force on the load cell. Load cells are often used for weighing. Heavy-duty types can be used for weighing hundreds or thousands of kilograms. In a weighbridge, they are used to weigh heavily loaded vehicles. Smaller versions are made which weigh masses of a few kilograms.

PRESSURE PAD

A pressure pad is a sensor for detecting the force due to the weight of a person or animal. The pad is enclosed in plastic and measures about 15 cm × 30 cm. Inside are two layers of conductive material, which are forced together and make electrical contact when a heavy object is on the pad. In other words the pressure pad acts as a switch, turned on when the pressure on it exceeds a given value. Usually a pressure pad is concealed beneath the carpet so that intruders are not aware that it is there.

QUANTUM TUNNELLING COMPOSITE

This is a new material which shows a large change in its resistance as the force on it is varied. It is supplied as a small pill or pad measuring about 3.6 mm square and 1 mm thick. Under zero pressure the resistance of a QTC pill is greater than 10 MΩ, so it is, in effect, an insulator. If force or pressure is applied to the pill, its resistance drops to less than 1 Ω. It is then a very good conductor.

This large change of resistance with pressure makes QTC pills highly suitable for use as force and pressure sensors, in speed control circuits and sensing circuits. It has many applications in control systems, including robots. The QTC pills are inexpensive and fit easily into a wide range of electronic circuits.

QUESTIONS ON TEMPERATURE, SOUND, AND FORCE SENSORS

1. Describe the action of a temperature-sensitive switch, mentioning three possible applications.
2. What is a thermistor? What is meant by the initials 'ntc'? What are the three shapes in which thermistors are manufactured? what shape would you choose for measuring the temperature in a small confined space?
3. What is the main advantage of thermistors as sensors for measuring temperature? What are the chief disadvantages?
4. Name two types of microphone. Which type would you use for recording a musical instrument on a CD? For what reason? Which type would you use to provide input to a clap triggered switch?
5. What is the piezo-electric effect? List three applications of this effect in electronic systems.
6. What is the usual operating frequency of an ultrasonic transmitter?
7. Explain how an ultrasonic transmitter and receiver are used to measure distance.
8. How does a strain gauge work? Name three ways in which you could find a use for a strain gauge.
9. Draw a simple circuit to show how you would use a pressure pad to detect an intruder entering a room through the doorway. Explain what happens as the intruder enters.
10. Describe three ways in which you could use a QTC pill in an electronic system.

DESIGN TIME

Design a switching circuit that turns a lamp on or off alternately each time it detects a handclap.

A car sounds an alert if a person is sitting in the passenger seat without their seatbelt fastened. Design the system, including a pressure pad and a microswitch.

Expanded PVC is a useful material for building projects. It is manufactured in 3 mm thick sheets. The material is easy to cut with a junior hacksaw, or with a craft knife and steel edge. It is light in weight and rigid enough for most applications. It has a certain amount of compressibility which means that nuts and bulkheads sink a fraction of a millimetre into the surface when tightened. This makes it less likely to be shaken loose by vibration. The seat is manufactured in a range of attractive strong colours, which improves the appearance of a unit built from this material.

Foam board is another useful building material. It consists of 5 mm thick sheet of solid plastic foam coated on both sides with plastic film, white on one side and coloured on the other. It is not quite as strong as PVC board, but is just as easily worked.

When your project is in the early design stages it is often a good idea to build a mockup — especially if you need to experiment with the design. There are various constructional sets that can help you this stage, including the well established favourite, *Meccano*®. At this stage the circuit can be breadboarded so that amendments and additions may be made.

Design a weighing scale based on a slightly flexible, horizontal beam. This carries a scale pan. When the object is placed in the pan the beam is bent slightly down. There is a strain gauge on the beam to measure the stress. Calibrate the scale in grams. Draw a system diagram of this device.

Magnetic Field Sensors

A **Hall effect sensor** responds to the intensity of the magnetic field around it. It has three terminals. If there is no magnet close to it, its output voltage is about half the supply voltage.

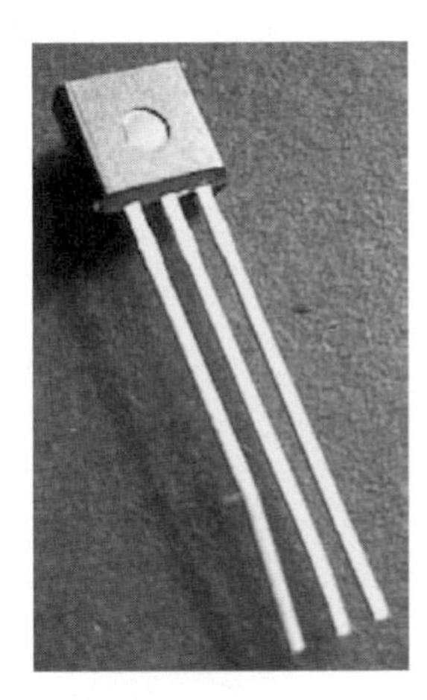

FIGURE 33.1

If the South pole of a magnet is brought close to it, the output voltage rises. The amount of rise is in proportion to the strength of the field. If the North pole of a magnet is brought close to it, the output falls.

The UGN3503 is typical of Hall effect sensors. It requires a supply voltage of between 4.5 V and 6 V (Figure 33.2).

It can be used as a switch, simply by moving a magnet towards it or away from it. It has the advantage that it changes state more quickly than a mechanical switch, and needs less force to operate it. It can operate at high speed, switching on and off thousands of times per second.

In this application it can be used as a door or window switch in a security system. The sensor is mounted on the door frame, and a small ferrite magnet is glued near the edge of the door. The magnet is close to the sensor when the door is shut. If the door is opened, the magnet is moved away from the sensor and the voltage across the sensor changes.

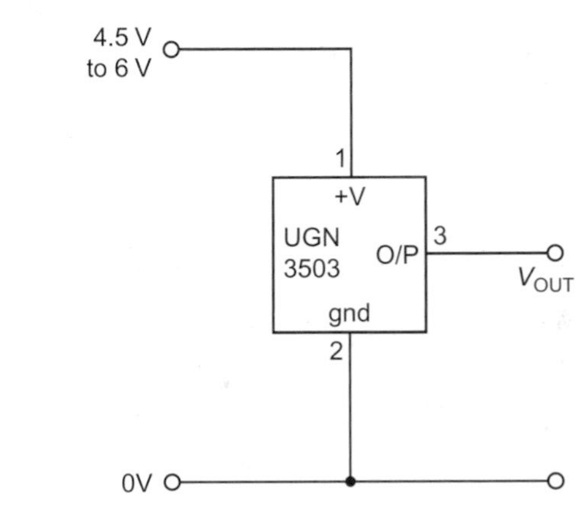

FIGURE 33.2

The device can also be used to measure the speed of rotation of an axle (Figure 33.3). There is a toothed iron wheel on the axle. A permanent magnet produces a field through the Hall effect device. This field changes as each tooth passes by the device. The output alternates at a frequency depending on the rate of rotation and the number of teeth on the wheel. A circuit measures the frequency and from this we can calculate the rate of rotation.

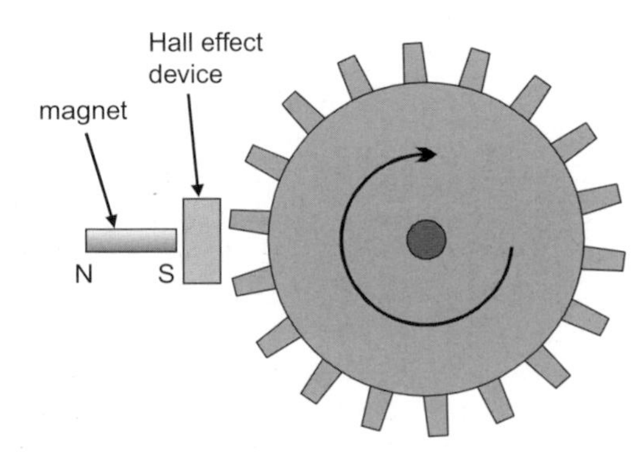

FIGURE 33.3

A similar device can be used to detect the direction of the Earth's magnetic field. The robot in

FIGURE 33.4

Figure 33.4 has the magnetic sensor mounted at the top of a mast, to keep it as far away as possible from the magnetic fields generated by the motors of the robot. The robot is programmed to take its magnetic bearings at frequent intervals and make the necessary corrections to its course. In the movie that is available on the companion website, the robot is programmed to follow the four sides of a square. Its heading is set successively to 0°, 90°, 180°, and 270° clockwise of magnetic North.

Position and Vibration Sensors

Specialised position sensors, such as linear inductive position sensors (LIPS) are obtainable, but are very expensive. They also need complex circuits to process the signals they produce. However, cheaper devices often can be used instead. For example, a microswitch (p. 105) can be used to detect whether an object is in position or not. It can tell us whether a cupboard door is open or closed. Better, use *two* microswitches, one for 'door open', and another for 'door closed'. When the door is halfway open neither switch is closed.

If you want to determine the exact position of an object, it may be possible to connect it to the wiper of a slider pot (p. 47). As the object moves, the wiper moves. The resistance changes between the wiper and one end of the pot. In this way position is converted to current. The information can then be processed by a logic or amplifying circuit. A rotary pot can be used in a similar way to sense rotary position or angle.

Another way to sense position is to use a beam of light. Preferably use infrared. Direct the beam so that it is broken when the object reaches a certain position. The voltage output from an infrared photodiode sensor (p. 112) falls when the object is in position. Below, a beam is being used at a supermarket checkout. It moves the belt until the first item reaches the end near the operator, the pickup position.

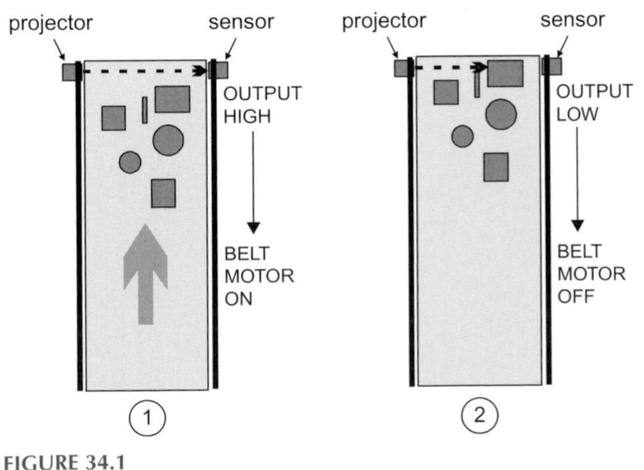

FIGURE 34.1

The mechanism halts with the object to be picked up by the checkout person. When the object is removed, the belt starts again and brings the next object to the pickup position.

Things to do

Try to think of an unusual sensing application for a switch or a sensor. For instance, how could we use a photodiode to sense force? Or how could we use a microswitch to sense water level?

Set up a temporary version of your sensor design and test. What are its good points? What are its bad points?

The examples above are methods of **sensing position** when, by the word 'position', we mean 'location'. Another meaning of the word 'position' is 'attitude'. By this we mean 'which way up?'. If there is just the right way up and the wrong way up, then a tilt switch is usually the solution. It can be set to turn on when the switch is more than a given number of degrees from vertical. This can be particularly useful for mobile robots, which are liable to overturn when travelling on rough terrain.

VIBRATION DETECTORS

A tilt switch can often be used to detect vibrations. If the vibrations are large enough, they will register as excessive tilt, breaking the electrical contact of the switch.

Vibrations that are of relatively high frequency are better registered by a microphone. The output of the microphone is amplified, and rectified. The rectified output increases in voltage (or possibly decreases, depending on the direction in which the rectifier operates). This changing voltage can then be fed to logic or other circuit for further processing. Amplification of the original signal must not be too great, otherwise the circuit will respond to sounds, as well as to actual vibration. After all, sounds are vibrations, so we must take care that the circuit can distinguish them.

Electronics: A First Course.

Moisture Sensors

Moisture sensors are generally home-made. A **water level sensor** has two probes of thick copper wire mounted close together.

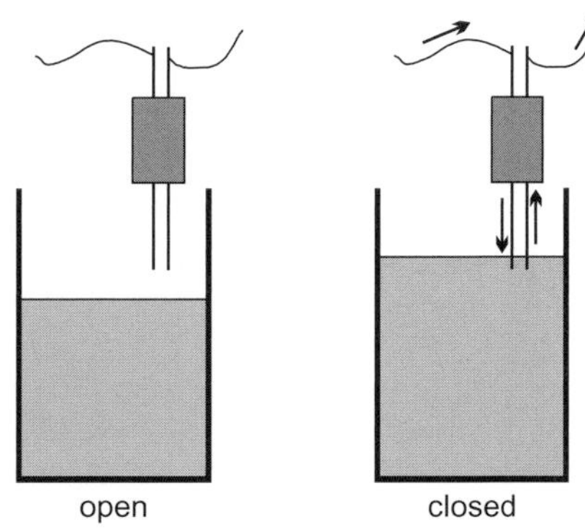

FIGURE 35.1

Normally the resistance between the wires is so high that it is like an open switch. When water partly covers the probes, conduction occurs and the resistance between the probes is low. Current flows, as through a closed switch.

Another type of moisture sensor can be made from a small rectangle of stripboard.

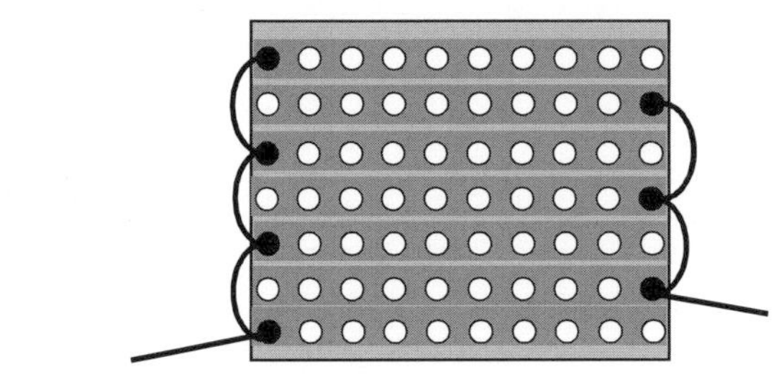

FIGURE 35.2

Connect alternate strips by soldered wires. This type of sensor is useful for detecting rain or water sprays.

Interfacing Sensors

An **interface** is a connecting link. In this Topic we look at the ways of linking a sensor to the rest of the system. The interface may be just a transistor. Or it may be more complicated.

TRANSISTORS

There are several instances of transistor interfacing in earlier Topics. The transistor may be a BJT or an FET. Usually, the transistor is either off, or it is on and saturated.

SIGNALS

Many sensors are resistive. Their resistance changes with temperature, for example, or with light level or position. We use a potential divider to produce a voltage signal from this change of resistance.

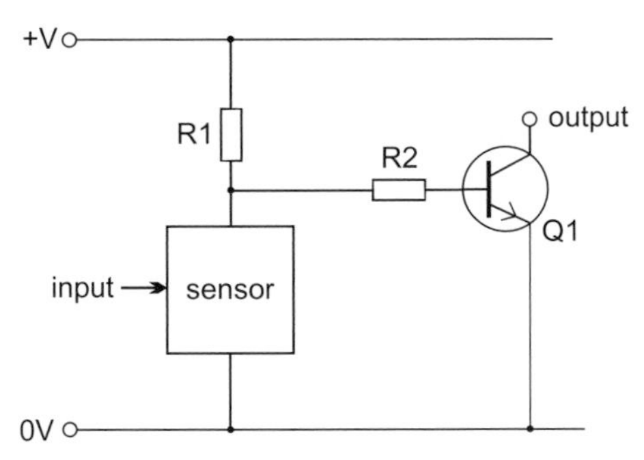

FIGURE 36.1

The output is at the collector terminal of the BJT (or at the drain terminal if it is an FET). The output is a varying *current*. Often we connect a load such as a lamp or a relay coil between the output and the positive supply line. There are several examples of this, such as the transistor switch circuits in Topic 24.

In some systems we need a varying **voltage**. Maybe the signal has to be amplified, as in an audio system. In this case we connect a resistor between the collector and the positive supply line. When current flows through the resistor a voltage appears across the resistor. This voltage signal appears at the output.

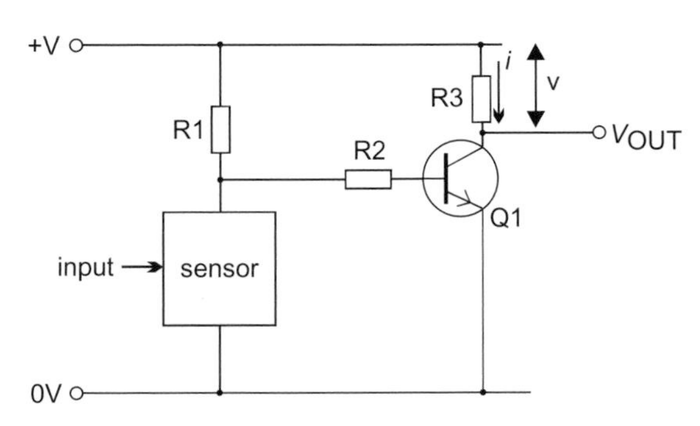

FIGURE 36.2

This is an example of:

Using a resistor to convert a current into a voltage.

Example

A transistor (Note: *not* saturated) has a collector current of 3.5 mA. The supply voltage is 9 V and the collector resistor (R3) is 1 kΩ.

What is V_{OUT}?

The voltage across the resistor is:

$$v = i \times R3 = 0.0035 \times 1000 = 3.5\text{ V}$$

One end of the resistor is connected to the positive supply line, so it is at 9 V. If there is a voltage drop of 3.5 V, the other end is at 9 − 3.5 = 5.5 V.

As the current increases, the voltage drop increases and V_{OUT} *falls*. As the current decreases, the voltage drop decreases and V_{OUT} rises. Summing up:

The voltage signal is proportional to the inverse of the current signal.

Self test

1 In the circuit above, $h_{fe} = 160$, and the base current is 30 μA. What is V_{OUT}?

2 If the base current falls to 20 μA, what happens to V_{OUT}?

Electronics: A First Course.

Two basic variations on this interface circuit are to:

- exchange the sensor and R1, to give the opposite action.
- include a variable resistor in the potential divider to vary the output for a given input.

DARLINGTON PAIR

A Darlington pair consists of two BJTs joined as in the diagram below.

You can wire two single BJTs together or you can buy a ready-made Darlington pair. The ready-made type has the two transistors in a single package with three terminal wires.

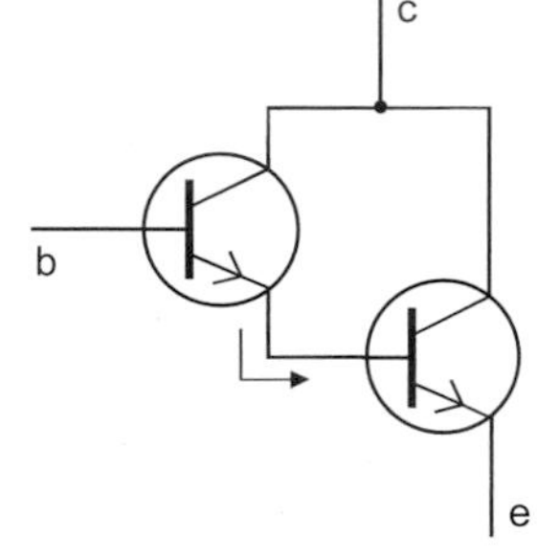

FIGURE 36.3

The advantage of a Darlington pair is its gain. This is because the emitter current of one transistor becomes the base current of the other. Assuming that the gain of each transistor is 100, the gain of the pair is 100 × 100, which equals 10 000.

In operation, there are voltage drops of 0.7 V between the base and emitter of both transistors. This gives a base-emitter voltage of 1.4 V for the pair.

Using a Darlington pair instead of a single BJT greatly increases the input sensitivity of a system.

SCHMITT TRIGGER INPUT

When experimenting with some of the transistor switch circuits (examples, pp. 78 and 90), you may have thought that the switch-on action is too gradual. As the LDR is shaded, for example, the LED *gradually* brightens. It would be better if it came on suddenly as the light level falls just below the set level. A Schmitt trigger input can produce this effect.

There is another advantage in using a Schmitt trigger. Suppose a system is designed to switch on a porch lamp at dusk. At this time of day, the light level falls very slowly. It may not fall steadily because of clouds occasionally moving in front of the Sun and then clearing again. Or there may be the shadows of leaves moving across the LDR. The effect in a system with a simple transistor switch is to make the porch lamp continually flicker on and off at dusk. This is annoying. Also, if the transistor is switching a relay, the chattering of the contacts will shorten its life.

The performance of the system can be improved by using a Schmitt trigger, as shown below.

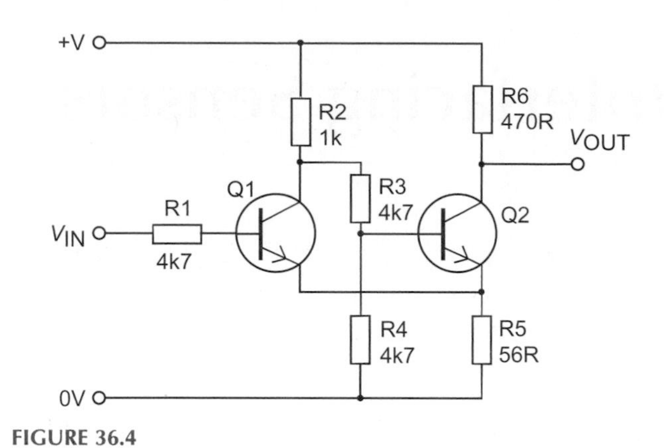

FIGURE 36.4

Without going into the theory, the action of the trigger is as shown in the graph below.

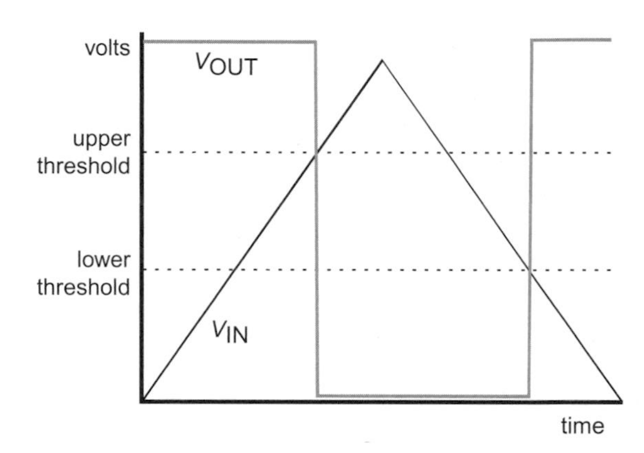

FIGURE 36.5

As the input rises from 0 V, the output is first of all high (supply voltage). As the input exceeds the *upper threshold*, the output very rapidly decreases to 0 V.

As the input decreases, the output does not become high again until the input has fallen below the *lower threshold*. Once the input is below the lower threshold, the output does not change again until the input has exceeded the upper threshold.

This graph shows a Schmitt trigger operating on a very irregular input.

Small reversals of the direction of change of input have no effect on the output. The effect of the trigger is to remove irregularities and to 'square up' the waveform. Note that the trigger inverts the waveform.

The output from the trigger may be taken from the V_{OUT} terminal, as seen in the graphs. It is also possible to replace R6 with a load, such as a lamp, LED or

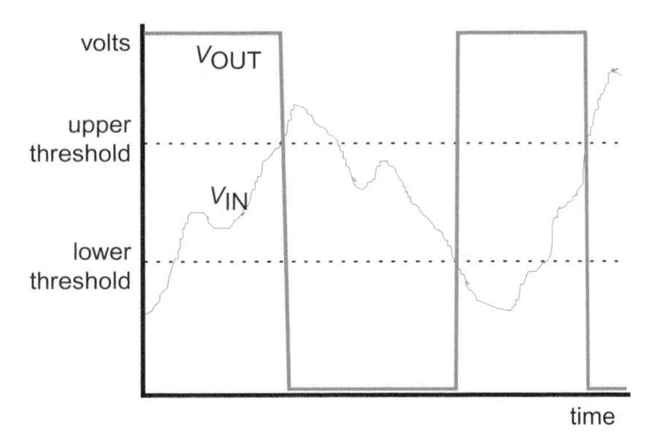

FIGURE 36.6

motor. R2 must have a higher resistance than the load. If a large driving current is needed, use a power BJT for Q2. If the sensor can supply only a small current to the trigger, Q1 can be an FET instead of a BJT.

The exact voltage levels of the thresholds and the difference between them can be altered by changing the values of R2, R5 and R6.

Things to do

Set up the Schmitt trigger circuit on a breadboard. Use a PSU with variable voltage to find out the upper and lower threshold voltages.

Experiment with different resistor values to alter the thresholds.

COMPARATORS

A comparator is an amplifying circuit with two inputs. Its output voltage is proportional to the *difference* between the two input voltages. Its gain is about 200 000, so an input difference as small as 100 μV is enough to swing the output well toward 0 V or the supply voltage.

FIGURE 36.7

Comparators are manufactured as **integrated circuits**. In one type, there are 23 transistors, 2 diodes, 19 resistors.

These and all the connections between them are built up on a tiny silicon chip. This is housed in an 8-pin package. You do not have to know how many components there are inside the package, or how the comparator works. But you need to know what it does and how to use it.

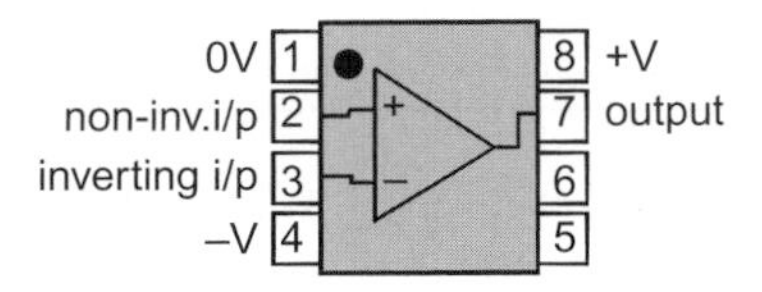

FIGURE 36.8

The drawing above shows the 8-pin package. Pin 1 is identified by the circular 'dot' (see photo also). The other pins are numbered as shown.

The two inputs to the amplifier are the non-inverting input (+) and the inverting input (−). The output swings to the positive supply rail if the (+) input is greater than the (−) input. The output swings to 0 V if the (+) input is less than the (−) input.

This device requires a dual power supply. Pin 1 is connected to 0 V. Pin 8 is connected to the positive supply line. Pin 4 is connected to the negative supply line. The positive and negative supplies must be of equal but opposite voltages.

The transistor at the output is like the one in the first diagram on p. 129. It has no collector resistor. At the output, we need to add a resistor connected to a positive supply line.

An example of a complete comparator circuit is:

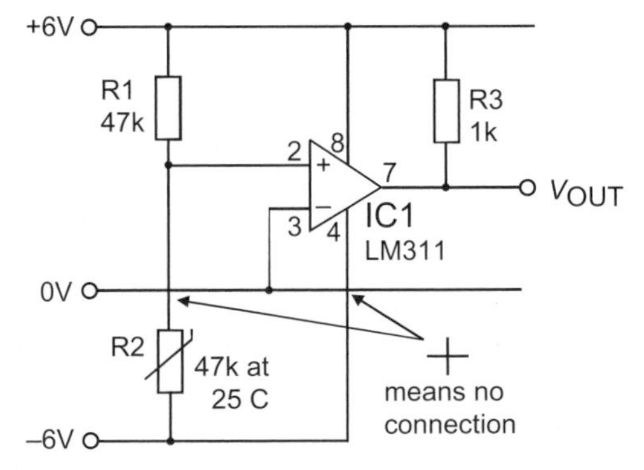

FIGURE 36.9

When you read this diagram, note where wires are connected.

The inverting input (−) is connected to the 0 V supply line, so the inverting input is at 0 V. The non-inverting input (+) is connected to a thermistor voltage divider. If the input from the divider is 0 V or less, the output is 0 V. If the input is a fraction of a millivolt more than 0 V, the output swings very rapidly to +6 V. The output changes from 0 V to +6 V when R2 precisely reaches the set temperature. In this circuit the output changes at about 25°C. Instead of connecting the (−) input to the 0 V line we can connect it to a two-resistor voltage divider that produces a different voltage. The output then changes at a different set temperature.

DESIGN TIME

Build the touch switch circuit on p. 89, but with a Darlington pair instead of a single transistor. Does this have greater sensitivity?

Then try to drive a heavier load, such as a torch lamp, a motor or a siren. You may need to use a power transistor for the second transistor.

A touch switch with a siren could make a useful panic button in a security system.

A moisture sensor and Schmitt trigger could be the basis of a rain detector. When the washed clothes are out on the line, this could warn someone to bring them in. Design and build a system. Select a suitable warning device.

Design and build a wind detector. Use it as part of a system that flashes an LED when it is windy.

Describe how this circuit works.

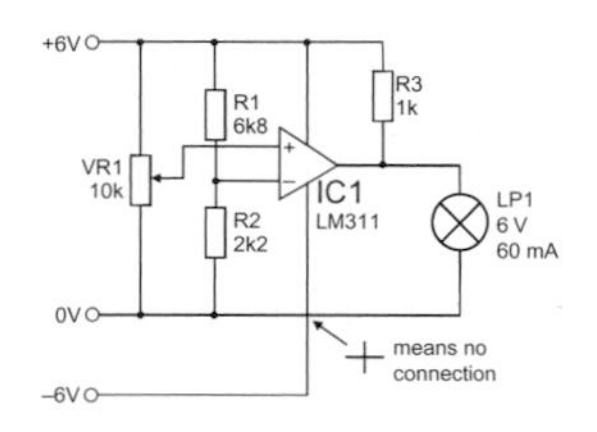

FIGURE 36.10

Draw its system diagram. Suggest an application for this circuit.

Adapt the moisture sensor to sense when a pot plant needs watering.

Try to think of an unusual sensing application for a switch or a sensor. For instance, how could we use a photodiode to sense force? Or how could we use a microswitch to sense water level?

Set up a temporary version of your sensor design, and test it. What are its good points? What are its bad points?

Amplifying Signals

In earlier Topics we have seen how to use a BJT for switching. We can use a BJT on its own, or use two BJTs as a Darlington pair or as a Schmitt trigger. A comparator has a similar switching action. In all these uses, the transistor is used in an all-or-nothing way. It is either off or saturated. All of these are amplifiers because a small change in input results in a large change in output.

Such amplifiers are no use when we need to amplify an audio signal. The waveforms are complicated. When we amplify them, we must maintain their shape as exactly as possible. The aim of the audio amplifier is to produce a varying output voltage that is an exact copy of the input varying voltage, except that the voltages of the output signal are much larger. We turn signal V_{IN} into signal V_{OUT}:

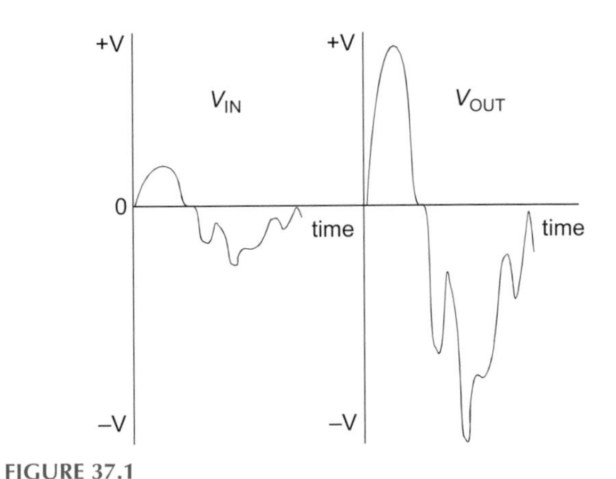

FIGURE 37.1

There are other signals that need similar processing. When a doctor is recording an electrocardiogram (ECG), the electrical signals from the muscles of the patients's heart are amplified before being fed to a chart recorder. A seismologist needs accurately amplified signals in order to analyse tremors during an earthquake. At the other end of the frequency scale, we need to be able to amplify ultra-high-frequency signals in a microwave receiver.

VOLTAGE GAIN

The **voltage gain** of an amplifier is given by:

$$G_V = v_{OUT}/v_{IN}$$

where v_{OUT} and v_{IN} are the output and input voltages at any instant. An amplifier may also have **current gain**, defined in a similar way. Putting these two gains together, and remembering that $P = IV$, we can see that an amplifier increases the **power** of a signal.

Voltage gain is not fixed. It depends on the frequency of the signal. This is mainly due to the effects of capacitance in the circuits. Try measuring this effect in a low-quality audio amplifier, as below.

Things to do

Connect a signal generator to the input of the amplifier. Connect an oscilloscope to the output of the amplifier.

1. Select an input signal of medium frequency, such as 1 kHz. Select the sine waveform, and a suitable amplitude. Use the oscilloscope to view the amplifier output.
2. Set the amplitude of the input signal to several different values. Measure frequency and amplitude of the corresponding output signals. Calculate the gain of the amplifier. Does the amplifier alter the frequency?
3. Repeat (1) and (2) at lower and higher frequencies, such as 100 Hz, 10 kHz, 100 kHz, 1 MHz and the highest frequency available from the signal generator. Calculate the voltage gain at each frequency.
4. Plot a graph to relate gain to frequency. To plot all the frequencies on one graph, it is best to use a logarithmic scale for frequency, like the one in Figure 37.2.

Figure 37.2 shows a typical amplitude–frequency graph taken on a single-transistor general purpose amplifier.

Frequency is plotted on a logarithmic scale, so that each step toward the right represents a tenfold increase in frequency. Amplitude is also plotted on a logarithmic scale. At the top level (0) is the

Electronics: A First Course.

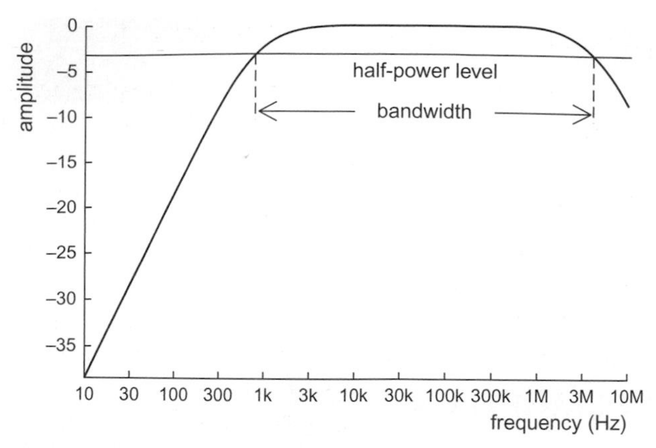

FIGURE 37.2

maximum amplitude. Negative values show decreased amplitude. At the level minus 3, the **power** of the amplitude is half the maximum power.

The graph shows that the output signal is below half power at frequencies less than about 800 Hz. It also falls below half power at frequencies greater than about 5 MHz. There is a range of frequencies, from 800 Hz to 5 MHz, in which the signal is above its half-power level. This range is defined as the **bandwidth** of the amplifier. In this case, the bandwidth is a little under 5 MHz.

AUDIO AMPLIFIER ICs

Audio amplifiers are widely used in audio equipment such as inexpensive radio receivers, music systems, and intercoms. The amplifiers are manufactured as integrated circuits. They are packaged as 8-pin or 14-pin devices at an average cost of £1. The amplifier is complete except for the addition of a dozen or fewer external components such as resistors and capacitors. The most popular audio amplifiers are the LM380, LM386 and TBA820. For data and constructional details, refer to the data sheets or the suppliers' catalogue.

OPERATIONAL AMPLIFIERS

These integrated circuits, known as **op amps** for short, are widely used for general purpose amplification because of their special features:

- Very high input resistance. They take very little current from the device that supplies them with input.
- Very low output resistance. They are able to supply a large output current without serious drop in output voltage.
- Very high gain.

Like comparators, they have two inputs, an inverting and a non-inverting input. Their pin arrangement is slightly different from that of the comparators:

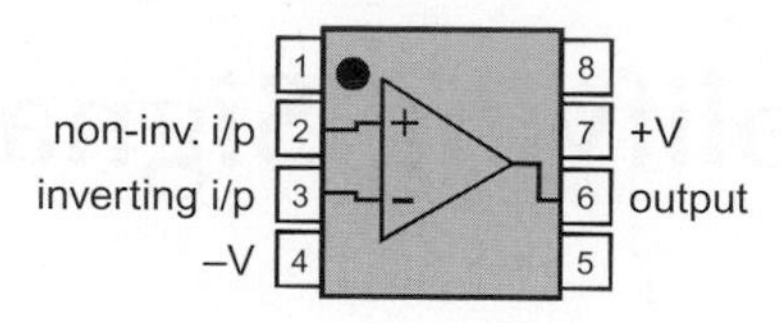

FIGURE 37.3

They operate on a dual power supply, +V and −V. They do not need a connection to the 0 V line. However, many op amps can operate on a single supply, in which case, pin 4 goes to the 0 V line.

The output of an op amp swings towards the +V line when the voltage at the (+) input is greater than that at the (−) input. This is the same as for comparators, but op amps do not need a pull-up resistor at the output. The output swings towards the **−V line** when the voltage at the (+) input is less than the (−) input. This is another difference from comparators.

OP AMP COMPARATOR

Because of their similarity, op amps can be used as comparators. The circuit (Figure 37.4) is similar, except for the power connections and the lack of pull-up resistor.

The output of this circuit swings toward the positive supply as temperature falls.

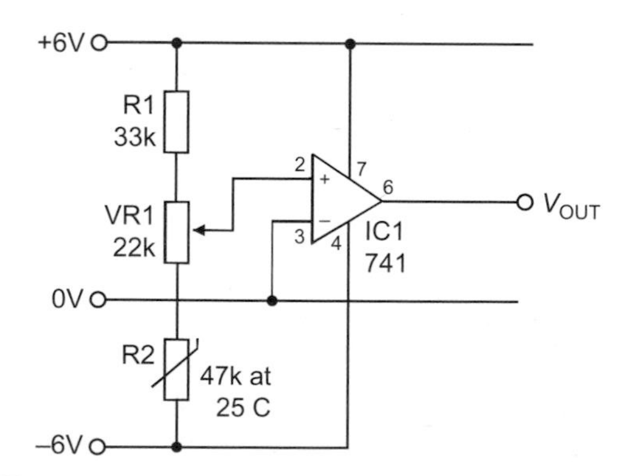

FIGURE 37.4

It swings toward the negative supply as temperature rises. The exact temperature at which the output changes is set by adjusting VR1.

The output of an op amp may not swing all the way to the supply voltages. In the 741, for example, on a ±15 V supply, the output can swing only to ±13 V. In other types it may swing closer.

INVERTING AMPLIFIER

An inverting amplifier is intended to amplify signals without the output swinging too far in either direction. The connections are like this:

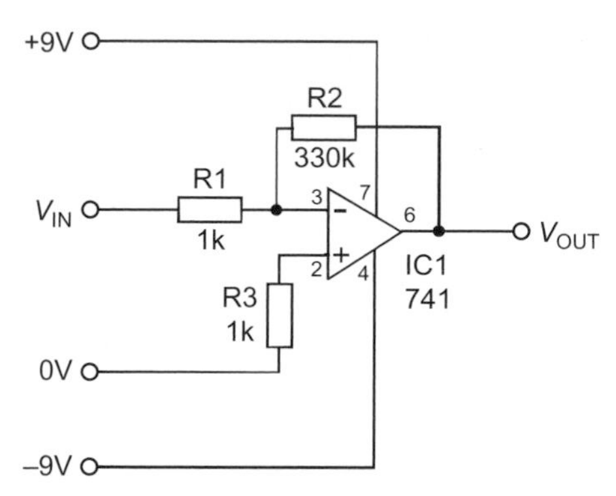

FIGURE 37.5

Note that in *this* diagram (and in some others) the amplifier is drawn with its (−) *above* the (+) symbol. This is to simplify the layout of the drawing. Always check which terminal is which.

In the inverting amplifier, part of the output signal is fed back to the inverting input. The effect of this **negative feedback** is to reduce the voltage gain to:

$$G_V = -R2/R1$$

The negative sign indicates that the amplifier *inverts* the input signal as well as amplifying it. In the diagram, the gain is:

$$G_V = -330\,000/1000 = -330$$

If the input is +17 mV, for example, the output is $-0.017 \times 330 = -5.61$ V. This is assuming that the voltage of the power supply would let the output swing that far. A ±6 V supply is too small to allow this. The amplifier needs at least ±8 V to amplify a voltage as big as 17 mV.

The resistor R3 is there to compensate for voltage drops across R1 and R2. It has a value equal to R1 and R2 in parallel. Here, R2 is so much larger than R1 that the nearest E24 resistor is 1 kΩ, the same as R1. For lower precision, R3 can be omitted and the (−) input connected directly to the 0 V line.

Self Test

1. An op amp wired as an inverting amplifier has R1 = 2.2 kΩ and R2 = 820 kΩ. What is its voltage gain?
2. Select resistors to give voltage gains of **(a)** −220, **(b)** −12, **(c)** −3, and **(d)** −1200.

BOOSTING OUTPUT CURRENT

The output of a typical op amp has a resistance of 75 Ω. This means that there is the equivalent of a 75 Ω resistance between the amplifier circuit and the output terminal. If the load is drawing a current of 1 mA, for example, the voltage drop across this resistance of $0.001 \times 75 = 75$ mV. The output of the op amp is 75 mV less than it should be.

If the op amp is to provide more than about 10 mA, its output must be boosted by using a transistor (BJT or FET), as in the circuit in Figure 37.6.

The load is an 8 Ω loudspeaker. A BC639 transistor has been used, so as to be able to carry up to 1 A.

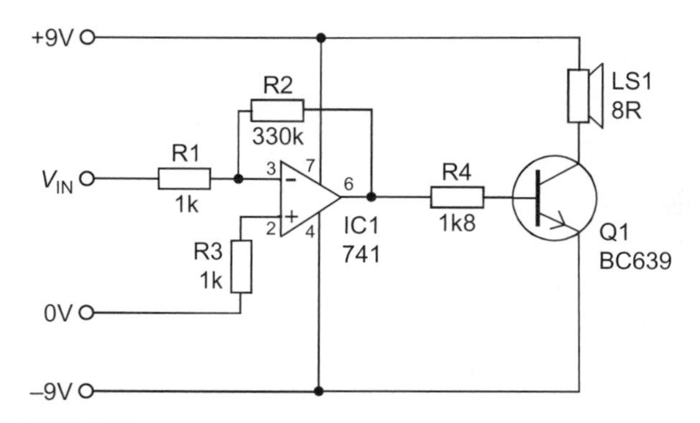

FIGURE 37.6

If necessary, a BJT of higher power can be used. For loads that require more than 1 A, the output of the op amp is fed to a Darlington pair with a power transistor as the second transistor, or to a power Darlington.

SINGLE SUPPLY OPERATION

An op amp is operated on a single supply by using a potential divider to supply the reference line. This is normally the 0 V line, but here the line is at 3 V.

This kind of circuit is useful for battery-powered

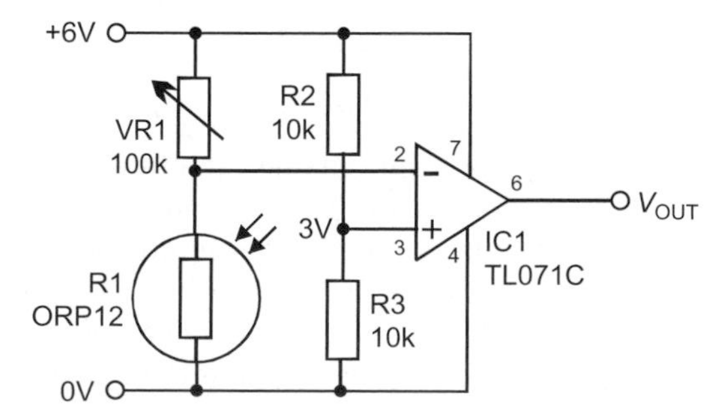

FIGURE 37.7

devices. This one runs on 6 V. However, not all op amps will run on such a low voltage. We have used the TL071C, which runs on a dual ±2 V supply, or on a single 4 V supply. The TL071C has FET inputs, which means that it requires much less input current than the 741, which has BJT inputs.

Another point about the TL071C is that it is a low-noise amplifier.

The output of this circuit is normally at 3 V and swings either side of this. It can not swing fully to 0 V or 6 V.

The LM358 op amp is specially intended for single supply operation, and will run on 3 V (or ±1.5 V). It accepts inputs that are close to the 0 V rails and its output will swing down to this level too. It has BJT inputs like the 741, but the input resistance is higher. The LM358 requires less power to operate it than the 741. The LM358 has two independent amplifiers in the same package, which means that amplifying circuits take up less board space. These features make it ideal for battery-powered portable equipment.

NOISE

Noise is an unwanted signal that has become added to the signal that is being processed.

Noise comes from various sources, either from inside the circuit or from outside. Noise is generated in several kinds of components, including resistors and transistors. It is generated in circuits such as amplifiers that contain transistors. In general, the amount of noise generated is related to the size of the currents passing through these components. One way to avoid noise is:

Keep currents as small as possible.

Transistors and op amps can be designed so as to generate less noise. There are three examples of low noise transistors in the data sheet on p. 100. So another rule for noise reduction is:

Use low-noise versions of transistors and op amps.

Noise that comes from outside the circuit includes voltage spikes on the mains power supply. A refrigerator that is constantly switching its motor on and off produces spikes on the mains supply that affect mains powered circuits in the same house. The spikes may pass through the transformer and into the low-voltage supply. To avoid this:

Place chokes in the power supply circuit.

A similar thing can happen within a circuit where, for example, a relay is continually being switched on and off. Chokes can help. Another technique is to:

Use capacitors for decoupling the supply lines.

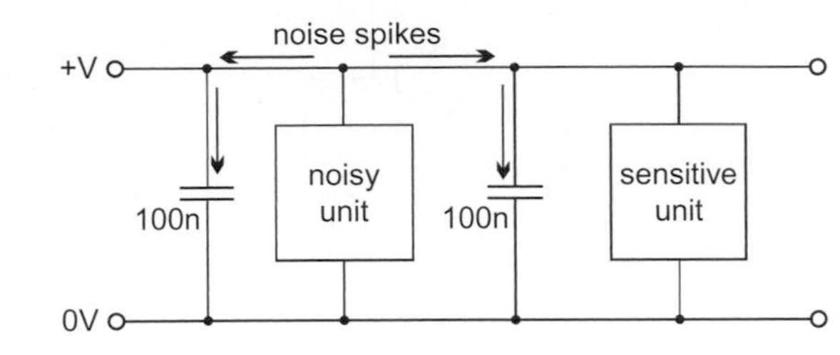

FIGURE 37.8

Spikes and pulses from a noisy part of a circuit pass along the supply lines but are absorbed by the capacitors. This is very important in large logic circuits.

Electromagnetic interference (EMI) is caused by magnetic fields from electrical appliances and equipment. Makers of equipment are required by law to keep EMI down to strict limits.

EMI is reduced by metal screening, connected to ground or the 0 V line.

This is the reason for grounding the aluminium case of a microphone. It is important for signal cable to be screened. Usually the screening consists of a flexible woven sheath of copper wires surrounding the conducting wire or wires. One end of the screening is connected to ground or 0 V line.

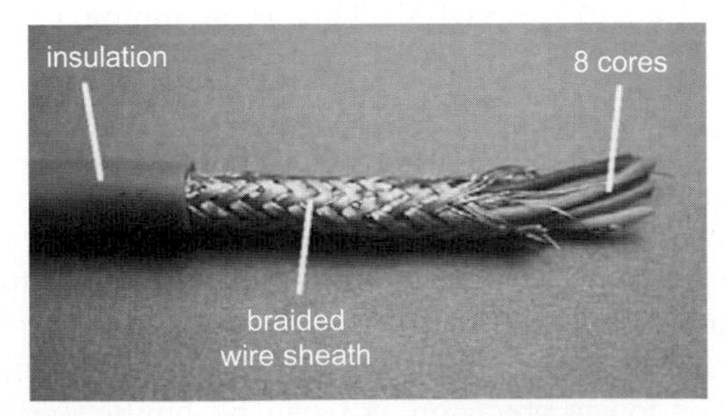

FIGURE 37.9

This 8-core computer cable is shielded by a braided wire sheath, with an outer layer of insulating plastic.

Non-inverting amp

The output of a non-inverting amplifier rises and falls as the input rises and falls. The circuit for a non-inverting amplifier based on an op amp is shown below.

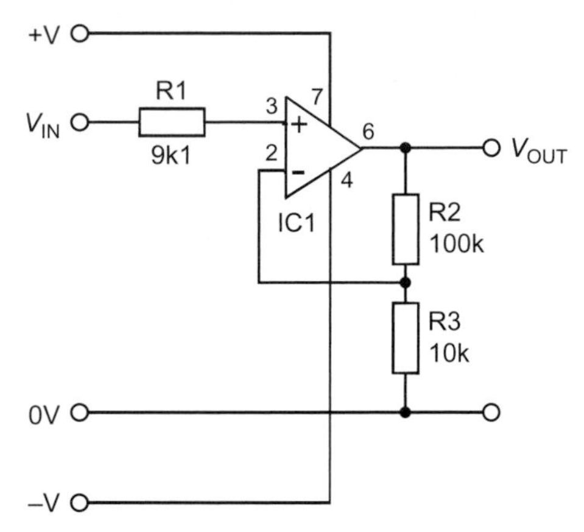

FIGURE 37.10

R1 equals R2 and R3 in parallel. The voltage gain of this amplifier is:

$$G_V = \frac{R2+R3}{R3}$$

For example, with the values in the diagram:

$$G_V = 110/10 = 11$$

In this amplifier, the input current flows through R1 to the (+) input. There, in the 741, it meets an input resistance of 2 MΩ. With such a high input resistance, the op amp draws very little current from the source circuit. The input resistance of op amps with FET inputs is even higher, and they draw almost no current.

Clipping

If the gain of an amplifier is too high, the output voltage can not swing far enough. Then the waveform becomes **clipped**. A clipped audio signal has a distorted sound.

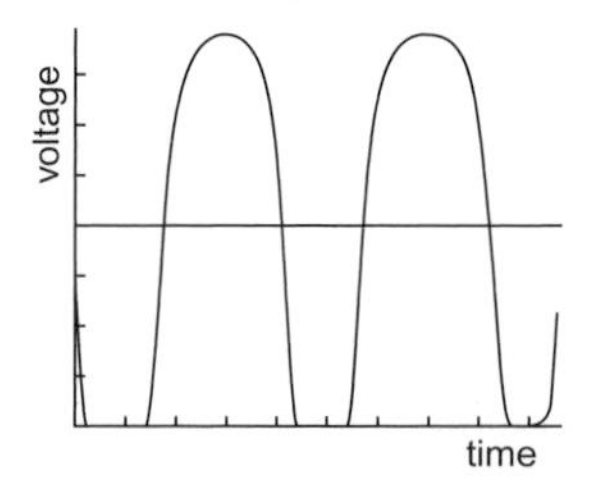

FIGURE 37.11

Op amp mixer

An op amp can be used to mix two or more signals. This is a variation on the inverting amplifier. The gain of each signal is:

$$G_V = -R/R_{input}$$

Where R_{input} is the input resistor for the signal. The output is the sum of these amplified signals.

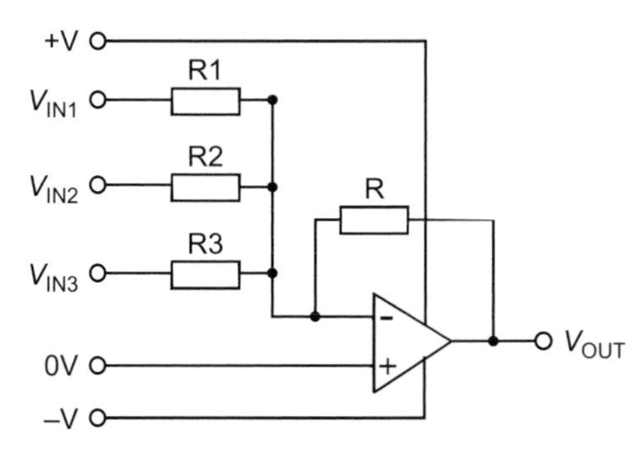

FIGURE 37.12

If all resistors are equal, the gain for each is −1. The signals are simply inverted and mixed. If R1, R2 and R3 are unequal, the signals are mixed in different proportions. An audio mixer has variable resistors for R1, R2 and R3, so that the volumes of each channel can be adjusted. A variable resistor for R allows the amplitude of the mixed signal to be increased or decreased.

Self test

An op amp mixer has R1 = R2 = R3 = 10 kΩ and R = 22 kΩ. V_{IN1} = 10 mV, V_{IN2} = 15 mV, and V_{IN3} = 5 mV. What is V_{OUT}?

Transistor pre-amplifier

The BJT amplifier below is intended to amplify the signals from a source such as a microphone or photodiode sensor.

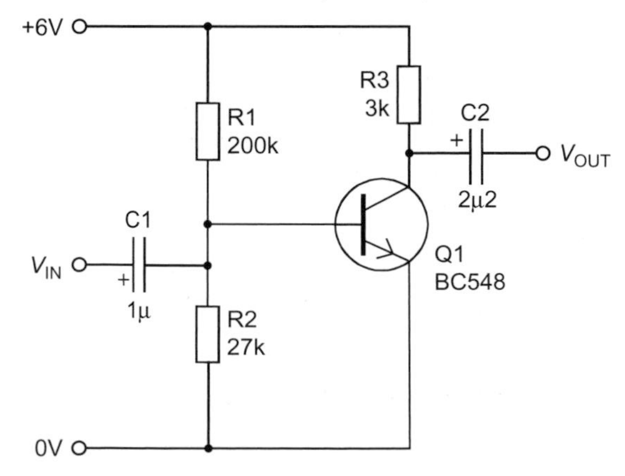

FIGURE 37.13

(continued)

Its features are:

- Resistors R1 and R2 are a voltage divider. They bias the base of Q1 to just over 0.7 V, so the BJT is just switched on but not saturated.
- It may be necessary to alter R1 or R2 slightly depending on the gain of Q1.
- With the biasing resistors as shown and a gain of 400 for Q1, the current through R3 is about 1 mA. This is a small current to minimise noise.
- A current of 1 mA through R3 causes a voltage drop of 3 V. This puts the voltage at the collector about half way between 0 V and the positive supply. This allows the collector voltage to swing freely up or down.
- C1 is a **coupling capacitor** that passes the signal from the microphone to the base of Q1. Its value is selected so that it will pass all frequencies in the audio range (30 Hz to 20 kHz).
- C2 is another coupling capacitor that passes the amplified signal at the collector to a further amplifying stage.
- The varying base current produces a larger and varying collector current. This produces a varying voltage across R3. The collector voltage is an amplified and inverted signal. The gain of this amplifier is about −100.

See the rest of this page and page 139 for more about this and other amplifiers.

Building and testing the pre-amplifier

Things to do

1. Build the pre-amplifier (p. 129) either on a breadboard or on stripboard (see p. 139 and p. 86).
2. Measure the voltages at the base and collector of Q1 to check that they are suitable. If not, replace R2 or possibly R1.
3. As input, use a signal generator, set to give a sine wave at 1 kHz and with amplitude 25 mV. Observe the output signal with an oscilloscope. Measure its amplitude. What is the voltage gain of this amplifier?
4. Check that the gain of the amplifier is constant over the whole audio range.

Figure 37.14 shows how to lay out the components on a piece of stripboard. This is a top view. The black dots at A1, E1 and J1 are 1 mm terminal pins. The diagram shows the copper strips, but these are *under* the board. You can see where to cut one of the strips at E8. A wire link runs from G13 to J13.

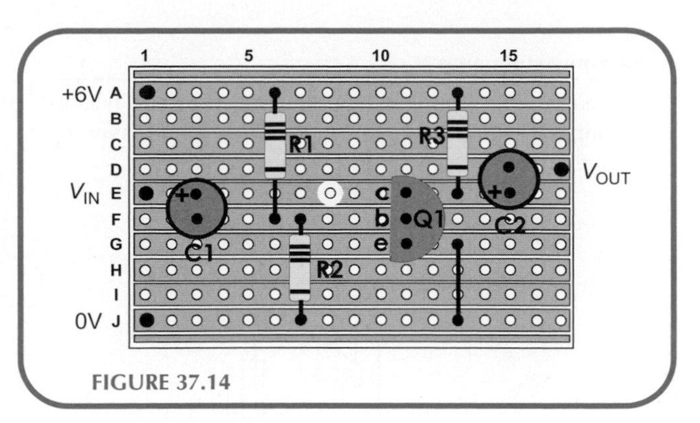

FIGURE 37.14

A stabilised pre-amplifier

The addition of an emitter resistor (R4) with a large-value capacitor (C3) reduces the gain of the amplifier but improves its stability. It makes the gain of the amplifier independent of the gain of the transistor. Because the gain of the transistor depends on temperature, the modified circuit is less dependent on temperature. It is stabilised.

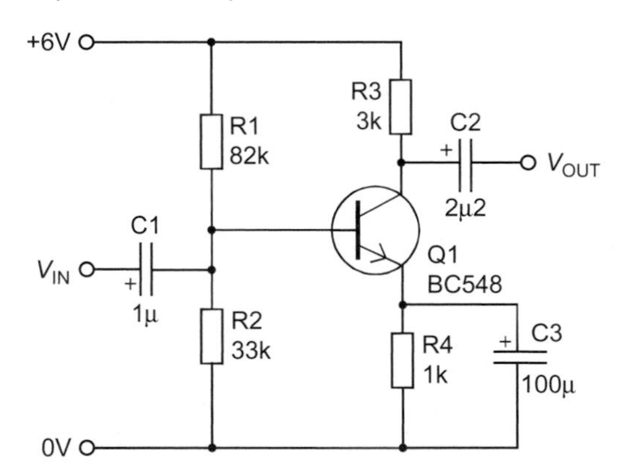

FIGURE 37.15

When the 'no signal' collector current of 1 mA is flowing through R3, an approximately equal current is flowing through R4. This generates a voltage of 1 V across R4. This puts the emitter of Q1 at 1 V. The base-emitter voltage is 0.7 V, as usual. The base must therefore be biased at 1.0 + 0.7 = 1.7 V. The values of R1 and R2 are selected to do this.

Things to do

1. Build the preamplifier, either on a breadboard or on stripboard (p. 86). For the stripboard layout follow the diagram (Figure 37.14). Add R4 and C3 in the space left vacant to bottom right. Do not include the wire link from G13 to J13.
2. Test the amplifier as described in the previous box.

Op amp inverting amplifier

This circuit uses almost any op amp, that has a single amplifier in the package. Instead of a double supply, with 0 V line and positive and negative supply lines, the op amp has a single supply (0 V and +V), with R2 and R3 forming a voltage divider. The divider provides a +3 V reference line.

Things to do

Build the amplifier on a breadboard or on stripboard (p. 86). Note that the copper strips are cut beneath the op amp, at C10, D10, E10 and F10.

1 Select a suitable op amp from data sheets.
2 Decide what gain you require and select the appropriate resistors.
3 Test it as described on p. 138.

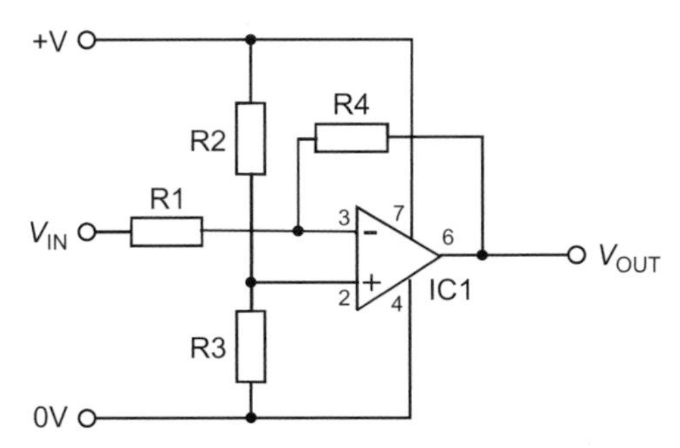

FIGURE 37.16

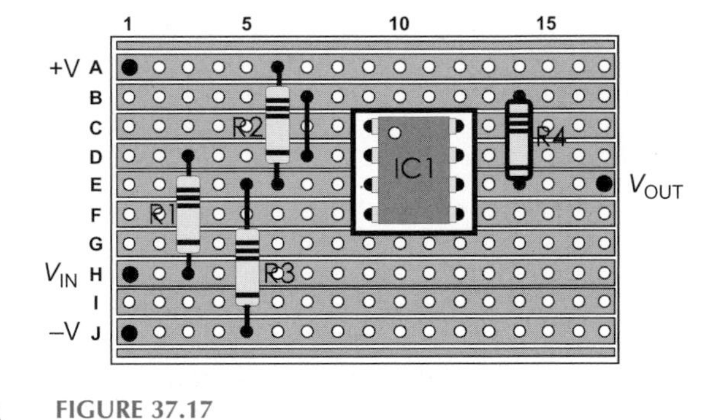

FIGURE 37.17

QUESTIONS ON SYSTEMS, SENSORS AND INTERFACING

1 Draw a system diagram of a electronic system that you might find in an office. Identify the three main stages of the system.
2 Describe a strain gauge, how it works, and one application in which a strain gauge is used.
3 Describe how you would obtain an amplified signal from a named type of microphone.
4 What is a Hall effect device? Describe an application in which a Hall effect device is used. Draw a system diagram to explain how it is used in this application.
5 Describe a simple moisture sensor and how to use it to switch a relay.
6 Why are position sensors important in industrial machinery?
7 Describe three position sensors used in industrial applications.
8 What is a Darlington pair? How does it work? If the voltage gains of the transistors are 120 and 220, what is the gain of the pair?
9 Describe the action of a Schmitt trigger? Draw a diagram to show how it is used.

QUESTIONS ON AMPLIFIERS

1 What is meant by the bandwidth of an amplifier?
2 List the main features of an op amp and explain why they are important.
3 Draw the circuit of an op amp comparator circuit. How would you use the op amp as a comparator to process the signal from a thermistor sensor?
4 Describe how the output of an op amp comparator differs from that of a regular comparator IC (such as a 311).
5 Show how to operate an op amp on a single supply.
6 Draw a circuit diagram of an op amp inverting amplifier. How do we calculate its voltage gain? Give an example.
7 What is noise? What are its causes? How may noise be avoided?

Timing

The circuits that we have looked at in previous topics have all operated instantly. At least, this is how it seems to us. They take a few nanoseconds to respond but, in effect, a change of input immediately produces a change of output. In this Topic we introduce noticeable periods of time into circuit operation.

DELAY

Charging a capacitor takes time (p. 51). The time taken to charge a capacitor introduces a delay into the operation of a circuit. The sequence is:

- Discharge the capacitor completely.
- Let current flow into the capacitor through a resistor.
- Wait until the voltage across the capacitor reaches a set level.
- The delay is the time taken to charge to the set level.

The problem is to monitor the voltage across the capacitor. We need to do this without drawing current from the capacitor. The solution is to use an op amp which has *high-resistance* inputs.

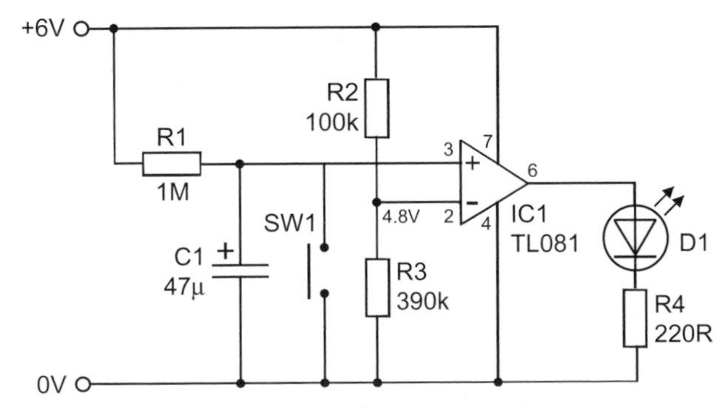

FIGURE 38.1

In the circuit above, current flows through R1 and charges C1. We press SW1 to discharge C1 at the start of the delay.

R2 and R3 form a voltage divider. The voltage at the (−) input of IC1 is 4.8 V.

When S1 is pressed briefly, the voltage across C1 and at the (+) input becomes zero. IC1 is acting as a comparator. As C1 charges, the voltage at (+) rises. At first, it is less than the voltage at the (−) input, so the output of IC1 is close to 0 V. The LED is not lit.

When the voltage at (+) reaches 4.8 V and above, it is greater than the voltage at (−). The output of IC1 swings up to nearly 6 V. The LED is lit. The time taken for charging C1 from 0 V to 4.8 V is approximately 75 s. When S1 is pressed, the LED goes out and there is a delay of 75 s before it comes on again.

This circuit works because IC1 has:

- **High input resistance**. It takes very little current from C1.
- **Low output resistance**. It can provide plenty of current to light the LED.

When a device is being used in this way, we say that it is a **buffer** between the capacitor part of the circuit and the LED part of the circuit.

Things to do

Set up the delay circuit on a breadboard.

1. Measure the delay.
2. Use a digital multimeter to measure the voltage across C1 as it charges. Why is an analogue meter of no use for this measurement?
3. How can you increase or decrease the delay time?
4. Draw a system diagram of the delay circuit.
5. Redesign the delay circuit to switch a lamp *on* when the button is pressed. This could be used to light a lamp in a dark corridor for a period, then switch it off automatically. Think of other applications.

PULSE GENERATOR

Instead of thinking of the previous circuit as a delay, we can think of it as a pulse generator. It produces a low pulse for 75 s, which turns off the LED for that period of time.

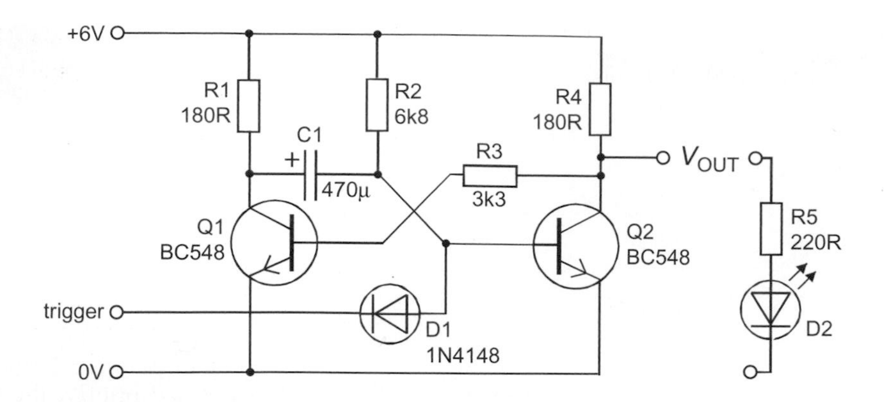

FIGURE 38.2

Figure 38.2 is a circuit for a pulse generator (or delay) that is based on two BJTs.

The diagram shows a possible output to an LED that indicates the state of the circuit.

The circuit consists of two transistor switches. The output from each switch is the input to the other. A system diagram shows that the connection from Q2 to Q1 is direct (through R3), but the connection from Q1 to Q2 passes through a delay stage. The delay stage is provided by C1 and R2 — another example of charging a capacitor through a resistor.

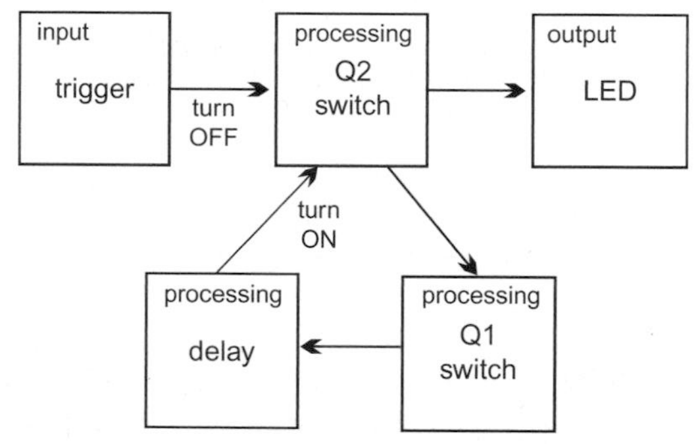

FIGURE 38.3

The circuit is triggered by briefly connecting the trigger input to 0 V. Triggering the circuit turns Q2 off. This puts the LED on (remember, transistor switches are inverting switches).

Without the delay, Q1 would turn Q2 on again immediately. You would not see even a flash of light from the LED. With the delay unit present, there is a delay of a few seconds while C1 charges through R2. Q2 is turned on when C1 has charged to the right level.

This circuit is stable when Q1 is off and Q2 is on. It remains indefinitely in that state.

It is unstable in the reverse state, with Q1 on and Q2 off. After the delay, it goes back to the other state. A circuit that is stable in only one of two states is called a **monostable**. There are several kinds of monostable. They are used to generate a single pulse when triggered.

Things to do

Build the monostable on a breadboard and check voltage levels as it operates.

When you have finished, leave the monostable on the breadboard ready for the addition of a sensor trigger.

FIGURE 38.4

TRIGGERING THE MONOSTABLE

The monostable is triggered by briefly bringing the base of Q2 to 0 V to turn it off. Instead of doing this by hand we can use a sensor. Figure 38.5 shows one way of interfacing the monostable to an LDR.

The interface is a transistor switch that takes its input from a voltage divider R6/R7. The switch turns on when a shadow falls across R7. This causes a fall in voltage at the collector of Q3. This is passed across the coupling capacitor C2 to the base of Q2 in the monostable circuit. We do not need the diode D1 when using this interface.

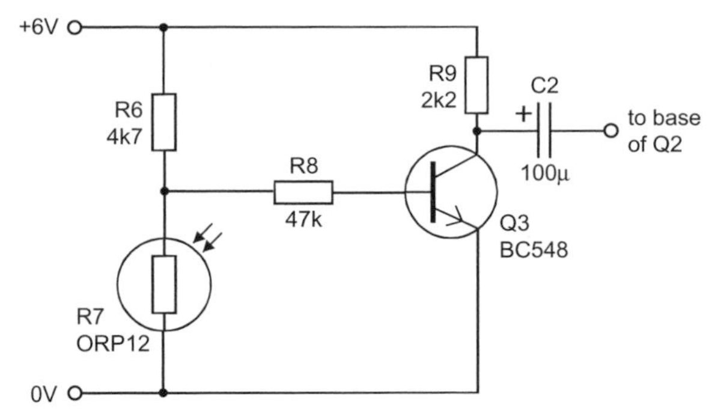

FIGURE 38.5

This is a good example of using a capacitor to couple two circuits together. Normally, the collector of Q3 and one side of the capacitor are at 6 V. Normally, the base of Q2 is at 0.7 V. The capacitor has a voltage difference of 5.3 V between its two plates. When a shadow falls on R7 the voltage at the collector drops suddenly to zero, a fall of 6 V. The voltage on the other plate of C2 falls by an equal amount, from 0.7 V to −5.3 V, maintaining for a while the 5.3 V difference across C3. The fall of voltage at the base of Q2 turns it off, triggering the monostable action. The system diagram is:

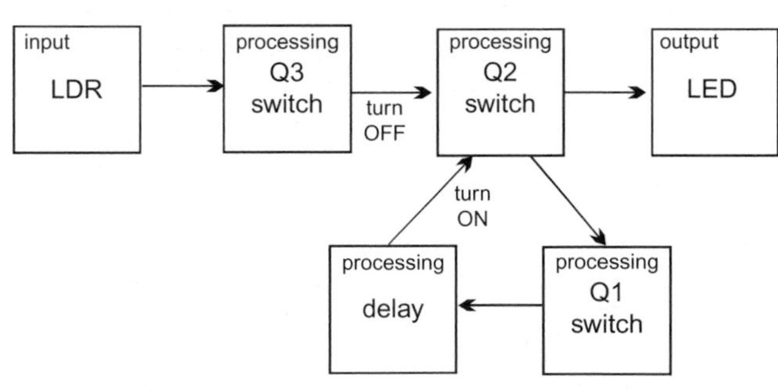

FIGURE 38.6

Things to do

Build the LDR interface on a breadboard and couple it to the monostable (remember to remove D1).

Test the combined circuits. Try to increase the length of the pulse that puts the LED out.

THE 555 TIMER IC

This IC can be used as the basis of a monostable circuit. The advantages of the 555 are:

- Greater precision in the length of the pulse.
- Pulse length is not affected by variations in the supply voltage.
- A long pulse length (up to 1 hour) is obtainable.
- Only a few additional components are needed.
- Output current is up to 200 mA, which is enough to light a lamp or power a relay.

A low-power version of the 555 is known as the 7555. This can generate even longer pulses, with output current up to 100 mA.

Below is the standard monostable circuit for the 7555.

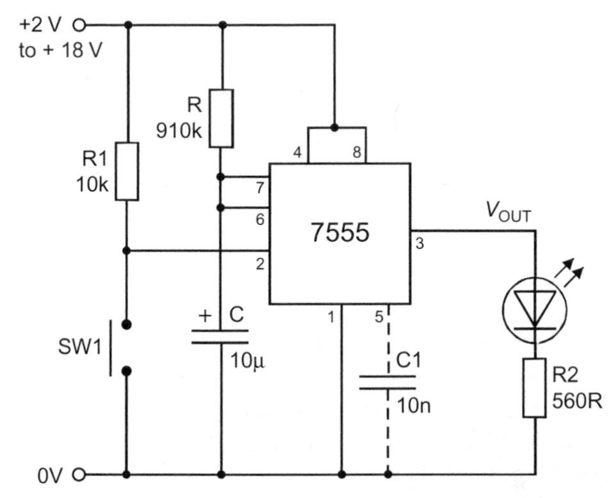

FIGURE 38.7

The 555 operates on a supply of 4.5–16 V. It requires the capacitor C1 connected between pin 5 and the 0 V line. This capacitor is not needed with the 7555.

The length of the pulse depends on the timing resistor R and the timing capacitor C. The length, *t*, of the pulse is:

$$t = 1.1RC$$

t is in seconds, *R* is in ohms and *C* is in farads. With the values in the diagram, the pulse lasts for 10 s.

The trigger input at pin 2 is normally held at the positive supply voltage. In the diagram we show a pull-up resistor R1 that does this. The timer is triggered by a brief low pulse. Here we have a push-button that, when pressed briefly, connects the trigger input to the 0 V line. The input can be triggered in other ways by interfacing various sensors to the timer.

The output (pin 3) is normally at 0 V. It rises instantly to the supply voltage when the timer in triggered. It falls instantly to 0 V at the end of the pulse.

Things to do

Build and test the monostable circuit, using a 555 or 7555 timer IC.

1. Try replacing R with resistors of other values and see what effect this has on the pulse length.
2. Try replacing C with capacitors of other values and see what effect this has on the pulse length.

(continued)

(continued)

3 Using the equation given on page 143, try to build timers with pulse lengths of 20 s, 60 s, 5 min, and 0.1 s. Use an oscilloscope to check the 0.1 s pulse.
4 Interface the timer to an LDR sensor. Find values for R and C so that the LED lights for 30 s when a shadow falls on the LDR.

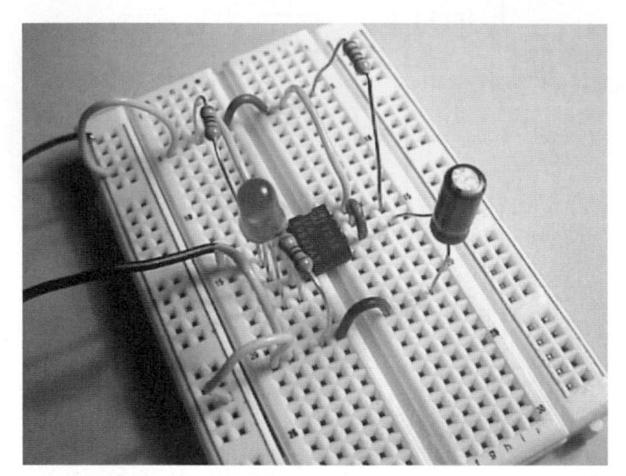

FIGURE 38.8

Pin 4 is the reset pin. Often this is permanently wired to the positive supply line, as in Figure 38.8. If this pin is briefly connected to 0 V while the timer is generating a pulse, the output of the timer at once goes to 0 V.

Things to do

Connect a resistor and push-button to reset the timer.

ASTABLE CIRCUITS

This kind of circuit is like a monostable that triggers itself to start again at the end of pulse. The result is a circuit that is *not* stable in any state. It runs continuously, producing pulses indefinitely. It is **astable**.

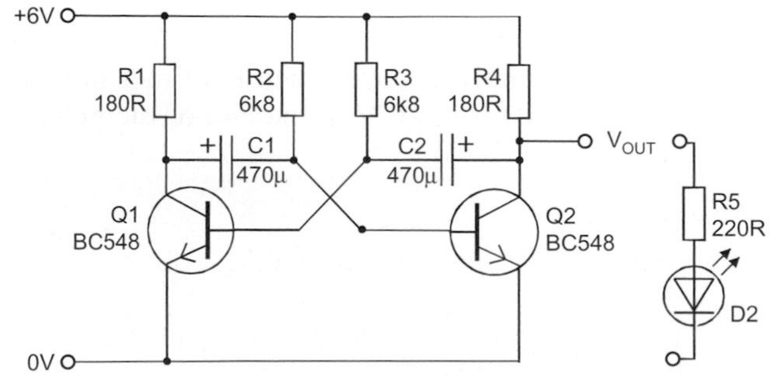

FIGURE 38.9

A clear example of this is the circuit in Figure 38.9. This is similar to the BJT monostable but with a delay circuit for each transistor switch.

The system diagram of the astable is:

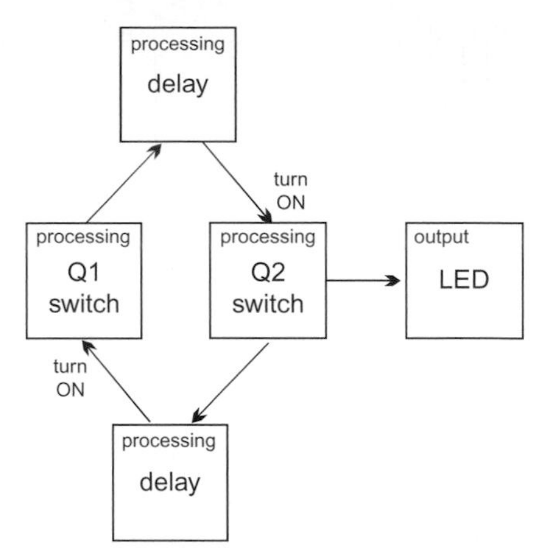

FIGURE 38.10

This is a system without a special input stage. The astable runs for as long as the power is switched on. However, the power switch (not shown in the drawing) can be thought of as an input. Another possible input is a switch that closes to connect the base of Q2 to the 0 V line. This holds Q2 off and prevents the astable from running. It runs again as soon as the switch is opened.

Things to do

Build the astable on a breadboard.

1 How long are the pulses? How long are the gaps (LED off) between the pulses?
2 Try the effect of replacing C1 and C2 with capacitors of other values, such as 47 μF. If the LED flashes very quickly, use an oscilloscope to measure the frequency.
3 Try the effect of replacing R2 and R3 with resistors of other values, such as 680 Ω.
4 Investigate what happens when C1 and C2 are not equal in value, or when R2 and R3 are not equal.
5 Try to get the astable running at several hundred hertz. Replace the LED with some other more suitable output stage.

THE 555 ASTABLE

The 555 or 7555 timer IC can be used to build an astable circuit:

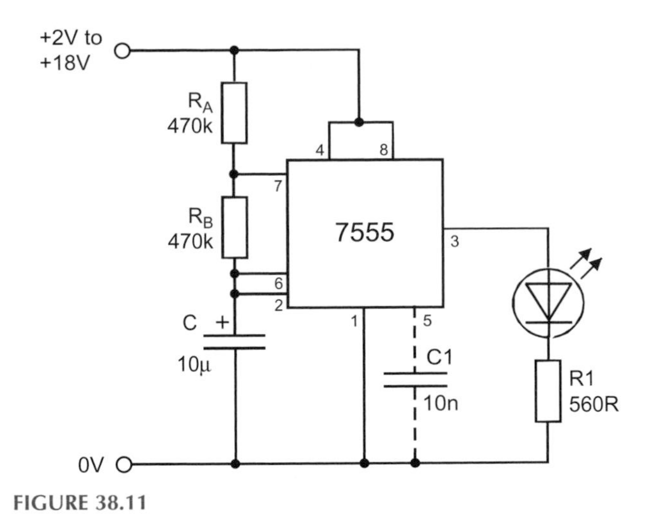

FIGURE 38.11

In the astable, the timing capacitor C is charged by current flowing from the positive supply through the timing resistors R_A and R_B. While this is happening, the output at pin 3 is equal to the supply voltage. The time taken for C to charge depends on the values of R_A, R_B and C.

As soon as the capacitor is charged, it is immediately discharged through R_B and pin 7. While this is happening, the output at pin 3 is at 0 V. The time taken depends on the values of R_B and C, but not on R_A.

One end of C is connected to the trigger input (pin 2, see p. 134). As C fully discharges, the falling voltage at pin 2 triggers the IC to start charging C again. With the values shown in the diagram, the IC takes about 10 s to run through a complete charge/discharge cycle.

The time, t, for one complete cycle is:

$$t = \frac{(R_A + 2 \times R_B)C}{1.44}$$

The time, t_1, for which the output is high is:

$$t_1 = 0.69(R_A + R_B)C$$

The time, t_2, for which the output is low is:

$$t_2 = 0.69R_BC$$

It is clear from the equations that t_1 is greater than t_2. We can make them *almost* equal by making R_A small compared with R_B. If we must have the periods exactly equal, or if t_1 must be smaller than t_2, we use a circuit with diodes, as in Figure 38.12.

For heavier loads, the output of a 555 or 7555, either in monostable or astable mode, can be used to drive a transistor switch. The circuit is the same as is used for boosting the output of an op amp (p. 135).

Things to do

Build and test a 555 or 7555 astable.

Duty Cycle

The duty cycle of an astable is:

$$\text{Duty cycle} = \frac{t_1}{t_1 + t_2} \times 100\%$$

A 555 astable circuit such as that in Figure 38.11 always has a duty cycle greater than 50%. With a circuit like the one below, the lengths of t_1 and t_2 can be set independently.

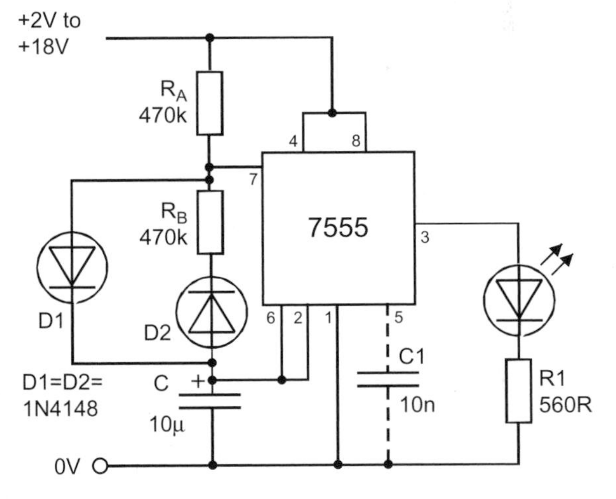

FIGURE 38.12

The capacitor is charged through R_A and D1. It is discharged through D2 and R_B.

$$t_1 = 0.69R_AC$$

$$t_2 = 0.69R_BC$$

Variable Frequency

To obtain variable frequency, make R_A small (say, 1 kΩ) so that the duty cycle is always close to 50%. For R_B, use a variable resistor with a small fixed resistor in series with it. The fixed resistor (say, 470 Ω) is there to prevent a zero resistance between pins 6 and 7 when the wiper of the variable resistor is at one end of its track.

Variable Duty Cycle (1)

The duty cycle can be varied by replacing resistors R_A and R_B by a variable resistor. The wiper is connected to pin 7. In this diagram, R_A is replaced by R1 and the 'top' part of VR1. R_B is replaced by the 'bottom' half of VR1 and R2.

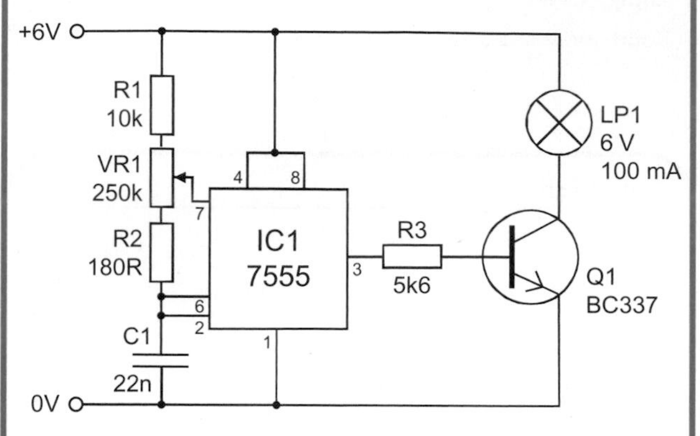

FIGURE 38.13

The duty cycle can be adjusted from about 70% to almost 100%. The astable runs at about 170 Hz. The lamp is turned on and off at this rate. It flickers too rapidly for the eye to see. It appears to be on continuously.

This is a lamp dimming circuit. The brightness can be varied smoothly over the range of 70% to 100%.

When the duty cycle is less than 100%, the lamp is being supplied with power for only part of the time. The shorter the 'on' period and the longer the 'off' period, the dimmer the lamp.

Variable Duty Cycle (2)

The disadvantage of the circuit in Figure 38.13 is that the duty cycle must always be 50% or more. This means that it is not possible to dim out the lamp completely. To give full dimout, we need to be able to reduce the duty cycle to almost zero. We use diodes as in the circuit in Figure 38.12.

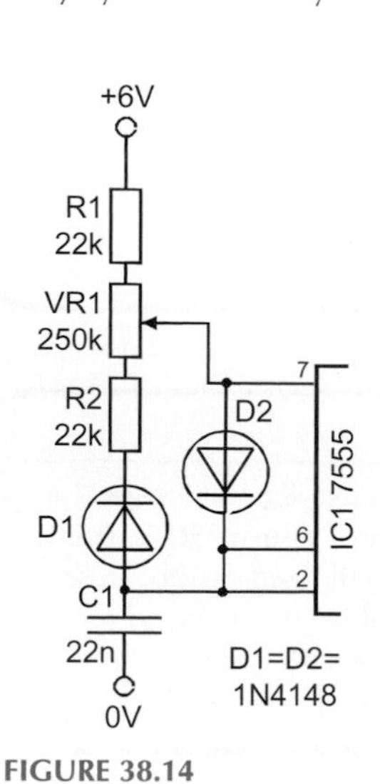

FIGURE 38.14

Rl and the 'top' part of VR1 represent R_A. The 'bottom' part of VR1 and R2 represent R_B. Now the duty cycle can range from 3.7% to 96.3%. The brightness of the lamp ranges from off to (almost) full brilliance.

This circuit is also useful for controlling the speed of a motor. It is better than simply putting a variable resistor in series with the motor. This is because a typical DC motor does not run well on a small voltage. It does not start well and, when running has a tendency to stall. This is particularly likely to happen if the mechanical load on the motor is suddenly increased.

Using the 7555 astable control, variation in the duty cycle causes variation in the *length* of the pulses, not variations in their voltage. The turning action of the motor is always at full strength. The result is that the motor can be run slowly without stalling.

Provided that the astable runs fast enough (a few hundred Hertz), there is no jerkiness in the action of the motor.

EXTENSION QUESTIONS ON AMPLIFIERS

1 Describe the circuit of an op amp non-inverting amplifier. What value resistors of the E12 series could be used to produce a gain of 23? What is the nearest value for the input resistor in the E12 series?

2 The amplifier of Q. 1 is running on a supply of ±15 V. The input is a sine wave of frequency 250 Hz, amplitude 600 mV. Describe the waveform of the output voltage.

3 This is the circuit of an op amp mixer with three inputs:

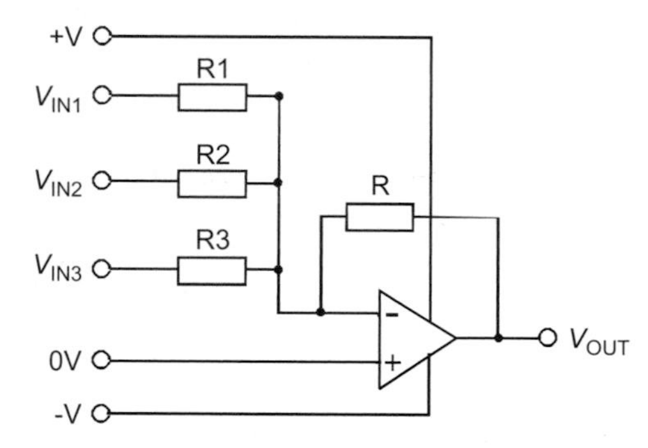

FIGURE 38.15

R=Rl=20 kΩ R2=10 kΩ, and R3=5 kΩ. If all three input voltages are 200 mV, what is V_{OUT}?

4 Given a single BJT pre-amplifier, what additional components are used to stabilise it? Where are they connected in the circuit. What is their effect on the output of the amplifier?

5 What is the purpose of the coupling capacitors in a BJT amplifier?

6 Using data sheets from manufacturers or other sources, make lists of three op amps with BJT inputs and three op amps with FET inputs. For each type, list its input resistance and its output resistance. What is the advantage of having (a) a high input resistance, and (b) a low output resistance?

7 The collector resistor of BJT preamplifier is chosen so as to make the output voltage about about half the supply voltage. What is the reason for doing this?
8 What is a suitable value for the collector current of a BJT pre-amplifier, and why?

QUESTIONS ON TIMING

1 Describe a delay circuit built from a resistor, a capacitor and an op amp. Why is the op amp needed?
2 Draw the circuit diagram of a monostable based on two BJTs.
3 Draw the system diagram of a BJT monostable. What are the advantages of a monostable circuit based on a 555 or 7555 timer IC?
4 With the help of a circuit diagram, describe a practical application for the 555 or 7555 IC when used as a monostable.
5 Given the values of the timing resistor and timing capacitor, calculate the pulse lengths produced by a 555 monostable:
   (a) $R=47\ k\Omega$ and $C=100\ nF$.
   (b) $R=10\ k\Omega$ and $C=2.2\ nF$.
   (c) $R=2.2\ M\Omega$ and $C=470\ \mu F$.
6 Given that the timing capacitor of a 555 monostable is 39 nF, what is the nearest E24 value of resistor required to produce a pulse of:
   (a) 1.15 ms.
   (b) 5 μs.
   (c) 22 ms.
7 With the aid of a circuit diagram and system diagram, describe the action of a BJT astable.
8 A 555 astable has $R_A=22\ k\Omega$, $R_B=47\ k\Omega$ and $C=100\ nF$. What are the lengths of (a) its cycle, (b) its high output pulses, and (c) its low output between pulses?

EXTENSION QUESTIONS ON TIMING

1 What is meant by duty cycle? If the output of an astable is high for 45 ms and low for 5 ms, what is its duty cycle?
2 Draw a diagram to show how to obtain duty cycles of 50% or less from a 555 astable.
3 Describe a circuit that uses a 555 astable as a motor speed control. What is its advantage over a variable resistor in series with the motor?

MULTIPLE CHOICE QUESTIONS

1 The processing stage of a typical electronic system contains:
   A a loudspeaker.
   B an amplifier.
   C a sensor.
   D an interface.
2 The arrows in a diagram of an electronic system represent:
   A the flow of information.
   B electric currents.
   C interfaces.
   D connecting wires.
3 A Hall effect device is sensitive to:
   A magnetic field.
   B moisture.
   C infrared.
   D ultrasound.
4 An example of a sensor that changes its resistance is a:
   A photodiode.
   B capacitor microphone.
   C Hall effect device.
   D strain gauge.
5 The device shown in Figure 38.16 is:

FIGURE 38.16

   A an infrared diode.
   B a crystal microphone.
   C a photodiode.
   D an ultrasonic sensor.
6 The circuit in Figure 38.17 is a:

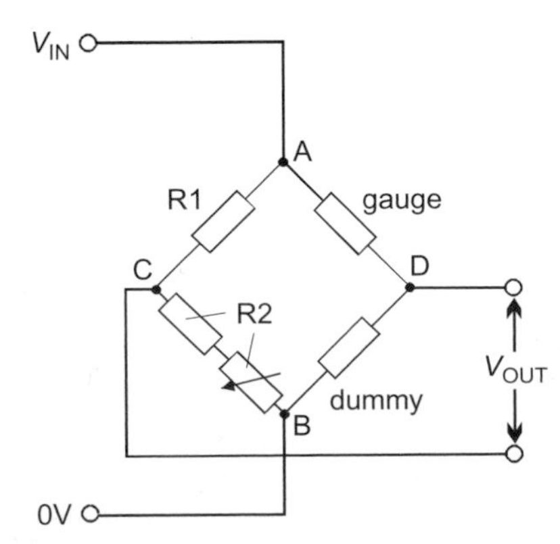

FIGURE 38.17

A voltage divider.
B Schmitt trigger.
C Wheatstone bridge.
D Load cell.

7 In the circuit Figure 38.18, the sensor is a thermistor. When a current of hot air is blown at the sensor, the voltage V_{OUT}:

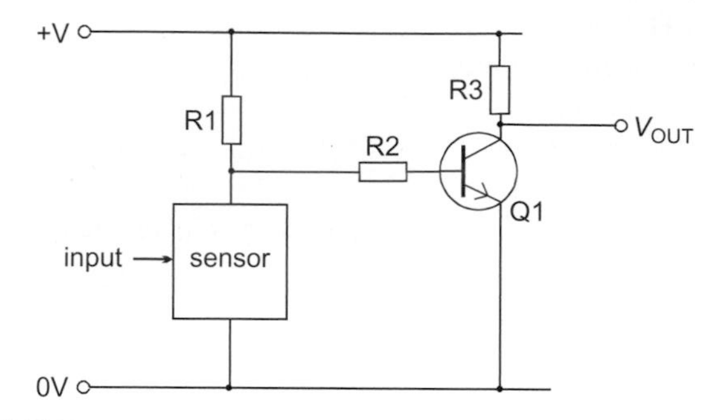

FIGURE 38.18

A rises.
B falls.
C does not change.
D becomes zero.

8 In an inverting op amp amplifier circuit, the input resistor is 15 kΩ and the feedback resistor is 180 kΩ the power supply is ±9 V. If the input voltage is 120 mV, the output voltage is:
A +1.44 V.
B + 8 V.
C −1.44 V.
D −7.2 V.

9 An appliance such as a large refrigerator can put voltage spikes on the mains when it switches on or off. To reduce the effects of these spikes getting into an amplifier circuit connected to the same mains supply, we:
A use a choke.
B use a shielded mains lead.
C keep currents in the amplifier below 1 mA.
D use low-noise transistors.

10 In the circuit Figure 38.19, c is several volts positive of e. A current of 50 μA flows into b. The gain of Q1 is 150 and the gain of Q2 is 100. The current flowing into c is:

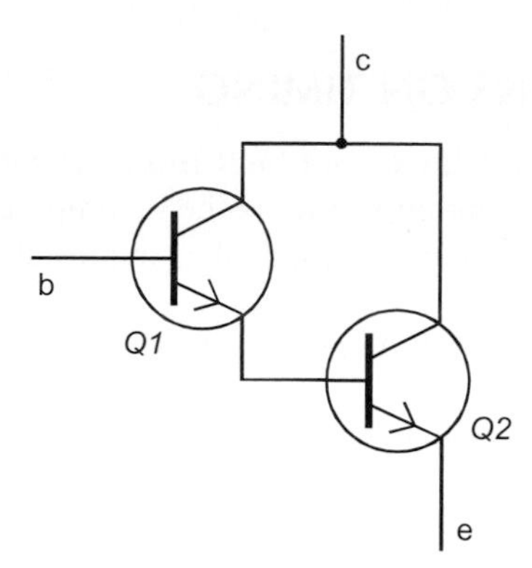

FIGURE 38.19

A 50 μA.
B 5 mA.
C 750 mA.
D 7.5 mA.

11 The circuit in Figure 38.19 is called:
A a Darlington pair.
B a Schmitt trigger.
C an inverting amplifier.
D a non-inverting amplifier.

12 A circuit that produces one pulse when triggered is called:
A a pulse generator.
B a monostable.
C an astable.
D a delay circuit.

13 One of the features of the 555 timer IC is that:
A the pulse length is not affected by the supply voltage.
B it can supply up to 1 A from its output.
C it is triggered by a positive-going input pulse.
D its output goes low when the IC is triggered.

DESIGN TIME

This page has some ideas for projects based on timing. Design and build!

A BJT monostable is used as a 'pulse stretcher' in an electronic version of the twisty wire game. When the loop touches the wire, the monostable is triggered for about 5 seconds, ringing a bell or flashing a lamp for that time.

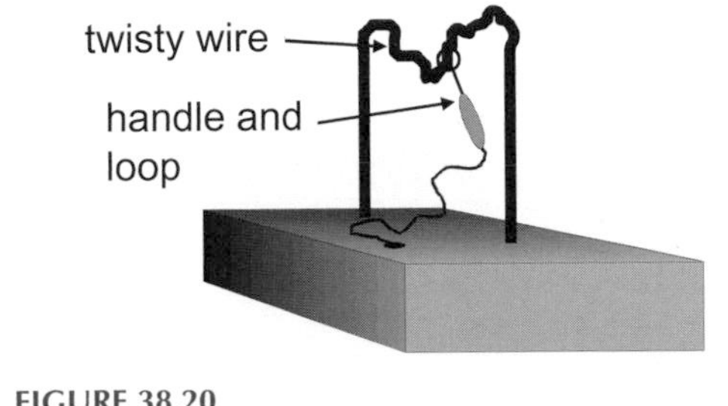

FIGURE 38.20

A porch lamp or a lamp for a dark corridor is to be switched on for 2 minutes by pressing a button briefly. Use a 555 timer in monostable mode.

An egg timer is based on a 555 monostable that switches on an LED. Press a button when you put the egg in the boiling water. The LED comes on until the egg is ready, 5 minutes later.

Try to improve the design by adding a switch to select cooking periods of 4, 5 and 6 minutes.

A simple buzzer or single-note siren (p. 188) is cheap. But the continuous sound it gives out is easy to ignore. Build an astable (either BJT or 555) that makes the siren 'beep' about once a second. This makes the sound more attention-catching.

A metronome circuit flashes an LED at a rate that can be set to several different musical tempi, from *largo* to *presto*. Preferably make the duty cycle low, so that the LED flashes crisply on the beat.

A 555 astable can be the basis for an audio signal generator. You can drive the speaker with a transistor switch. The frequency of the generator should be adjustable over the range 50 Hz to 1 kHz.

Finally, a rather tricky one. The panic button, when pressed, makes this project emit a note at about 1kHz for a period of 3 minutes.

To make it even trickier, let the note sound for half-second 'bleeps' with a half-second gap between them.

There are lots of other timing projects that you could design. Think up some for yourself!

Logic

Logic circuits are used for processing **binary** information. By 'binary', we mean that the information has only two possible states. For example, a switch is open or it is closed. It can not be half-open or half-closed.

There are two switches in the circuit below. There is one lamp. The circuit has two binary inputs and one binary output.

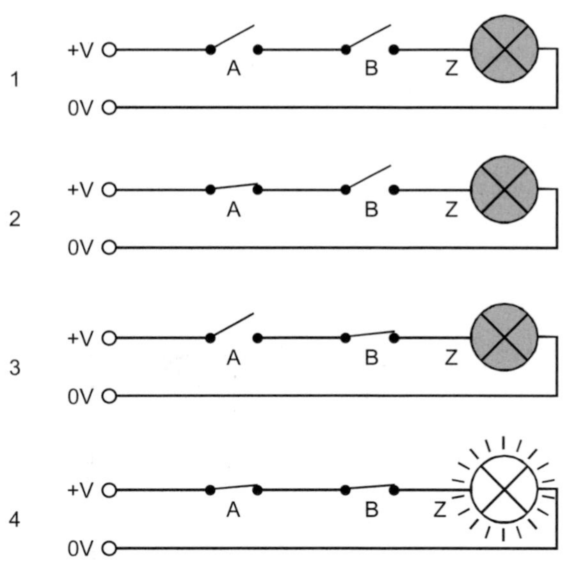

FIGURE 39.1

There are four possible ways in which the two switches can be set:

1 A open and B open: lamp off.
2 A closed but B open: lamp off.
3 A open but B closed: lamp off.
4 A closed AND B closed: lamp ON.

There is only one way to light the lamp — close A AND B.

The circuit performs a **logical operation**, the **AND** operation. The circuit works only for closing the switches. Leaving A and B open has no distinctive result. Note the binary nature of the inputs. The switches are open or closed. Note the binary nature of the output. The lamp is on or off.

This circuit can have practical applications. The switches could be microswitches that detect the positions of two safety grids on a power drill. The switches close when the grids are locked in position. The lamp comes on when *both* grids (A AND B) are in position. This signals 'OK to drill' to the operator.

The action of the circuit can be summarised if we represent the binary states of inputs and output by '0' and '1'. For the switches, 0 = 'switch open' and 1 = 'switch closed'. For the lamp, 0 = 'lamp off' and 1 = 'lamp on'.

Now we can set out the four states of the switches in a **truth table**:

Inputs		Output
B	A	Z
0	0	0
0	1	0
1	0	0
1	1	1

The table shows the lamp on only when both A AND B are closed.

Logic is not limited to two inputs. There may be any number. For example, there could be four switches in series. The lamp lights only if A AND B AND C AND D are all closed.

Another example of switched logic is illustrated in Figure 39.2. By connecting the switches in parallel we obtain logical OR. The lamp lights when any one or more of the switches is closed.

A system such as this provides control over a device from one of a number of stations. A domestic example is the lamp on an upstairs landing. This is controlled from switches at the top and bottom of the stairs.

Another example is a fire alarm system in which the alarm is switched on by closing any one of a number of switches located in different parts of a building.

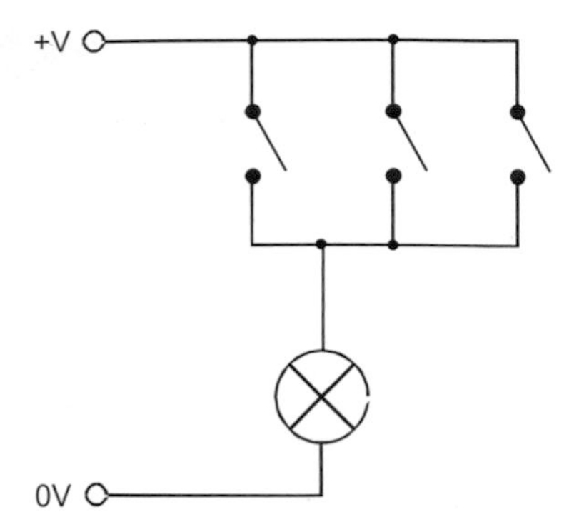

FIGURE 39.2

There are many instances when a few switches wired together can perform logic, as in the examples given above. In some systems we use relays instead of ordinary hand-operated switches, but the principle is the same.

However, there is a limit to what can be done with mechanical switches. The next example shows how we use electronic logic circuits for the same purpose.

Electronic logic is faster and cheaper than mechanical switching, and we can build complex logic functions in a very small space.

ELECTRONIC LOGIC

Electronic logic circuits work with two levels of voltage:

- **Low:** 0V or close to 0V.
- **High:** The positive supply voltage, or close to it. In some types of logic circuit, 'high' is always 5 V. In others, it may have other values.

Usually the low voltage level corresponds to logical'0' and the high level to logical '1'.

To see how logic circuits operate, we will consider a practical example of a logical system. This is part of a security system that controls a floodlight in the garden of a house. Intruders are detected when they break a beam of infrared light that is directed at aphotodiode. The floodlight is to be switched on when the beam is broken. However, there is no point in turning on the floodlight during the daytime, so there is an LDR set to tell whether it is day or night.

The system diagram in Figure 39.3 shows the two sensors and the processing of their outputs by two interface circuits. These are a Schmitt trigger and a transistor switch. The diagram shows which logic levels correspond with which conditions at the inputs.

The logic signals from the interfaces both go to the same next stage. This is a logic circuit that performs the AND operation. It works according to the truth table early in this chapter. It accepts two inputs, A and B, and produces a single output, Z. According to the table, the floodlights are switched on only when it is night AND an intruder is detected.

LOGIC GATES

Building logic circuits is simple. All the logic gates and other more complicated logic circuits that you might need are available as integrated circuits. There are two commonly used 'families' of logic ICs:

- **TTL**, which is short for transistor–transistor logic. This runs on 5 V, so it needs a regulated power supply. TTL type numbers all begin with '74' so this is sometimes known as the 74XX series. There are various types of TTL, of which the Low Power Schottky type has almost replaced the original 74XX series. 74LSXX ICs need less power than the 74XX type.

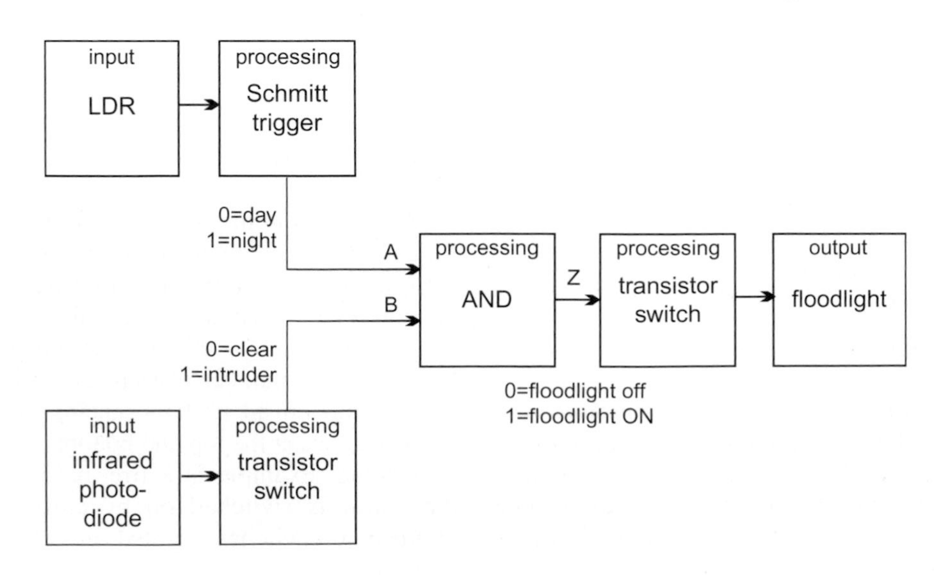

FIGURE 39.3

- **CMOS**, which is short for complementary MOS. These have type numbers ranging upward from 4000, so are sometimes known as the '4000' series. There is also the '4500' series. Members of both series run on any voltage between 3 V and 15 V.

CMOS is slower than TTL but requires less current. It has the additional advantage that it does not require a regulated power supply.

Many of the 74XX series are also available as CMOS ICs. Their type numbers begin 74HC. They operate on 2 V to 6 V, require less current than TTL and are faster than CMOS.

LOGIC ICs

Both TTL and CMOS are packaged as double-in-line ICs (see Figure 39.4). They usually have 14 or 16 pins, but sometimes more. There are four AND gates to the 7408, 74LS08, and 4081 ICs. The four gates share the power supply pins.

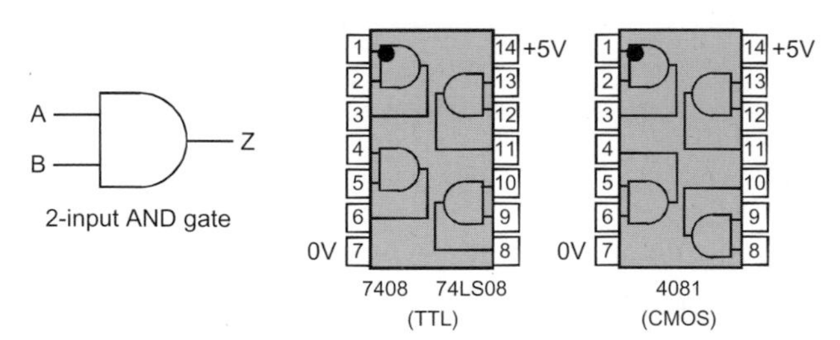

FIGURE 39.4

Note that the connections to the gates are different in the two types of IC.

Design Tip

The two types of logic IC differ in the way they must be used.

1. TTL must have a 5 V regulated supply. CMOS can run from batteries.
2. All inputs of a CMOS IC must be connected to 0 V, the positive supply, a CMOS output, or another point in the circuit. If you leave any input unconnected while testing a circuit, the IC will not work properly. It may overheat and destroy itself.
3. TTL inputs may be left unconnected. An unconnected input behaves as if it has a high (= '1') signal applied to it. For best operation, the unused inputs should be connected to the positive supply through a 1 kΩ resistor. Several inputs can share the resistor.

BOOLEAN NOTATION

On previous pages we have represented logical operators by writing out the name in full. For example, the action of a two-input AND gate is written:

$$Z = A \text{ AND } B$$

In this equation, A and B are the values of the two inputs. They may be 'low ('0') or 'high' ('1'). Z is the value of the output, which may also be 'low'or 'high', depending on the entries in the truth table.

The equation can also be written using a 'dot' (.) for 'AND'. The equation is written:

$$Z = A.B$$

The OR operation is represented by a 'plus' symbol. Thus, the action of a three-input OR gate is:

$$Z = A + B + C$$

The NOT or INVERT operation is indicated by placing a bar or line over the variable. So the action of a NOT gate is written:

$$Z = \overline{A}$$

OR GATE

This performs a common logical operation. In words, the output of an OR gate is high (1) if any one OR more of the inputs is high.

Figure 39.5 shows the symbol for a 2-input OR gate. Any larger number of inputs is possible.

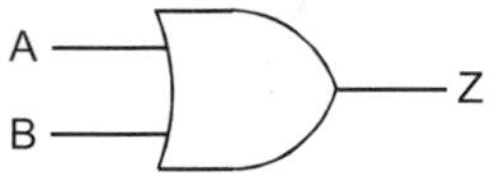

FIGURE 39.5

The Boolean equation representing the action of this gate is:

$$Z = A + B$$

Remember that this is a Boolean equation, so it is read as 'Z equals A OR B', not as 'Z equals A plus B'.

The truth table is:

Inputs		Output
B	A	Z
0	0	0
0	1	1
1	0	1
1	1	1

OR gates are available in 14-pin ICs, with the same pinouts as shown in Figure 39.4 for AND gates.

NOT GATE

This gate is unusual because it has only one input. The output is always the inverse of the inputs. This is why it is also called the INVERT gate.

Here is its symbol. The small circle indicates that the output is inverted.

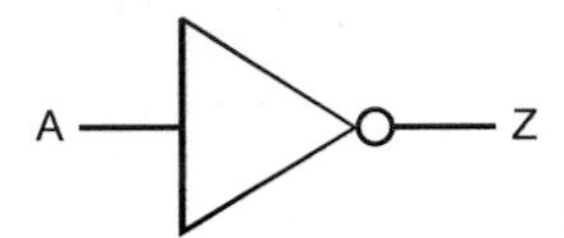

FIGURE 39.6

The Boolean equation that represents the action of this gate is:

$$Z = \overline{A}$$

The truth table is:

Input	Output
A	Z
0	1
1	0

SYSTEMS OF GATES

The output of a gate can be fed to the input of one or more other gates. In this way we can create more complex logic functions. As an example, take this system of two AND gates:

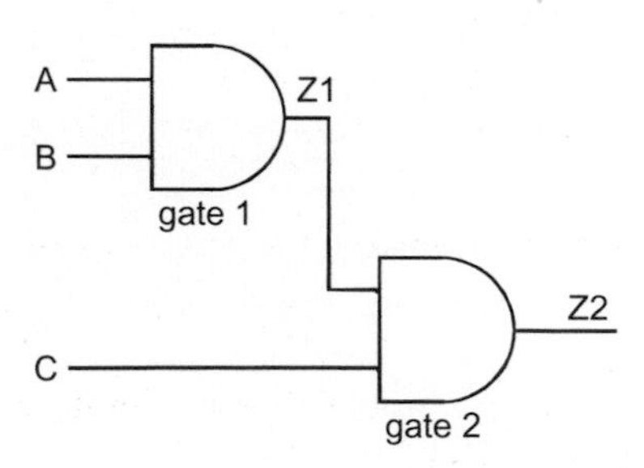

FIGURE 39.7

There are eight possible combinations of the three inputs. Their truth table is shown below. The stages of working out the table are:

1 Draw a blank table with the headings listing system inputs (C, B, A), the intermediate output (Z1) and the system output (Z2).
2 Fill in the eight combinations of inputs.
3 Work out the logic at gate 1. Looking only at inputs A and B, work down the column Z1. Enter the value of A AND B for each combination of inputs. Use the truth table to help with this.

Inputs			Z1	Output
C	B	A		Z2
0	0	0	0	0
0	0	1	0	0
0	1	0	0	0
0	1	1	1	0
1	0	0	0	0
1	0	1	0	0
1	1	0	0	0
1	1	1	1	1

4 Now we can go on to work out the logic at gate 2. Looking only at columns C and Z1, work down column Z2. Enter the value of C AND Z1 or each combination of inputs to gate 2.
5 Z2 is the output of the system. This is '1' only when all three inputs are '1'. In other words, these two gates are the equivalent of a 3-input AND gate.

Here is another example:

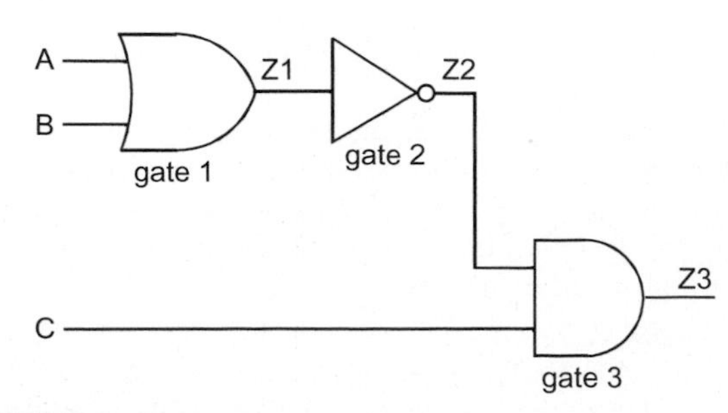

FIGURE 39.8

As before, there are eight possible combinations of inputs (see the table below). This time there are two intermediate outputs (Z1, Z2).

The stages of working out the table are:

1 Draw a blank table with the headings given above.
2 Fill in the eight combinations of inputs.

Inputs			Z1	Z2	Output
C	B	A			Z3
0	0	0	0	1	0
0	0	1	1	0	0
0	1	0	1	0	0
0	1	1	1	0	0
1	0	0	0	1	1
1	0	1	1	0	0
1	1	0	1	0	0
1	1	1	1	0	0

3 Work out the logic at gate 1. Looking only at inputs A and B, work down the column Z1. Enter the value of A OR B for each combination of inputs.
4 In Z2 enter the inverts of the values in Z1 (0 for 1, and 1 for 0).
5 Looking only at columns C and Z2, work down column Z3. Enter the value of C AND Z2 or each combination of inputs to gate 3.
6 Z3 is the output of the system. This is '1' only when A and B are '0' and C is '1'. There is no single gate equivalent to this system.

Things to do

On a breadboard, set up systems of AND, OR and NOT gates. To provide the inputs, join the input pins to the 0 V line or the positive supply line. To read the outputs, use either a logic probe, or an LED driven by a transistor switch (see the circuits on p. 85, but without R1 and R2).

For each system of gates, run through all possible combinations of inputs. Fill in a truth table to show how the output varies with combinations of inputs.

Verify your practical results by drawing a diagram of each system and working out its action on paper.

Self Test

Write out the truth table of each of the systems of gates shown below. Where possible, draw the equivalent single gate that could replace them.

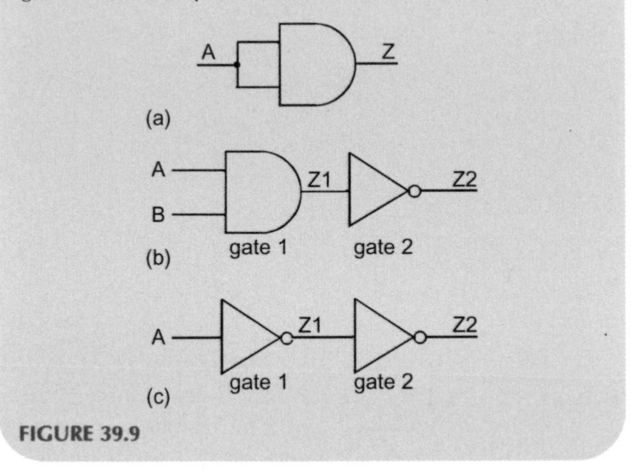

FIGURE 39.9

NAND GATE

The NAND gate or NOT-AND gate is one of the most useful gates for logical processing. The NAND gate is the equivalent of an AND gate followed by NOT, as in Figure 39.9 (b) above.

The symbol is an AND gate with a small circle at its output to indicate that the output is inverted.

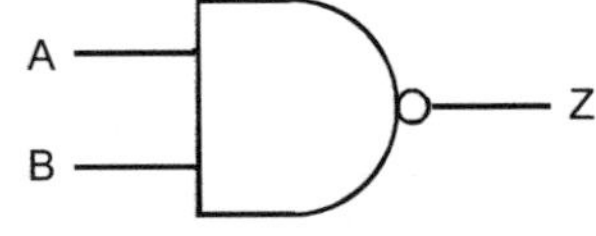

FIGURE 39.10

The output is '0' only when A AND B are '1', as shown in the truth table:

Inputs		Output
B	A	Z
0	0	1
0	1	1
1	0	1
1	1	0

As a Boolean equation, the action of the gate is:

$$Z = \overline{A.B}$$

Two-input NAND gates are available as ICs, with four gates to the package.

NOR GATE

This is the equivalent of an OR gate followed by a NOT gate. Its truth table is:

Inputs		Output
B	A	z
0	0	1
0	1	0
1	0	0
1	1	0

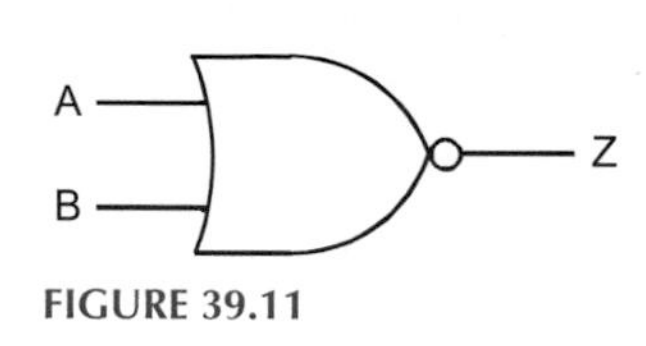

FIGURE 39.11

The output is '0' when A OR B are '1'.

The symbol is an OR gate with a small circle at its output to indicate that the output is inverted.

As a Boolean equation, the action of this gate is:

$$Z=\overline{A+B}$$

EX-OR GATE

An Ex-OR, or exclusive-OR gate is a relative of the OR gate. The output of an OR gate is '1' when inputs A OR B **OR BOTH** are '1'. The output of an Ex-OR gate is '1' when inputs A OR B but **NOT BOTH** are high. The truth table differs from that of OR in the last line:

Inputs		Output
B	A	Z
0	0	0
0	1	1
1	0	1
1	1	0

The Boolean equation of the gate is:

$$Z=A+B$$

The Ex-OR gate is sometimes known as the 'same or different' gate. Output is '0' if the inputs are the same and '1' if they are different. They can be useful in circuits when we want to compare two logical quantities to see if they are alike or not.

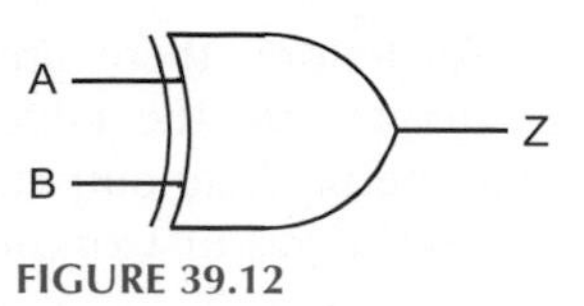

FIGURE 39.12

The ex-OR gate has only 2 inputs.

An **ex-NOR gate** is an ex-OR followed by NOT. Its output is the inverse of ex-OR.

Self Test

Write the truth tables of these circuits:

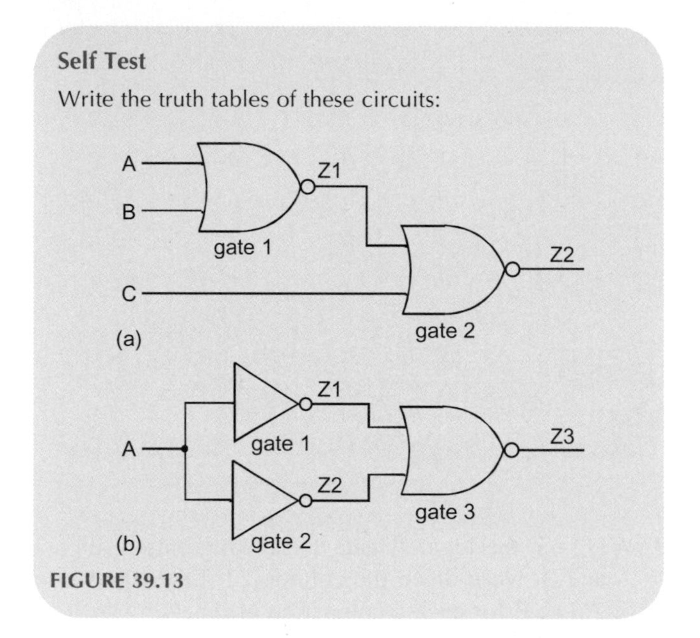

FIGURE 39.13

QUESTIONS ON LOGIC

Use this combined truth table for reference when answering these questions.

Inputs		Outputs					
B	A	A.B	$\overline{A.B}$	A + B	$\overline{A+B}$	$A\oplus B$	$\overline{A\oplus B}$
0	0	0	1	0	1	0	1
0	1	0	1	1	0	1	0
1	0	0	1	1	0	1	0
1	1	1	0	1	0	0	1

1. What is meant by 'binary'? Give some examples to illustrate this idea.
2. Describe a switched logic circuit to control the electric motor of a drilling machine and make the machine safer to use.

3 Write the truth tables of these logic circuits:

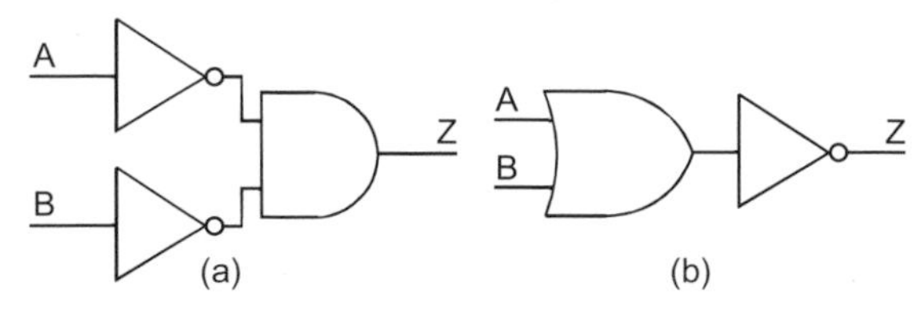

FIGURE 39.14

4 Use a system diagram to help you describe a system with two named inputs and one named output, that is based on AND logic.
5 Use a system diagram to help you describe a system with three named inputs and one named output, that is based on OR logic. The system may also include NOT logic.
6 Use a system diagram to help you describe a system with two named inputs and two named outputs, that is based on both AND and OR logic. The system may also include NOT logic.
7 Compare the properties of TTL logic ICs and CMOS logic ICs.
8 A system is designed to switch on a room heater if the temperature is below 15°C, but not if the door has been left open. Draw the system diagram, including the logic required.
9 A system is designed to flash a beacon lamp and sound a siren when any one of three windows or the door is opened. But the beacon is not to be flashed at night. Draw the system diagram, including the logic required.
10 Write the truth table for the logic circuit below. To what gate is it equivalent?

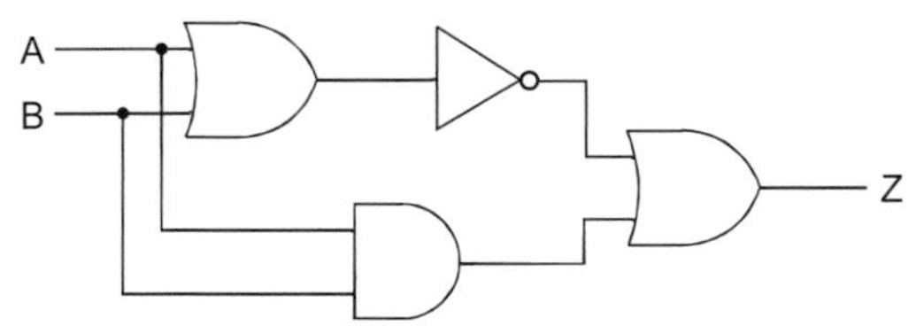

FIGURE 39.15

MULTIPLE CHOICE QUESTIONS

1 The logical operation performed by the circuit below is:

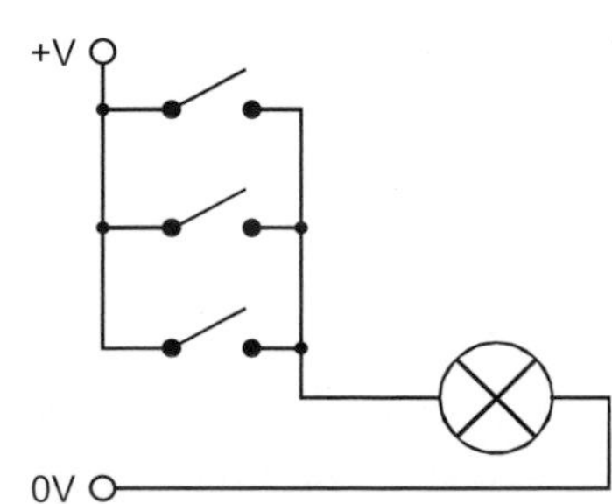

FIGURE 39.16

A NOR.
B NAND.
C OR.
D INVERT.

2 Given the values for switch positions and the state of the lamp listed on p. 141, the output Z is '1' in the system below when:

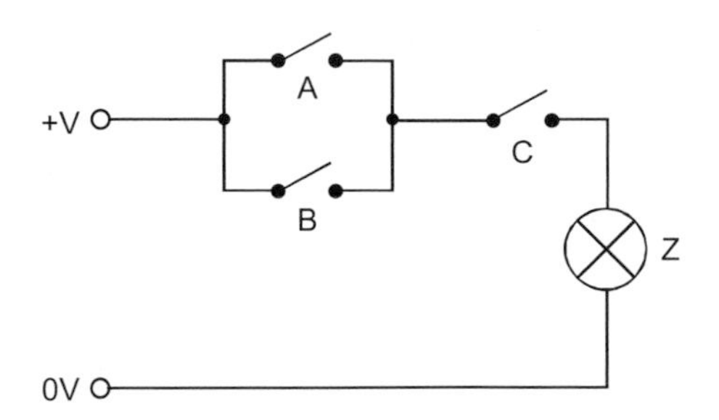

FIGURE 39.17

A A = 1, B = 1, C = 0.
B A OR B = 1, C = 1.
C A = 1, B OR C = 1.
D Z = A OR B OR C = 1.

3 When using TTL, the supply voltage should be:
A 5 V.
B 6 V.
C 3 V.
D 15 V.

4 The logical operation performed by this gate is:
A AND.
B NOR.
C NAND.
D Ex-OR.

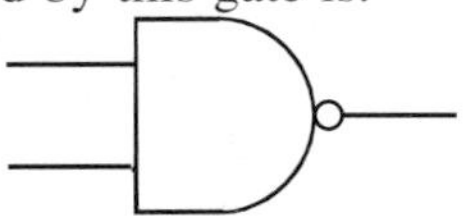

FIGURE 39.18

5 If a system has 3 input terminals the number of possible combinations of inputs is:
A 8.
B 3.
C 4.
D 2.

6 The output of a two-input NOR gate is high when:
A A is high and B is low.
B A is low and B is high.
C Both A and B are low.
D Both A and B are high.

7 The output from this logic system is low when:

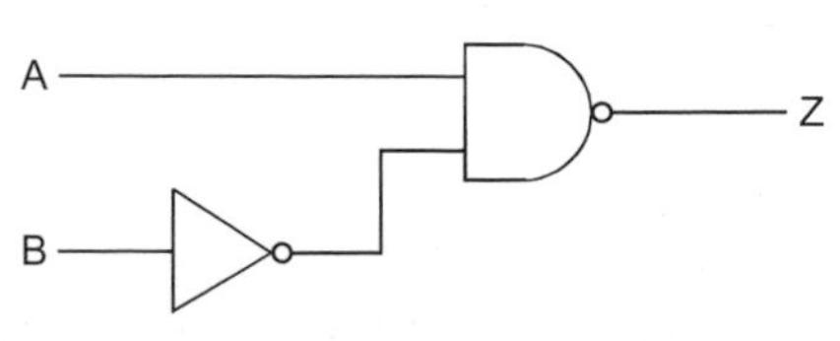

FIGURE 39.19

A A is high and B is low.
B A is low and B is high.
C Both A and B are low.
D Both A and B are high.

8 The output of an Ex-OR gate is '1' when:
A A is '1'.
B B is '0'.
C A and B are the same.
D A and B are different.

9 If $Z = \overline{A.B}$, then Z is low when:
A A and B are high.
B Never.
C A or B is low.
D A is low and B is high.

10 The combination of inputs that produces a low output from the system below is:

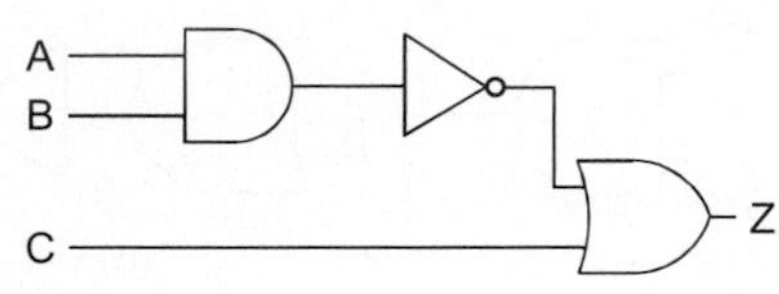

FIGURE 39.20

A All low.
B A high, B and C low.
C All high.
D A and B high, C low.

Logical Systems

The logic gates described in Topic 39 produce a fixed output for every combination of inputs. Each gate has a truth table that describes its action. We can connect several gates together in a logical circuit and write a truth table that describes the outputs for every *combination* of inputs. This is called **combinational logic**.

In this Topic we see how to design combinational logic circuits and look at some useful examples.

DESIGNING CIRCUITS

Follow this plan:

- Set out a truth table with a row for each combination of input states. With one input there are two states (0 and 1). With two inputs there are four combinations of states (00, 01, 10, and 11). With three inputs there are eight combinations (000 to 111).
- Fill in a column for each output. On each row enter '0' or '1', according to what you want to happen to that output for each combination of inputs.
  - Taking each output column in turn, scan down it and look for one of these:
    - An exact match with one of the input columns.
    - An exact inverse of one of the input columns.
    - A set of '0's and '1's that is identical with the output of one of the two-input gates: AND, OR, NAND, NOR ex-OR or ex-NOR.
  - If you find one of the above, write out the logical equation. If the system has two inputs and you do *not* find one of the above, you will probably find that inverting *one* of the inputs before sending it to a gate will produce the result you need.
- Use the logical equations to design the system, keeping the number of gates used to a minimum. Look for logical terms that appear in several of the equations. For example, you may find that $\overline{A}$ occurs in two or more of the equations. If so, you need to invert A only once and feed it to the inputs that need it.

DESIGN EXAMPLES

Half Adder

A half adder circuit takes two one-digit binary numbers and adds them together. There are only two different one-digit binary numbers — '0' and '1'. These are to be added in all four possible combinations.

The first step in designing the logic is to set out the truth table. This describes the inputs and what output we expect them to produce. Here is the table for a half adder:

Inputs		Outputs	
B	A	Carry	Sum
0	0	0	0
0	1	0	1
1	0	0	1
1	1	1	0

This is ordinary binary arithmetic, in which:

$$0+0=0$$

$$0+1=1$$

$$1+0=1$$

$$1+1=0, \text{carry } 1$$

Looking down the Carry (C) column, we can see that C is '1' only when both A and B are '1'. This is an example of the AND operation. As a logic equation, we write:

$$C = A.B$$

The column for the Sum (S) is recognisable as the output of an exclusive-OR gate. The Boolean equation is:

$$S = A \oplus B$$

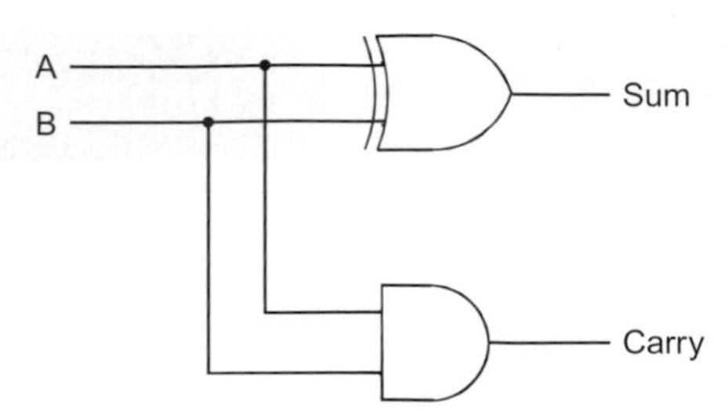

FIGURE 40.1

The symbol ⊕ means 'exclusive-OR'.

The logic circuit for these two equations is:

Sprinkler Control

A system is designed to turn on a garden water sprinkler when the soil is dry, but not when the sun is shining. A light sensor A has outputs 0 = dull, and 1 = sunny. A soil moisture sensor B has outputs 0 = moist, and 1 = dry. For the sprinkler S, 0 = off, and 1 = on. The truth table for the system is:

Inputs		Output
B	A	S
0	0	0
0	1	0
1	0	1
1	1	0

The sequence of '0's and '1's in the output column does not fit any of the regular patterns suggested opposite.

The third line down corresponds to 'dull day and dry soil', the ideal conditions for watering. For S = 1, we see that A = 0 and B = 1. As a logic equation, this is written:

$$S = \overline{A}.B$$

The equation shows that we need an AND gate, and that the A input must be inverted.

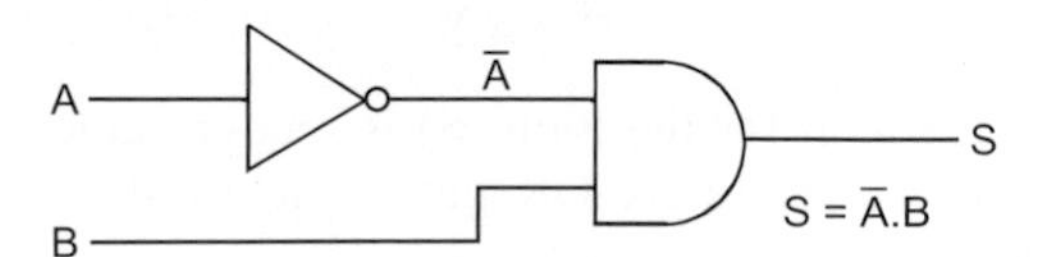

FIGURE 40.2

This is a case of inverting one of the inputs to a standard 2-input gate.

Decoder

You have probably noticed that the inputs to a 2-input system in the truth table are the binary numbers 00, 01, 10, and 11. These are equivalent to decimal numbers 0, 1, 2, and 3.

This system decodes the binary numbers to drive a 7-segment display. Each of the segments *a* to *g* has an output to drive it.

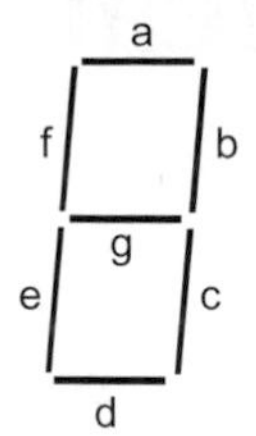

FIGURE 40.3

The truth table is:

Inputs		Outputs						
B	A	a	b	c	d	e	f	g
0	0	1	1	1	1	1	1	0
0	1	0	1	1	0	0	0	0
1	0	1	1	0	1	1	0	1
1	1	1	1	1	1	0	0	1

Scanning the table one row at a time, we enter output 'on' (1) or 'off' (0) for each segment.

Logic equations are:

$a = d = \overline{A.\overline{B}}$ (NAND of A and $\overline{B}$)
$b = 1$ (on for every numeral, connect to +V)
$c = \overline{\overline{A}.B}$ (NAND and $\overline{A}$ and B)
$e = \overline{A}$
$f = \overline{A + B}$
$g = B$

The decoder circuit (the circuit that produces the outputs corresponding to the given inputs) is shown below.

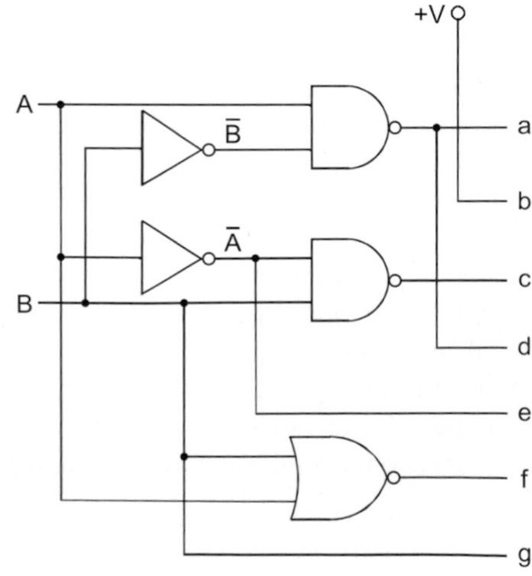

FIGURE 40.4

Priority Encoder

The system has three inputs, numbered 1, 2, and 3, according to their priority. Input 3 has the highest priority and input 1 the lowest.

The system has two outputs, Z0 and Z1, which correspond to the four binary numbers 00, 01, 10 and 11 (or 0, 1, 2 and 3 in decimal).

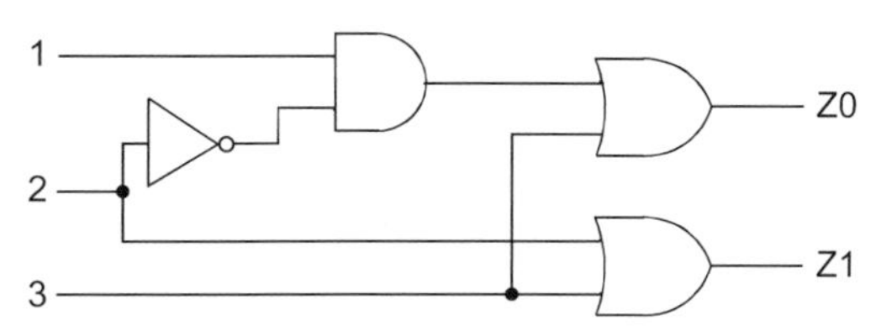

FIGURE 40.5

Output Z0 is the least significant digit, the rightmost digit of the pair.

Normally all inputs are low and the output is zero (00). When one *or more* of the inputs is made high, the output indicates which of the high inputs has the highest priority.

For example, if 1 and 2 are high, the output shows '2' (10 binary). If 3 and 2 are high, the output shows '3' (11 binary).

Data Selector

The system has two data input terminals, A and B. These could be receiving digital signals from two sources (for example, two digital radio receivers). The system has a third input terminal named SELECT, which is labelled 'S' in the diagram.

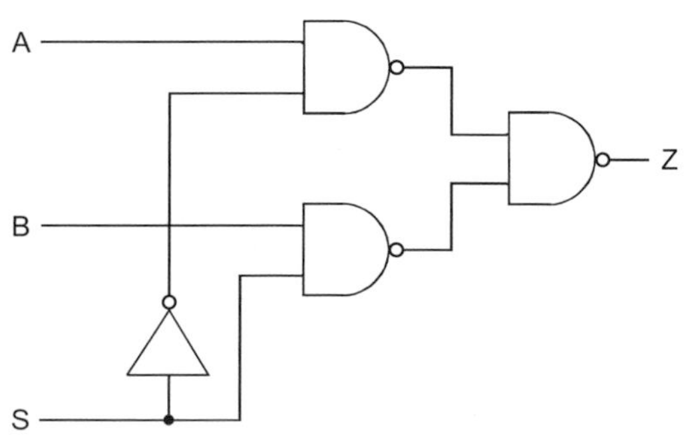

FIGURE 40.6

When S is low, data (a sequence of low and high pulses) that arrives at A is sent through to the output Z. Data arriving at B is ignored.

When S is high, data arriving at B goes through to Z, but data arriving at A is ignored.

By making S low or high we can select which set of data is sent on to the next stage.

A data selector system is often called a **multiplexer**.

Things to do

Draw the truth tables of the priority encoder and the data selector. Confirm that they operate as described above. Build them on a breadboard.

NAND LOGIC

One or more NAND gates can be wired together so that they are equivalent to any other kind of gate (see Figure 40.7).

The diagrams shown in Figure 40.7 show how to connect NAND gates to produce modules that have the same action as the other kinds of gate.

When you are building up systems from these modules, it is often possible to obtain two or more functions from a single module. For example, the module for NOR also produces OR and the inverts of A and B. An example is given on p. 162.

Things to do

Work out the truth tables of the diagrams in Figure 40.7 and confirm that they perform as stated.

Alternatively, using ICs on a breadboard or a ready-made logic board, connect up some of these NAND gates circuits, and check their truth tables.

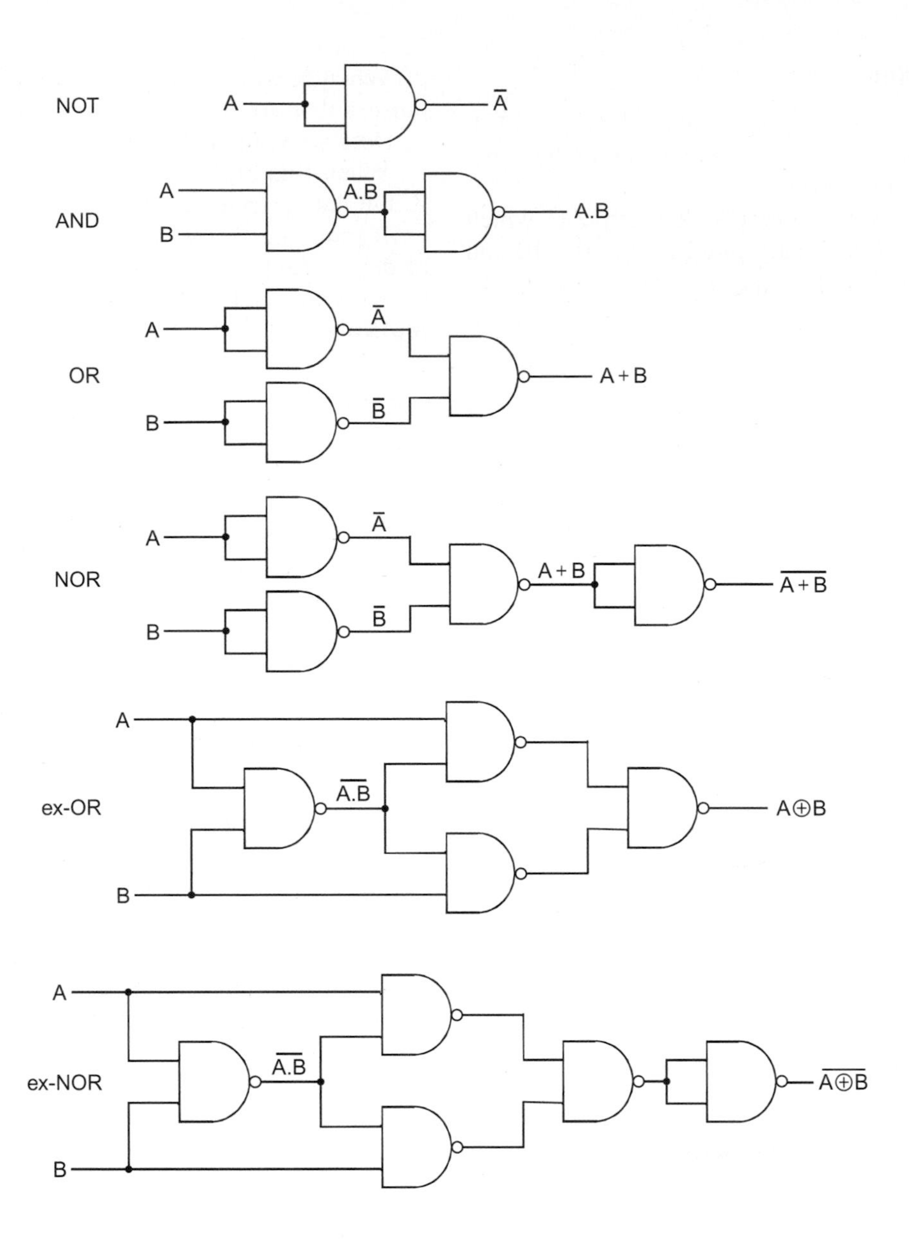

FIGURE 40.7

Replacing with NAND

Logic gates are manufactured as ICs, each of which typically contains four two-input gates. NOT gates are six to a package. To be most economical of cost, as many as possible of the gates in an IC should be used.

Look at the priority encoder circuit in Figure 40.5. This has a NOT gate, an AND gate, and two OR gates. As there are three different kinds of gate we have to have *three* ICs. to build the circuit. Using one NOT gate leaves five gates unused (but needing current). Using only one AND gate leaves three unused. Using two OR gates leaves two unused. The total is 9 unused gates out of 14, which is costly, and wastes power and board space.

In the upper diagram of Figure 40.8 the four gates of the decoder are each replaced with their NAND equivalents. The circuit has the same action as before but now needs eight gates.

However, as it consists wholly of NAND gates, this circuit can be simplified, as shown in the lower diagram. Compare the upper and lower diagrams and note the changes.

Gates *b* and *d* invert the signal twice, so replace them with a plain wire. Next, there is no need for two gates (*e* and *h*) to invert signal 3. Replace them with one gate and split its output to send to *f* and *i*.

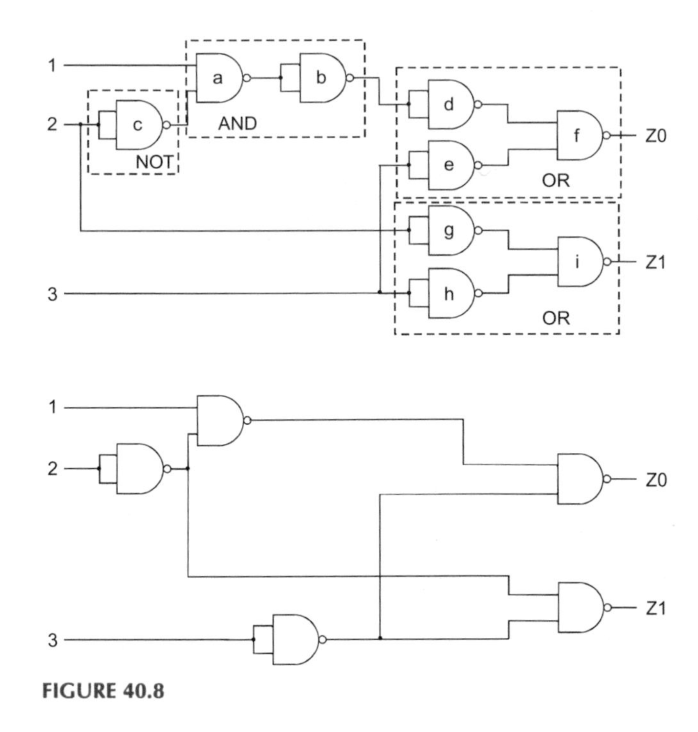

FIGURE 40.8

Finally, gate *i* needs the invert of signal 2, provided by gate *g*. But this signal is already available from *c*, so remove *g* and take the signal from the output of *c*. The encoder now needs only five gates, in two ICs. There are only three unused gates, which possibly might be used in other NAND logic.

Questions on Logical Systems

1. Write the truth table for this circuit. To what gate is it equivalent?

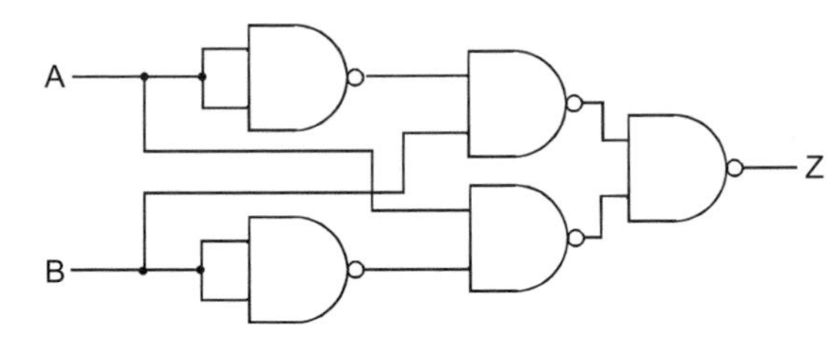

FIGURE 40.9

2. For the circuit below, write the value of Z as a Boolean expression, using A and B.

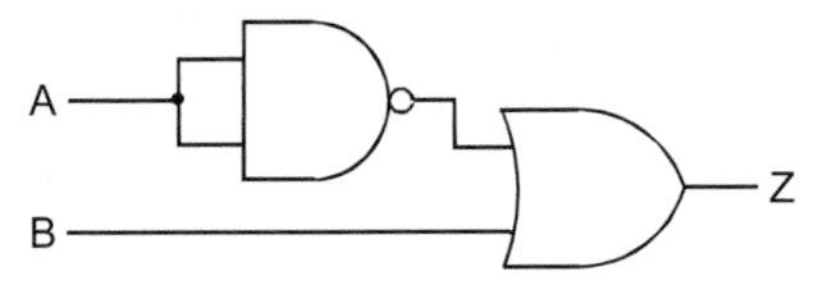

FIGURE 40.10

3. Draw a circuit that has output $Z = \overline{A + B + \overline{C}}$.
4. Convert the half-adder circuit into a NAND gate version.
5. Convert circuit (b) in 'Self test' Topic 35, p. 148 into a NAND gate version.

DESIGN TIME

Build the sprinkler control system (p. 160). Design your own soil moisture sensor. Use a solenoid-operated valve to turn on the water, or a small fountain pump driven by an electric motor.

Build a 2-line to 1-line data selector. Give it two outputs: Z and $\overline{Z}$. Design a display of 4 LEDs to show the state of the two inputs and two outputs.

Build the decoder system. Use logic that produces counts 0 to 3. Or you could try out the logic for your solution to Q. 3.

Alternatively, redesign the logic to give 1, 2, 3, and 4, or to give capitals A, B, C and D.

A model railway is on display in a shop window. The train runs continuously in the direction shown by the arrows. There are two sensors (A and B) that detect when the train is passing. For A and B, 0 = no train, 1 = train passing. The train is to circulate the loops alternately.

The junction can switch left or right. Its motor responds to two control signals, Z1 and Z2. For Z1, 0 = motor off, 1 = motor on. For Z2, 0 = go left, 1 = go right.

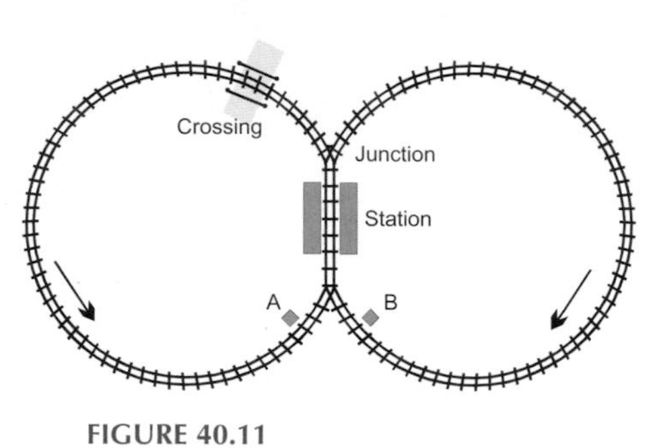

FIGURE 40.11

Draw a truth table of the operation of the junction. Note that one combination of inputs A and B is not possible. Design the logic circuit.

As extensions of the system, add logic and other circuits to (a) make the train halt in the station for 1 minute every time it goes through and (b) stop the train if it approaches the crossing and the gates are closed across the track.

Feed the output of a slow-running 555 astable to a NOT gate. What do you notice about the duty cycle of the output from the gate? A useful tip!

In an industrial process, two liquids are run into a tank through valves W and X and stirred together by stirrer Y. Then the mixture is run out of the tank through valve Z. There are two level sensors A and B which have output 0 when they are uncovered and 1 when they are covered. Starting with a full tank (A covered) and the stirrer running, valves W and X are to be closed and valve Z is to be opened. When the mixture has drained so that both sensors are uncovered, valve Z is to close and valves W and X are to open. The stirrer continues running. The tank then fills until both sensors are covered. The cycle repeats indefinitely.

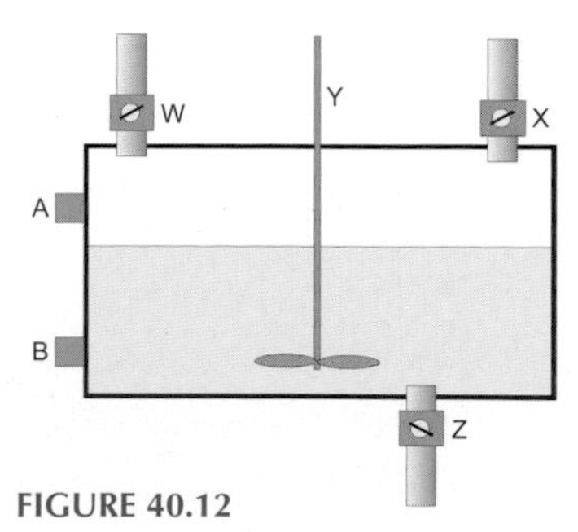

FIGURE 40.12

Design a logic system to produce the action specified above. Then add a manual over-ride switch that immediately closes W and X, and opens Z. The stirrer stops when the lower sensor is uncovered and the mixture is allowed to drain away completely.

Logical Sequences

In Topics 39 and 40 we looked at combinational logic. In this Topic we describe logic circuits that go through a *sequence* of changes. Their outputs depend not only on the present inputs but also on what inputs there have been in the past. This is called **sequential logic**.

The circuit below is a simple example of a sequential logic circuit.

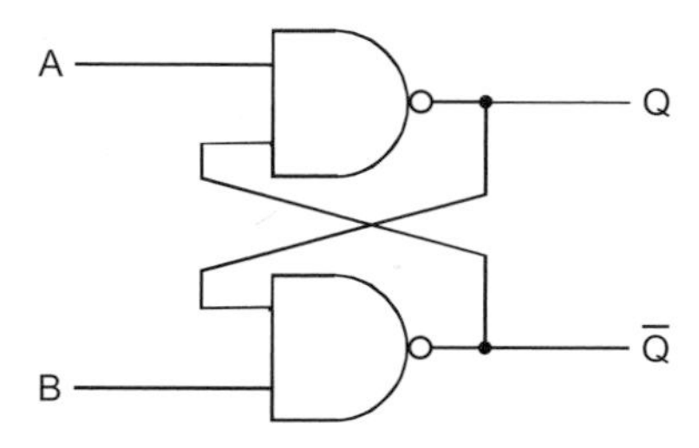

FIGURE 41.1

The output of each gate connects back to an input of the other gate. This is a reminder of the circuits. These circuits have two possible states, but are stable in only one state or neither state, respectively. This circuit also has two possible states, but it is stable in both of them. It is called a **bistable** circuit.

If you try working out the circuit using the NAND truth table, or if you try some practical runs on an actual IC, you find that the circuit is stable only when both inputs are high (=1). When it is stable, one of its outputs is low (=0) and the other is high (=1).

In one of its stable states the logic levels are as shown in Figure 41.2

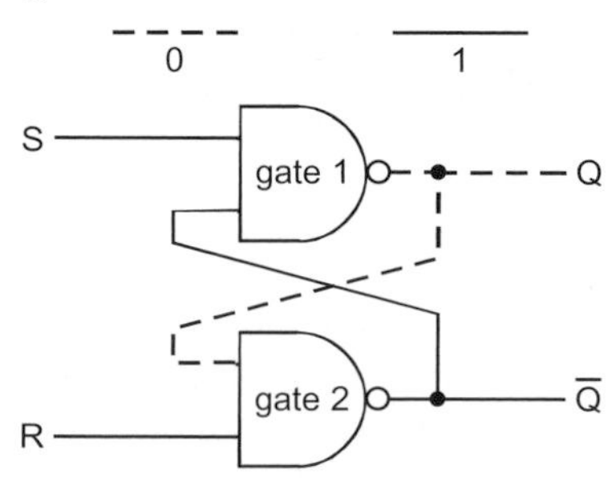

FIGURE 41.2

The inputs have been relabelled S (=set) and R (=reset). In Figure 41.3, the circuit is in the reset state, with its Q output low (0). Its $\overline{Q}$ output is the reverse of this. The bar over the $\overline{Q}$ indicates this relationship.

In Figure 41.4, the inputs appear not to be connected to anything. In practice, they must be connected to something that will hold them at a high logic level. A pull-up resistor connected to the positive supply line would do.

Watch what happens if we make the S input low for an instant:

1 Start with circuit in RESET state, Q low.

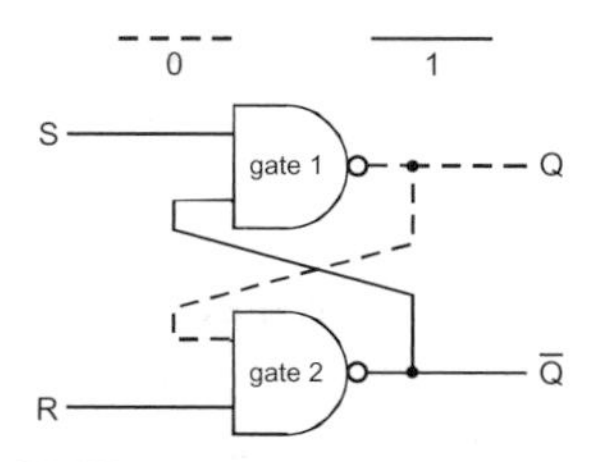

FIGURE 41.3

2 Make S input low. Q goes high.

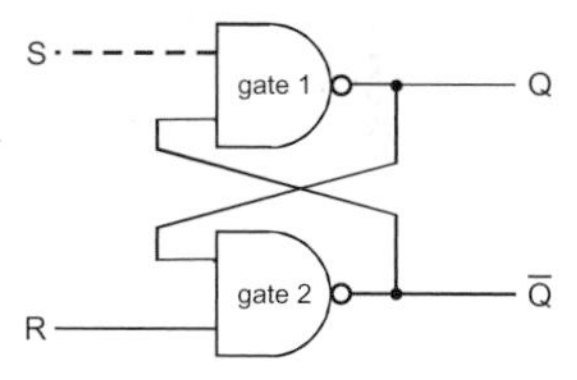

FIGURE 41.4

3 Gate 2 has two high inputs. $\overline{Q}$ goes low. Now Gate 1 has two low inputs, Q goes high.

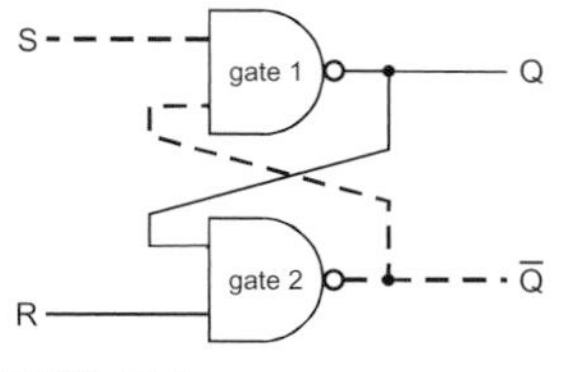

FIGURE 41.5

4 Make S high again. No change in Q or $\overline{Q}$. The circuit is in the SET state.

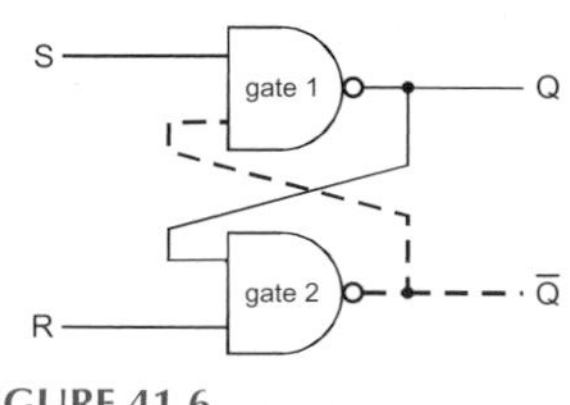

FIGURE 41.6

The circuit is stable in this state too. Making S low *again* has **no effect** on the output.

The only way to make the circuit change back to its reset state is to make the R input low for an instant. The resulting action is the reverse of that above.

S-R BISTABLES

The circuit described opposite is a **set-reset bistable**. It is also known as a set-reset **flip-flop**. It has the same type of action as the thyristor bistable (p. 93).

The result of a brief low pulse to one of its inputs may be 'no change' or a change of state from reset to set, or from set to reset. It all depends on the state the circuit is already in. It depends on what has happened to the circuit in the past. In this way, the circuit has a **memory**. Set-reset bistables are used as the units of certain types of digital memory.

Self Test

An S-R bistable can also be built from a pair of NOR gates. Draw the diagram and work out the logic levels as it is set and reset.

Design Tip

S-R bistables are a useful circuit unit because they can be set or reset by the output from sensors or from other logic. Also, they are easy to build using a pair of gates from a NAND or NOR IC. The next example shows you how to use a NAND bistable to 'remember' that there is an intruder around.

USING AN S-R BISTABLE

On pages 81 and 85 are circuits that switch on a lamp or LED whenever the light falling on an LDR is reduced. These circuits could be used with an electronic siren instead of LP1 or D1.

This could be the beginning of a security system. A beam of light shines on an LDR. If an intruder passes through the beam, the siren sounds. This is not quite good enough, for the siren stops sounding as soon as the intruder moves out of the beam. The system needs to 'remember' that the beam has been broken. This is where we use a bistable.

The system diagram looks like this:

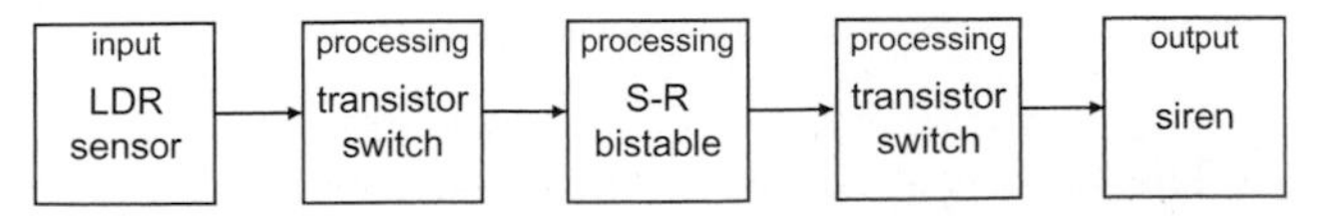

FIGURE 41.7

The circuit might be:

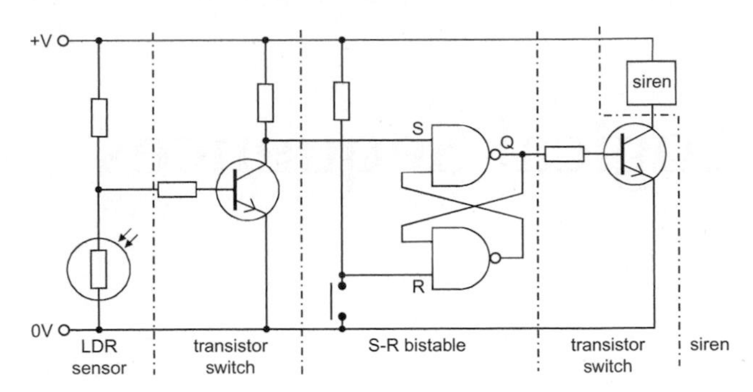

FIGURE 41.8

When the beam is broken the transistor switch is turned on. It delivers a low pulse to the S-R bistable. This makes the output (Q) go high. This turns on the second transistor switch, which switches on the siren.

Once the beam is broken, the siren sounds continuously until someone (a) switches off the power supply, or (b) presses the reset push-button. Pressing the button produces a short low pulse at the reset input. The bistable changes state. Q goes low and the siren is switched off.

CLOCKED LOGIC

In a big logic system there may be dozens, hundreds, or even thousands of gates. Any of them may change state when they receive a suitable input. Different gates take different lengths of time to change. The situation is very complicated. It is like an orchestra with all the players following their scores at their own speed. It needs a conductor. Watching the conductor's baton, the players all keep in step and the music sounds right.

The conductor of a logic system is the **system clock**. This might be a 555 astable circuit.

The system clock (or simply 'clock', as we shall call it) produces a series of pulses at a fixed rate. In a computer, which must perform many actions in a short time, the clock beats very fast. It may run at several hundred megahertz.

In a clocked logic system, the logic circuits do not act immediately there is a change in their inputs. Instead, they all wait. They do nothing until the clock tells them to act. Most act on the 'rising edge' of the clock. That is, the instant when the signal from the clock changes from low to high. However, some logic is clocked by a falling edge, so you must always check the data sheets for this point when designing a clocked logic circuit.

D-TYPE FLIP-FLOP

This is our first example of a clocked logic device. The device actually consists of several logic gates joined together, but we will just look at the action of the circuit *as a whole*. The circuit is available as a CMOS integrated circuit (4013). Each IC contains two identical flip-flops. Figure 41.9 shows the symbol for a D-type flip-flop.

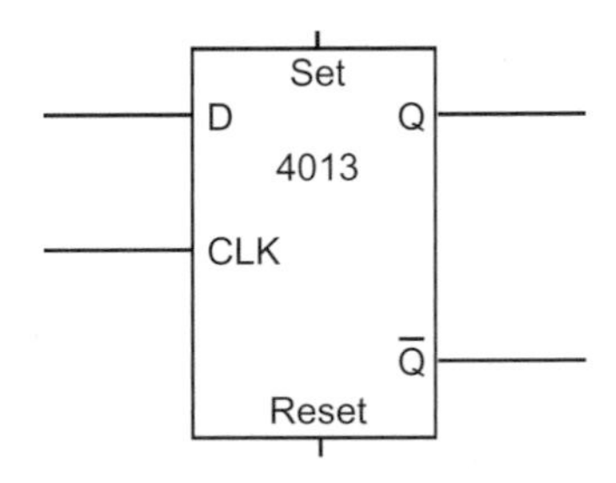

FIGURE 41.9

There are four input terminals:

- D, the data input.
- CLK, the clock input.
- Set.
- Reset.

There are two output terminals, Q and $\overline{Q}$. $\overline{Q}$ is the inverse of Q.

Data is fed to the D input. The data may be high or low and may be changing, but nothing happens to the outputs. Then, when the clock input rises from low to high, the data that is present at the D terminal *at that instant* appears at the Q output. If the input changes again, there is no change at Q. There are no further changes at Q until the *next* rising edge of the clock signal. At all times, $\overline{Q}$ is the inverse of Q.

The graph below plots typical changes in the inputs and outputs of a D-type flip-flop.

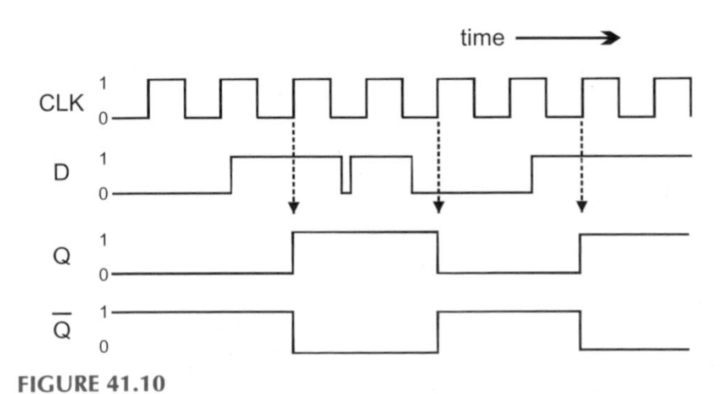

FIGURE 41.10

Although D changes, there is no change in Q until the next rising edge. Then Q changes to be equal to D at that time. Note that in the sequence above the second time D changes, it changes back again *before* the next rising edge of the clock. In this case, the change in D does not register as a change in Q.

A D-type flip-flop acts as a **latch**. It samples the data input at regular intervals. The value is held until the next sampling. This is useful if data is changing rapidly. It gives a logic circuit time to process data without it changing while it is being processed.

The Set and Reset inputs are used when we need to change the output immediately, without waiting for the clock. Normally these two inputs are held low and have no effect. If the Set input is made high, Q immediately goes high and $\overline{Q}$ goes low. The reverse happens if the Reset input is made high.

COUNTER/DIVIDER CIRCUIT

The D-type flip-flop can be built into several useful circuits.

The Set and Reset inputs are not used. They are connected to the 0 V line.

The $\overline{Q}$ output is connected to the D input.

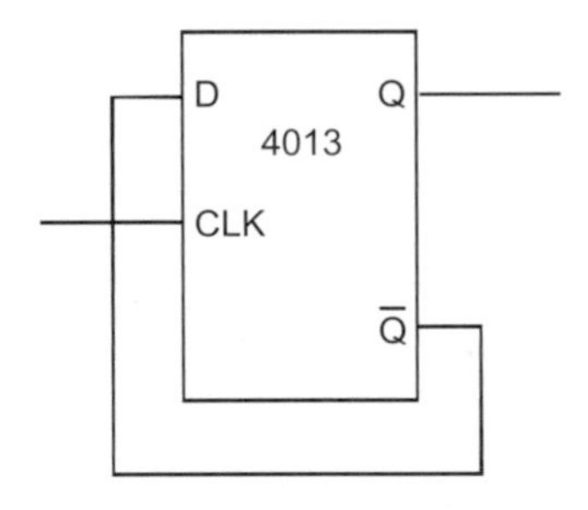

FIGURE 41.11

The stages of its operation are as follows:

1. Suppose the Q output is 0. The $\overline{Q}$ output is 1, and this is fed back to the D input.
2. At the next rising edge of the clock, Q takes the value of D and becomes 1. $\overline{Q}$ changes to the inverse, 0, and this is fed back to D.
3. At the next rising clock edge, Q becomes 0. $\overline{Q}$ changes to 1, and this is fed back to D. The circuit is now back to stage 1 and the cycle repeats indefinitely.

Look at this as a graph Figure 41.12:

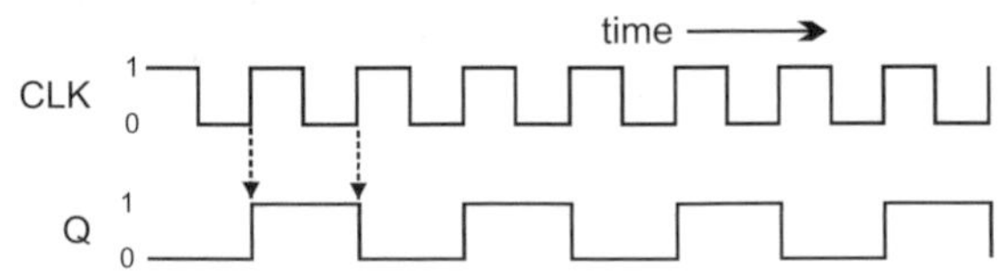

FIGURE 41.12

Q changes every *other* time that the clock changes. In other words, it changes at half the rate of the clock. Or, we can say that the frequency of Q is half the clock frequency. The flip-flop is acting as a **frequency divider**.

Design Tip

If the output of an astable is fed to the clock input of this circuit, the output Q of the circuit has half the frequency of the astable. The more important fact is that Q has a duty cycle of *exactly* 50%, no matter what the duty cycle of the astable. This is a simple way of obtaining a 50% duty cycle.

TWO-STAGE DIVIDER

If the Q output from the 4013 is fed to the CLK input of a second 4013 (perhaps the other flip-flop in the IC), we divide the frequency again. The output of the second flip-flop has a frequency that is one quarter of the original frequency. By setting up a chain of flip-flops, we obtain half-frequency, quarter-frequency, eighth-frequency, and so on.

An interesting action comes from feeding the second flip-flop with the $\overline{Q}$ output.

The circuit is:

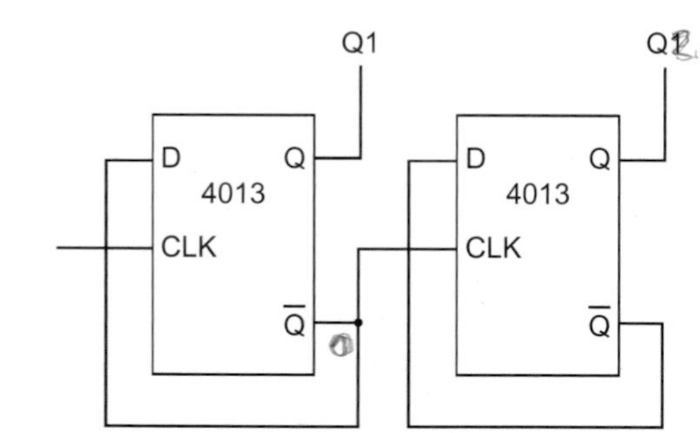

FIGURE 41.13

The graph of the outputs is:

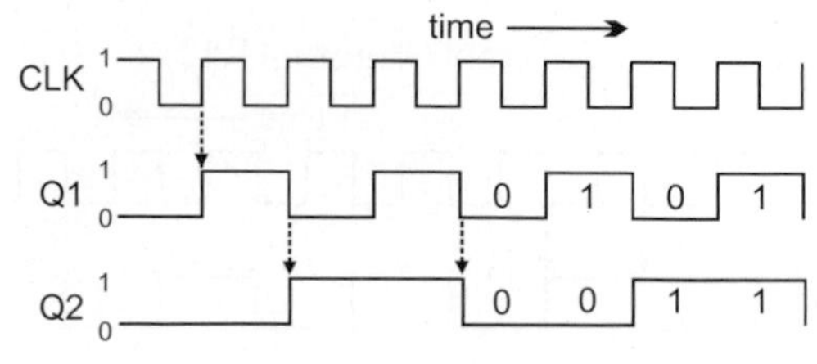

FIGURE 41.14

The frequency is halved at each stage, as before. However, the logic levels of Q1 and Q2 go through an interesting sequence. In the table below, we set them out more clearly:

Q2	Q1	Binary	Decimal
0	0	00	0
0	1	01	1
1	0	10	2
1	1	11	3
0	0	00	0
0	1	01	1
		And so on ...	

Because the CLK input for Q2 comes from the inverted output of Q1, Q2 changes on the *falling* edge of Q1. Taking the Q1 and Q2 logic levels as numbers written in binary, we see that the output from the flip-flops repeatedly runs through the decimal equivalents: 0, 1, 2, 3, 0, 1, ... and so on. The circuit is a **counter**.

We can add more stages to the counter:

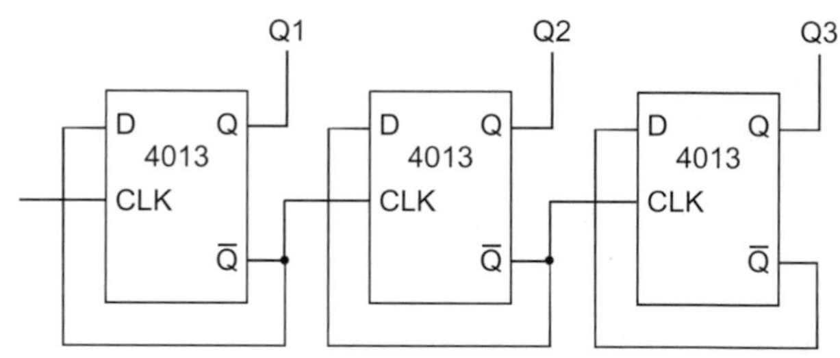

FIGURE 41.15

The output graphs are:

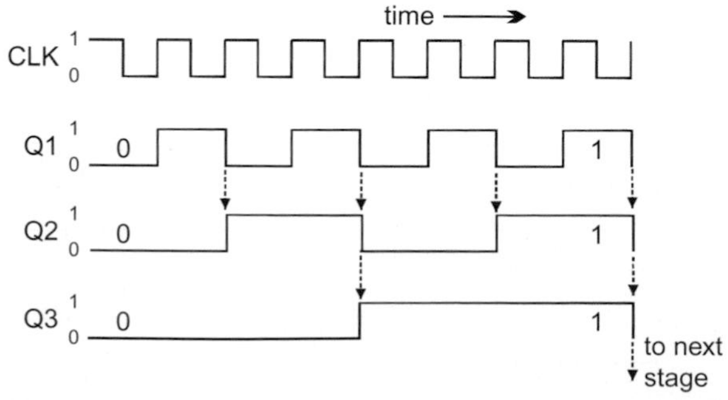

FIGURE 41.16

The falling edge of each stage triggers the next stage to change state. This produces the binary equivalents of 0 to 7, repeating. If we add a fourth flip-flop to the chain, the counter runs from 0000 to 1111, repeating. This is equivalent to 0 to 15 in decimal.

The graphs show that the counter circuit also acts as a frequency divider. It divides the frequency by two at each stage.

OTHER BINARY COUNTER/DIVIDERS

A four-stage binary counter/divider is available as a single IC, the 74LS93. The counting chain is in two parts, one containing a single flip-flop and the other containing three flip-flops already connected as in the Figure 41.15. This gives three dividing options: divide by 2, divide by 8 and (by connecting the two parts with an external link) divide by 16. In terms of counting, the IC can count from 0 up to 1, 7, or 15.

CMOS binary counter/dividers usually have more stages. The 4020, for example has 14 binary stages, so it divides by 2^{14}, which equals 16 384. As a counter, it counts from 0 up to 16 383. However, the outputs of stages 2 and 3 are not connected to terminals, so not all binary numbers can be obtained.

The 4040 IC has 12 stages, dividing by 2^{12}, or 4096. It counts from 0 to 4095. Outputs are available from all stages. The 4060 IC has 14 stages, like the 4020, though it does not have outputs from stages 1 to 3 and 11. Its advantage is that it has an astable circuit included. It needs only a capacitor and a pair of resistors to build the clock generator. For further details of these ICs, refer to the data sheets.

DECADE COUNTERS

In this topic, several ICs are identified by type number so that you know which ones to use when experimenting, or for building a project. The aim is to give you an idea of the range of counters that is available. However, apart from the 4013 and (later) the 4017, you are not expected to memorise type numbers and their distinctive features. All the details you may need are provided in data sheets or in retailers' catalogues.

Although digital circuits work most easily with binary numbers, humans are more familiar with decimal numbers. Several types of counter IC produce decimal output. The output runs through the sequence from 0000 to 1001 (0 to 9) and then returns to 0000 to repeat the sequence. The 74LS90 is an example of a decade counter IC.

Another decade counter is the 4518, which has two decade counters in one IC (see 'Cascading counters', below). The 4510 contains a counter that has the useful feature of being able to count up or down. If its up/down input is low, it counts up in the normal way (0, 1, 2, . . . , 9, 0, 1, . . .). If this input is high, it counts down (9, 8, 7, . . ., 1, 0, 9, . . .). Another feature of this counter is that it has four inputs that are used to load the counter with any starting value from 0 to 9. Probably the most versatile of the decade counters is the 4029, which has all the features of the 4510, and may also be set to count either in binary or in decimal.

CASCADING COUNTERS

A single decade counter runs from 0 to 9. If we want to count numbers larger than 9, we need to join two or more counters in series. This is called **cascading**

Things to do

Build a two-stage counter/divider, using the two flip-flops of a 4013 IC. Figure 41.17 shows how to connect the power supply and the Set and Reset terminals.

Take the input from a slow 555 or 7555 astable or a signal generator. Observe the counting action as indicated by the LEDs.

Extend the counter by adding two more stages and two more LEDs.

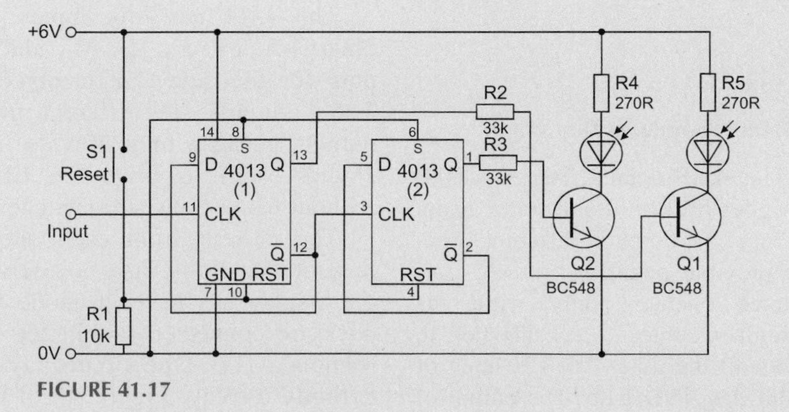

FIGURE 41.17

them. The 4518 is cascaded as shown in Figure 41.18.

Some counters have special outputs for use when cascading. The 4518 does not have such an output so we use an AND gate to detect when the output of the counter is '9', by ANDing the '1' and '8' outputs.

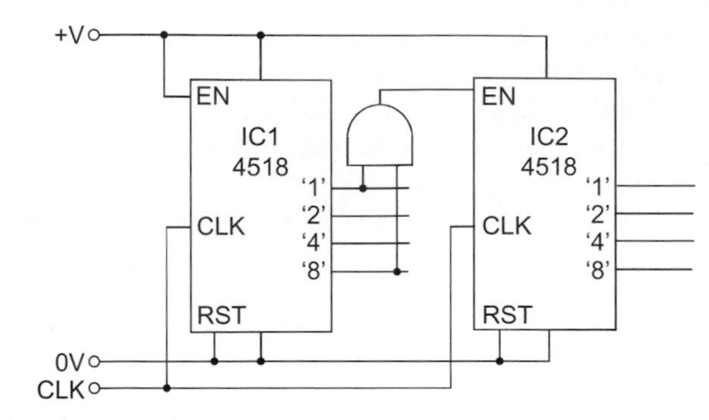

FIGURE 41.18

Note that the clock input drives *both* counters. The logic levels are:

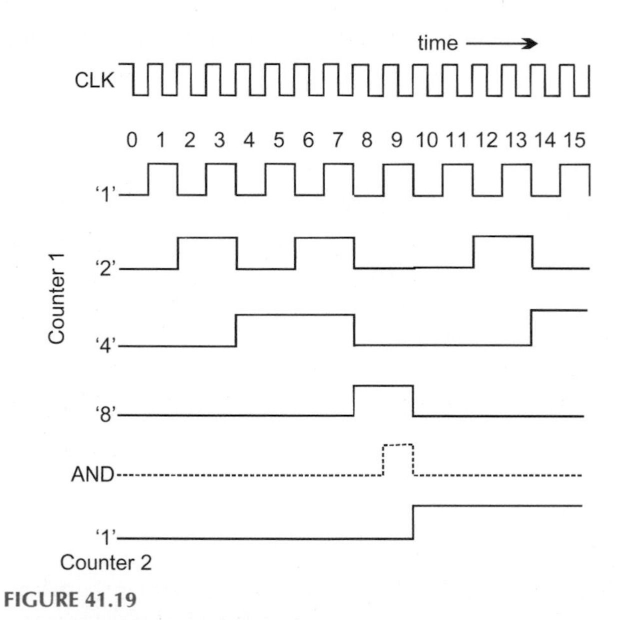

FIGURE 41.19

There are several points to note in this graph:

- **Decade counting:** The 4518 counts from 0 (0000) up to 9 (1001), then goes back to 0 and starts again. This gives a sequence of outputs different from binary counting (see previous page).
- **Carrying:** A counter operates only while its ENABLE input is high. Counter 1 has this tied to the +V line. It counts all the time. The EN input of Counter 2 is fed from the AND gate. The output of this is normally low, so Counter 2 ignores the clock pulses. However, when Counter 1 gets to the count of '9', the output of the AND gate goes high. With a high EN input, Counter 2 advances on the next clock pulse. Then the EN input goes low and Counter 2 does not count again until after the next '9' on Counter 1.
- **Binary coded decimal:** Numbers are represented on this counter by using a four-digit binary number for every digit of the decimal number. In binary, decimal 10 is 1010. In a counter made from two decade counters, the code for decimal 10 is '0001' and '0000'. These binary codes represent the digits '1' and '0' respectively. They thus represent a decimal number.

As another example of binary coded decimal (or **BCD**), take the decimal number 25. In binary, this is '1 1001'. In BCD, it is '0010' and '0101', representing '2' and '5' respectively. BCD is important for driving numeric displays.

Self Test

Write these decimal numbers as BCD:
(a) 84, **(b)** 33, **(c)** 29, **(d)** 362.

BCD DECODERS

Earlier we described a circuit that decodes the first four binary numbers to produce the logic levels for driving a seven-segment display. There are ICs that contain decoding circuits for the whole range of binary numbers from 0 to 9. The 4033 is a decade counter with a built-in seven-segment decoder. Instead of the usual '1', '2', '4', and '8' outputs, it has seven outputs that will drive a display. The outputs are not able to drive the LEDs directly, so you need seven transistor switches for that purpose. There are also decoders that can take their input from sources other than counters. One of the most popular of these is the 4511.

The 4511 has four inputs for the digits of the binary values '1', '2', '4', and'8'. It has seven outputs for the seven segments of a numeric display. These outputs are different from the usual CMOS outputs because they provide enough current (up to 25 mA each) to drive the LED segments directly without using transistor switches.

The current from each output passes through a resistor that limits it to a safe amount. Each LED of the display has its own anode terminal but the cathodes are connected within the display to a common terminal. This type of display is called a **common cathode** display. Figure 41.20 is a typical circuit for using the 4511.

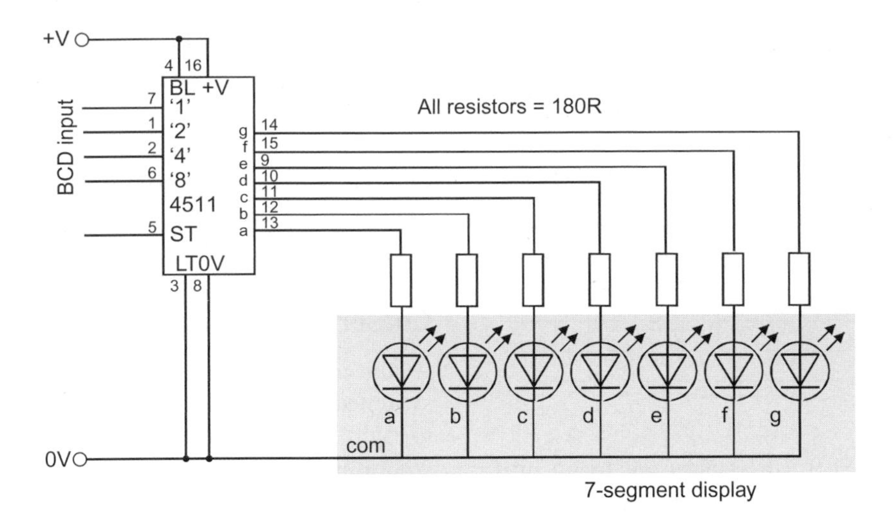

FIGURE 41.20

The 4511 has latches that store the input. When the STORE input is low, the output follows the input. If the STORE input is made high, the input data is held latched. The output shows the latched data, in decoded form. Any further changes that occur in the data arriving at the IC have no effect on the display. This is useful because it makes it possible to hold a rapidly changing input, to allow time to read the display.

The BLANK input is normally held high. If it is made low, the LEDs all go out. This happens also if the IC receives an input corresponding to numbers 10 to 15. The LAMP TEST input is normally low. If it is made high, all segments come on. This allows the display to be tested.

RESETTING THE COUNTER

Sometimes we want a counter that counts up to a number other than 10 or 16. For example, we might want to count up to 6, then continue from zero. This is easily arranged by external gates. The technique is to generate a pulse that resets the counter as soon as it reaches the number *after* the required maximum count. Here is a circuit for making a 4518 count up to 6:

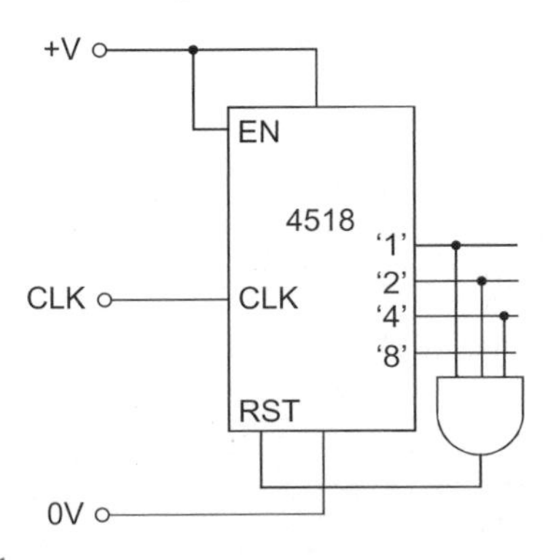

FIGURE 41.21

The counter counts normally from 0 to 6. During these stages there is no stage in which all inputs to the AND gate are high. Its output stays low. On the next clock pulse, the outputs are the binary equivalent of '7'. Outputs '1', '2', and '4' all go high. Immediately, the 3-input AND gate detects there are three highs and its output goes high. The high level at the RESET input resets the counter to 0000. There is in fact a brief count of 7, but resetting is too fast for this to be noticed.

Design Tip

Unless you are going to use several 3-input AND gates for other purposes in your circuit, you would probably not use a 3-input AND gate for resetting. Instead, you could use a 3-input NAND gate and follow this with an inverter. The inverter too might be another NAND gate with its inputs joined together (Figure 40.7).

Any surplus 3-input NAND gates can easily be converted to 2-input NAND gates by joining two of their inputs together.

THE 4017 COUNTER

The output stage of this counter is completely different from that of other counters looked at so far. It is a decade counter and its outputs are described as 'one of 10' outputs. The IC contains a special decoder to provide these outputs.

It has 10 output terminals all of which, except for the '0' output, go low when the counter is reset. Then, on each rising edge of the clock input, one of the outputs goes high, in order from 1 to 9 Figure 41.22.

This type of counter is useful for triggering a series of actions. As each output goes high, it triggers an action. An example is driving a display of changing coloured lights. A complex sequence of lighting effects

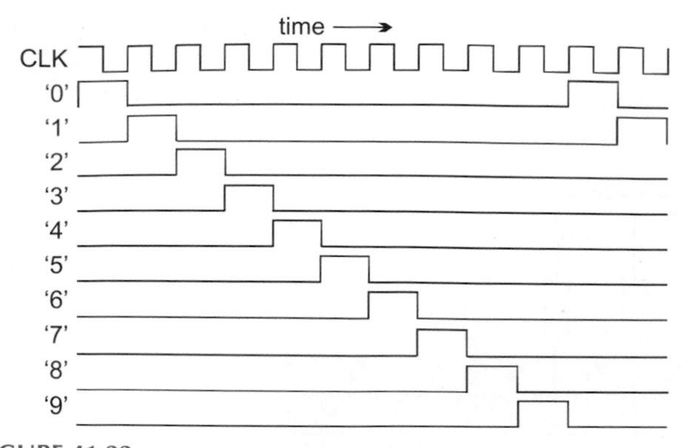

FIGURE 41.22

can easily be produced. If any stages are to go on for longer than the others, they may be triggered repeatedly by consecutive outputs fed to an OR or NOR gate.

It is simple to arrange for this counter to produce a lower count. If we want it to count to 7, for example, the output '8' is fed back to the RESET input. The counter has an OUT output. This goes high for counts 0 to 4 and low for counts 5 to 9. Two 4017 counters may be cascaded by feeding this output to the CLOCK input of the second counter.

The 4017 is useful for dividing a frequency by 10 or, if cascaded, for dividing by 100 or higher powers of 10. Note that cascading does not extend the 'one of 10' action to '1 of 20' or more.

TWO-DIGIT COUNTER

The single-digit 0 to 9 counter may be extended to make a two-digit 00 to 99 counter.

The first stage consists of two decade counters with a gate to decode the output of Counter 1. This enables Counter 2 on the count of '9'. The circuit for this is shown in Figure 41.18.

The second stage consists of two 7-segment displays. These are each connected as in Figure 41.20. A system diagram of the complete circuit is:

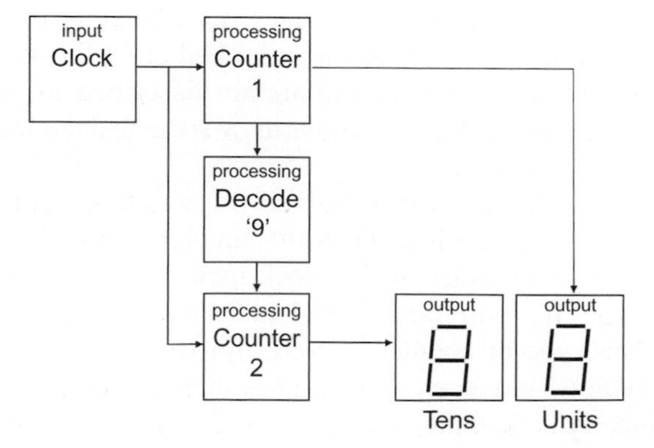

FIGURE 41.23

COUNTER-DRIVEN SYSTEMS

We have shown how counter circuits can be used for counting objects or events or for dividing frequencies. Another use for counters is to provide a sequence of inputs for multistage combinational circuits. We will look at two examples.

Chaser Lights Display

A fairground booth has a row of lamps that are switched on in pairs, the pattern travelling from left to right. At a given instant, the lamps look like this, where black = 'off' and white = 'on':

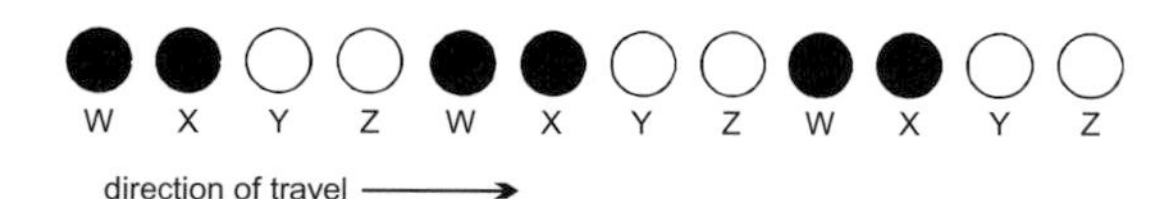

FIGURE 41.24

All lamps labelled with the same letter are switched on together. This means that we can wire the lamps in four groups (W, X, Y, Z).

Because the system needs only four outputs, W, X, Y, and Z, it needs only two inputs. A 2-bit counter circuit such as that described earlier produces the inputs 00, 01, 10 and 11 in a repeating sequence. To activate the display the counter is driven by a slow-running astable. This could be based on a 555 timer IC. We switch on the groups two at a time according to the four stages shown in this truth table:

Inputs		Outputs			
B	A	W	X	Y	Z
0	0	1	1	**0**	**0**
0	1	**0**	1	1	**0**
1	0	**0**	**0**	1	1
1	1	1	**0**	**0**	1

1 = 'lamp on' and this travels across the output side of the table from left to right.

Scanning the columns:

- Z is the same as B:

$$Z = B$$

- X is the inverse of B:

$$X = \overline{B}$$

- W has the same outputs as an ex-NOR gate:

$$W=\overline{A \oplus B}$$

- Y has the same outputs as an ex-OR gate:

$$Y=A \oplus B$$

The logic circuit is:

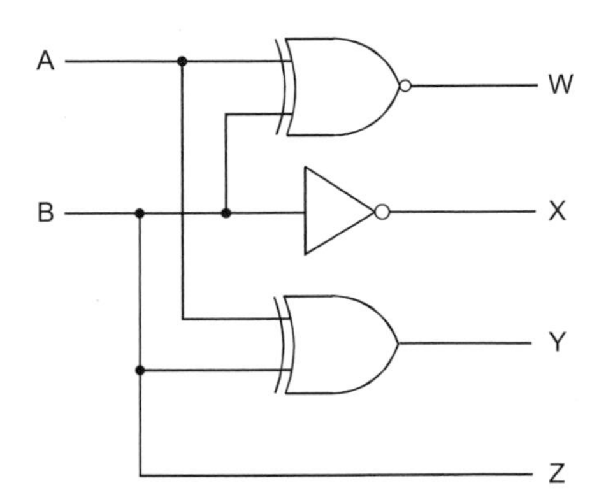

FIGURE 41.25

This circuit produces the required result. However, it might be possible to build it more economically, using NAND gates, as explained on pp. 157–8.

Traffic Lights

The sequence of lighting the red, yellow and green lamps goes through four stages, as shown in the truth table:

Inputs		Outputs		
B	A	R	Y	G
0	0	1	0	0
0	1	1	1	0
1	0	0	0	1
1	1	0	1	0

For four stages, we need two inputs A and B, which are easily provided by a 2-bit counter driven by a very slow astable. The output columns conform to the standard cycle. Scanning the columns, we note that:

- Y is the same as A, so:

$$Y=A$$

- R is the inverse of B, so:

$$R=\overline{B}$$

- With three '0's and one '1', G looks like the output of an AND gate, but the bottom two lines are in the wrong order. Invert the A input, and obtain:

$$G=\overline{A}.B$$

The logic circuit is:

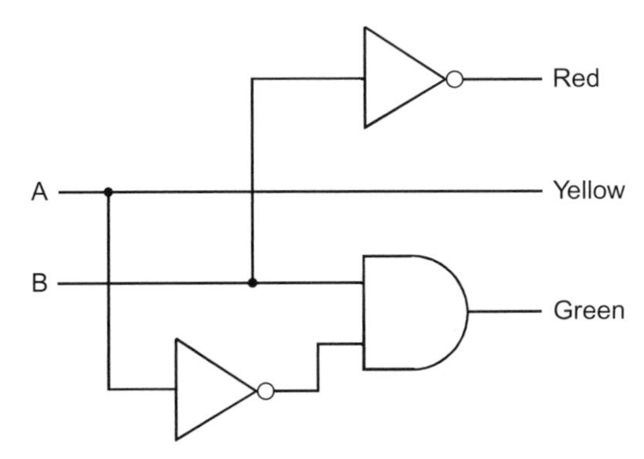

FIGURE 41.26

QUESTIONS ON LOGICAL SEQUENCES

1. Describe the action of a set-reset bistable built from two NAND gates. Suggest an application for this circuit.
2. Describe the 4013 D-type flip-flop. Name its input and output terminals and outline their functions.
3. Draw a graph to show the changes in input and output of a D-type flip-flop as the logic levels of the data change. What is the advantage of a system that is based on clocked logic?
4. Describe how a D-type flip-flop is used as a data latch.
5. Draw a circuit in which a D-type flip-flop is used to divide the frequency of a signal by two. Draw a graph of all input and output signals.
6. Draw a circuit in which two D-type flip-flops are used to build a two-stage counter. Draw a graph of all input and output signals.
7. Describe the features of three different types of counter IC and suggest a possible application for each type. Type numbers are not required.
8. Describe how two decade counter ICs are cascaded to count from 00 to 99.
9. Explain what is meant by 'binary coded decimal'.
10. Write the decimal number '69' (a) as a binary number, and (b) in BCD.
11. What is meant when we say that an IC decodes BCD into levels for a 7-segment display?
12. Taking any digit between '0' and '9', write it in BCD form and in the form in which it is used to drive a 7-segment display.

13 Describe a circuit that can be used to make a decade counter count up to 4 instead of to 9.

14 Describe a circuit that can be used to make a 4-stage binary counter count up to 12 instead of up to 15.

15 Describe the 4017 decade counter IC. Draw a graph of its inputs and outputs while it counts from 0 to 4. Suggest an application for this counter IC.

16 Design a system based on the 4017 decade counter that counts cars driving into a car park and turns on a 'Full' illuminated sign when eight cars have entered.

17 A car park has separate entry and exit gates. Design a system to count up when a car enters and to count down when a car leaves.

18 Design a circuit to implement your design for Q. 16 or 17.

19 Design a logic circuit for the chaser lights display (p. 172) using only NAND gates.

20 Design a 2-input decoder circuit (as on p. 160) to display letters a to d on a 7-segment display, as below:

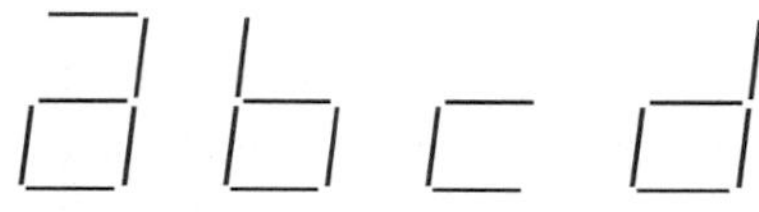

FIGURE 41.27

21 Using a system diagram, illustrate how to build a 2-digit display circuit that counts from 00 to 99.

22 In what way could the circuit of Q. 21 be modified to count from 00 to 45?

23 Draw a system diagram of a counter that runs from 000 to 999 and drives a display.

24 Design a circuit using a 14-stage binary counter/divider to produce two output signals running at 1024 Hz and 64 Hz, driven by an astable running at 8192 Hz.

MULTIPLE CHOICE QUESTIONS

1 Output Q of an S-R bistable made from two NAND gates changes from high to low when:
A the Set (S) input is made high.
B the Reset (R) input is made high.
C the Set input is made low.
D the Reset input is made low.

2 The Q output of a 4013 D-type flip-flop is high if:
A the Reset input is made high.
B the Data input is high and the Clock input goes high.
C the Data input is low and the Clock input goes low.
D The Data input is high and the Clock input goes low.

3 A 4013 D-type flip-flop acts as a divide-by-two frequency divider. The D input is connected to the:
A clock input.
B Q output.
C $\overline{Q}$ output.
D positive supply line.

4 A binary counter with 14 stages counts from zero to:
A $2^{14}-1$.
B 2^{14-1}.
C 14^2.
D 2^{14}.

5 The number 1000 0011 in binary coded decimal equals the binary number:
A 0011 0101.
B 38.
C 0101 0011.
D 53.

6 To make a four-bit binary counter count from 0 to 5, we feed back to the reset input the:
A outputs '4' and '1' ANDed together.
B outputs '4' and '2' ANDed together.
C output '4'.
D outputs '4' and '1' NANDed together.

DESIGN TIME

More circuits for you to design and (possibly) build.

An alarm sounds briefly when someone stands on a pressure mat. The alarm continues to sound after the person has stepped off the mat. It stops only when a reset button is pressed.

The system sounds an alarm, as on the left, but the sound continues for 5 minutes after the person has stepped off the mat, then stops automatically.

A metronome circuit with variable speed flashes a red LED on every beat. It also flashes a green LED on every fourth beat.

Design and build a system that counts people passing through a doorway and displays the number.

Next, try to modify the system so that it rings a bell when the seventh person passes through the doorway.

A clock running at 50 Hz or more drives a D-type flip-flop. The outputs of this switch on two LEDs alternately. One LED is labelled 'Heads', the other 'Tails'. There is a button which, when pressed, stops the clock. This leaves either 'Heads' or 'Tails' lit, virtually at random.

A simple reaction tester has a 4017 IC and a row of 10 LEDs which light up one at a time, in order. The counter is driven by a 10 Hz astable. Design the logic so that the counter starts when the operator presses a button. The subject watches for the first LED to light and presses a button as soon as they see it. Pressing this button stops the counter. The subject's reaction time to the nearest tenth of a second is measured by noting which LED is lit.

Increase the frequency of the astable to obtain greater precision.

Build the chaser light display circuit (see Figure 41.25) and use it to drive 12 LEDs arranged in a circle.

An alarm sounder unit that emits an intermittent (beep-beep-beep-...) sound, instead of a continuous sound.

Slightly more complicated is a unit that emits a two-toned sound.

Storing Data

The processing stages of some systems may need to store data. The data, or **information**, is stored in the system to be used later.

The set-reset bistable is an example of a circuit that stores data.

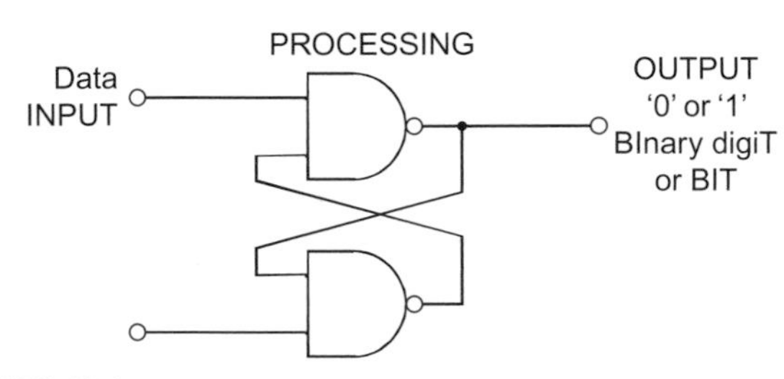

FIGURE 42.1

It 'remembers' if the Set input has been made low since the bistable was last reset. Its output is a low voltage or a high one. We may write its output as '0' (usually for 'low') or '1'. These are the two digits used for numbers in the binary system. The output of the bistable is a single **BI**nary digi**T**, or **BIT**.

The D-type flip-flop is another memory device with a one-bit output.

BITS AND BYTES

A bit can have one of two possible values, 0 or 1. The bit can represent facts, for example, 0 = 'It is not raining', and 1 = 'It is raining'. The facts represented by '0' and '1' must always be the *exact* opposites of each other. Alternatively a bit can represent a quantity, 'zero' or 'one'.

If we have eight bistables, their eight outputs are each represented by '0' or '1'. The collection of eight bits is called a **byte**. One possible set of outputs is shown in Figure 42.2. There, the byte is 10100110. This might represent eight different facts, such as 'It is/is not raining', 'It is/is not windy', 'It is/is not Monday', 'It is/is not a public holiday' and so on.

Alternatively, the byte might represent a quantity. '10100110' is the binary equivalent of the decimal number 166. This could be the output of an 8-stage counter and '10100110' (or decimal 166) is the number of cars in an automatic car park.

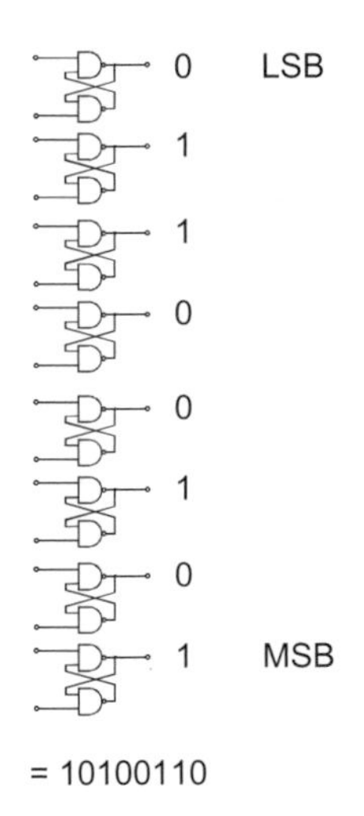

FIGURE 42.2

It is important to read the outputs in the right order.

In this example, we are reading from bottom to top. The bottom bistable outputs the most significant bit (MSB). The top bistable outputs the least significant bit (LSB).

A byte, when representing a number, can take any value in the range 0000 0000 (all bistables reset) to 1111 1111 (all bistables set). This is equivalent to 0 to 255 in decimal, a total of 256 possible values altogether.

MULTIPLES

The memories of most of today's systems need to store more than a few bytes of data. Larger amounts are rated in:

- **Kilobyte** (Kb): A kilobyte is 1024 bytes. Note that it is *not* 1000 bytes, as one might expect from the

Electronics: A First Course.

meaning of 'kilo' in the metric system. A kilobyte of memory is enough to store a short document, such as half a page of text. It could also store the instructions for operating a simple circuit such as an automatic dishwasher.

- **Megabyte** (Mb): A megabyte is 1024 kilobytes. It is *roughly* equal to a million bytes, or 8 million bits. A megabyte could store a small colour photograph, or the instructions for a simple computer game.
- **Gigabyte** (Gb): A gigabyte is 1024 megabytes, or approximately a thousand million bytes. The total storage capacity of a personal computer is usually a few gigabytes.

MEMORY

Many automatic systems such as dishwashers, microwave ovens, robots, and computers are based on logic and need memory for data storage. As suggested above, the data may be in the form of instructions to the machine, telling it how to perform its tasks, or how to respond to given inputs. Data of this kind is called a **program**. We shall say more about programs later. Data may also be numeric, as in an automatic weighing machine.

Both program data and numeric data are stored in binary form. The memory chip carries thousands or even millions of bistables, each storing a single bit of data, '0' or '1'. The bistables may each be similar to that shown opposite but a different kind of bistable is found in many types of memory chip.

A typical example of a memory chip is the 6116, which stores 2 kilobytes of data.

FIGURE 42.3

The IC is a large one, but not because the chip inside is particularly big. The main reason for its size is to provide space for the 24 pins that are needed for the power supply and for input and output signals.

Data is loaded into the chip or recovered from the chip a byte at a time. This means that the chip must have eight output terminals, one for each bit of the byte. When data is being loaded (or, as we more often say, **written**) into memory we have to be able to tell the chip which set of eight bistables to store it in. This is easily done because each byte has its own **address**. This being a 2 Kb memory, the addresses are numbered from 0, up to 1 less than 2 Kb.

2 Kb is a little more than 2000 bytes. It is 2 times 1024 bytes, or 2048 bytes. So the addresses run from 0 to 2047. These numbers need to be in binary form for the logic circuit of the chip to be able to decode them. The binary equivalent of 2047 is 111 1111 1111. This is an 11-bit number, so we need 11 input lines to specify the address of the byte in which the data is to be stored.

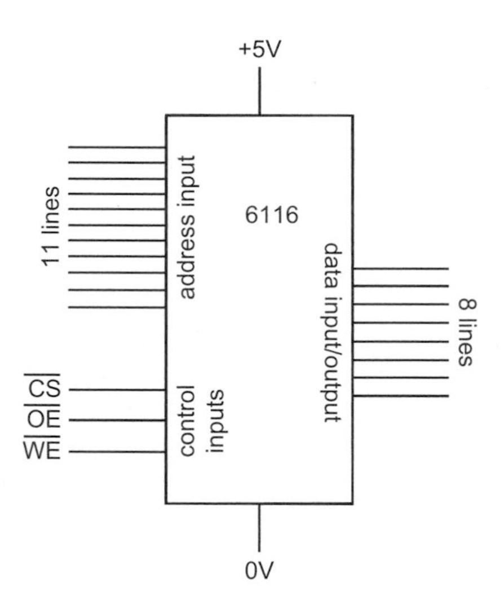

FIGURE 42.4

Two of the remaining five pins are used for the power supply to the chip. Finally there are three control inputs:

- **CHIP SELECT**: A system may have more than one memory chip. To make one particular chip operate, its $\overline{CS}$ line is made low.
- **OUTPUT ENABLE**: If this line is made low, the chip outputs data from an address in its memory.
- **WRITE ENABLE**: When this is low, data that is placed on the 8 data lines is stored at an address in the memory.

The bars over $\overline{CS}$, $\overline{OE}$ and $\overline{WE}$ mean that the lines are normally held at logic high, and have to be made low to act. We call this **active low**.

There are two kinds of operation, writing and reading. To write some data into memory, the CHIP SELECT input is made low. Then the byte to be written into memory is put on the data lines, and the address where it is to be written is put on the address lines. Finally, when the WRITE ENABLE line is made low, the data is stored in the eight bistables at the given address.

The diagram on the right shows logic levels at the moment when the byte 1001 1100 is being written into address 100 0110 0110. In decimal, the value 156 is being written into address 1126.

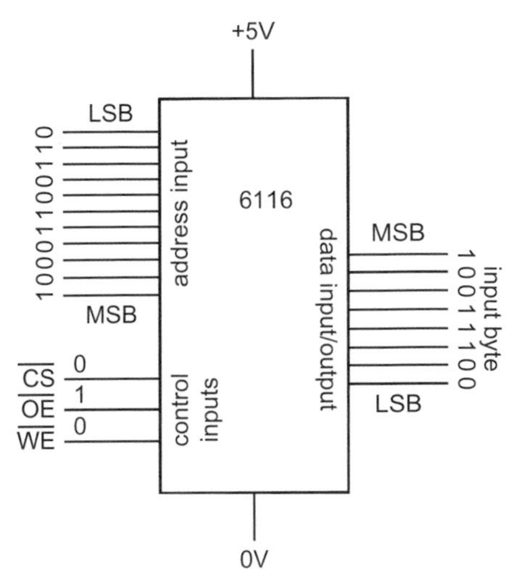

FIGURE 42.5

To **read** data that is stored in memory, make the CHIP SELECT input low. Then put on the address lines the address to be read from. The stored data appears on the data lines when the OUTPUT ENABLE line is made low. Below, we are reading from address 15.

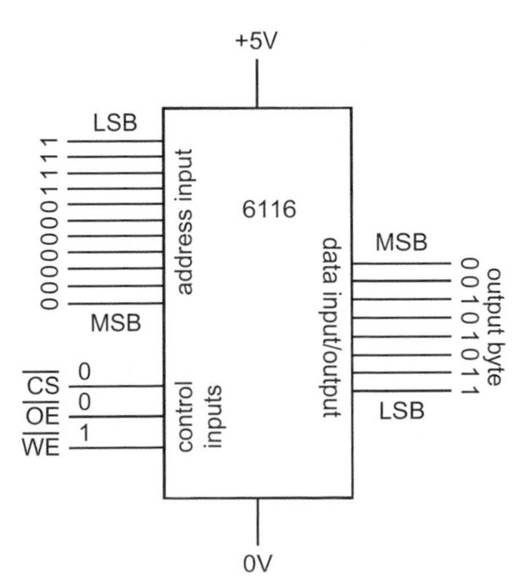

FIGURE 42.6

The 6116 and similar memory ICs are used for temporary storage of data. When the power supply to the circuit is switched off, the data is lost. This type is used as RAM in computers and other equipment. Other types of memory IC include ROM, used in computers and other equipment, which retains the stored data permanently. It remains there when the power is switched off. Some types of ROM are re-programmable, but retain the data when power is switched off.

FLASH MEMORY

A flash memory chip consists of arrays of MOS transistors each of which has two gates. One of the gates is completely insulated so that any charge of electrons placed on it remains there indefinitely. Once the gate is charged or discharged, the transistor remains on or off indefinitely. Its state is read by a technique that employs the other gate. The two possible states of each transistor are equivalent to the binary logic values '0' and '1', and in this way the chip stores arrays of binary data.

The reader might wonder how it is possible to alter the charge on a gate that is electrically insulated. A conductor would seem to be essential. However, there are ways of transferring electrons without conduction.

The advantages of flash memory are:

- It needs no power supply to keep the data in memory. This makes it ideal for portable equipment.
- It is removable. An owner of a digital camera can have several flash cards in use. Flash memory used in a computer can be removed and locked away for security.
- Writing and reading is fast, so it is ideal for memories holding large amounts of data. Flash memories holding hundreds of megabits are easily obtainable; one storing 8 gigabits has been developed.
- It is not damaged by mechanical shock, unlike hard disks.

The disadvantages are:

- Individual bytes can not be erased; flash memory is erased in blocks.
- They eventually wear out; typically, a memory can be erased 10 000 times.
- They are more expensive per stored bit than a hard disk.

The advantages outweigh the disadvantages, so flash memory has become very popular.

Flash memory is used for storing data in a digital camera. A typical flash card stores 256 Mb, which is enough for hundreds of high-resolution digital colour photos. Later, after the data has been transferred to a computer, or has been printed out on to paper as a photo, the data can be erased and a new set of photographs recorded.

The photo below is the flash memory card from a Personal Digital Assistant. The PDA is a hand-held computer, slightly larger than a mobile phone, and with a large (for the PDA's size) LCD screen and a built-in phone. It can exchange data with a remote computer through the mobile phone network. The card shown here stores 256 Mb. The text and illustrations of *Electronics: A First Course* need about 95 Mb so this tiny card can store over two books the size of this one.

FIGURE 42.7

A memory stick is a similar device which may contain as much as 1 Gb of storage. It is a convenient way of transporting data between computers as well as filing away text and images. It has a plug at one end to fit into a USB socket on the computer.

Flash memory is also widely used in MP3 players and mobile phones.

QUESTIONS ON STORING DATA

1 What is a bit and what values can it have?
2 How many bits are there in 2 bytes?
3 How many bytes are there in 5 Kb?
4 Describe a 1-bit memory circuit.
5 When a given bit is '1', it means 'The temperature is greater than 25°C'. What is meant when the bit is '0'?
6 What is the value of the most significant bit in the binary number 1 0011 0010?

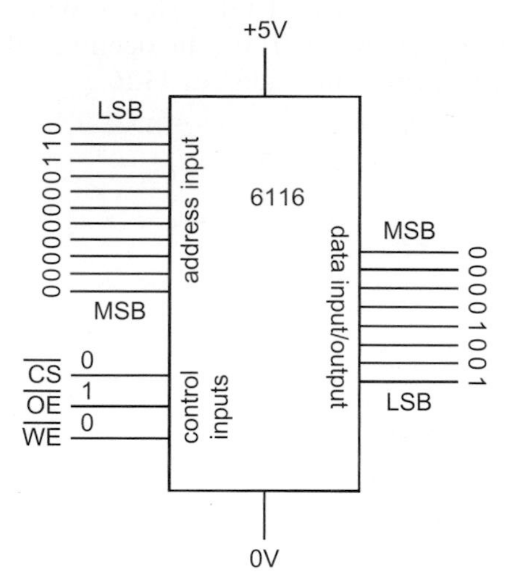

FIGURE 42.8

7 Given the logic levels shown in the diagram, what operation is being performed by this memory IC?
8 What are the three main groups of inputs or outputs of a typical memory IC?
9 What is the action of the $\overline{\text{CHIP SELECT}}$ input on a memory IC, and when is it needed?
10 What is the meaning of the bar over the names of input terminals such as $\overline{\text{WE}}$ and $\overline{\text{CS}}$?
11 Give three examples of equipment or appliances that have memory built in to them. In each case, say what kind of data is stored in the memory. If possible, state the size of the memory.
12 Explain what is meant by the terms 'write' and 'read' when applied to the action of a memory IC.
13 What is 'flash memory'? List three types of electronic equipment that have flash memory. Explain why flash memory is so suited to these applications.

Topic 43

Microcontrollers

Microcontrollers are not included in some Electronics specifications.

All kinds of electronic equipment, from mobile phones to microwave ovens and from dishwashers to digital cameras, have a microcontroller at the heart of their system. The microcontroller performs all the complicated processing needed to link the inputs of the system to its outputs.

A microcontroller is sometimes referred to as a 'computer on a chip'. This is a good description. A microcontroller is a single integrated circuit on which are combined most of the circuit blocks that we find as separate units within a computer.

FIGURE 43.1

There are hundreds of different microcontrollers available. It is difficult to say that any one type is 'typical'. The 28-pin integrated circuit in the photo is average in size and contains the items that are needed by all controllers:

- **Arithmetic-logic unit (ALU)**: Logic circuits that perform addition, subtraction, and many logical operations.
- **Memory**: Logic circuits for storing data. There are two kinds. **RAM** stores up to 72 bytes of data. This memory is used by the ALU for a temporary store of data that it needs while processing. **ROM** stores up to 3 Kb of data. This is where the program is stored that tells the controller what to do.
- **Clock**: The chip contains all the components of the system clock except for the crystal.
- **Input and output**: Of the 28 pins of this IC, 20 are used for the input or output of data. They may be connected to sensors and other input devices. They may be connected to lamps, displays, motors, loudspeakers and other output devices. There is more on the next few pages about interfacing these to the microcontroller.

As well as these essential units, the controller in the photo has special facilities that many, but not all, microcontrollers possess. These include a pair of digital timers and an 8-stage binary counter.

MEMORY

There are many different types of memory, and manufacturers are always inventing new types. Essentially, the two main types are known as RAM and ROM.

RAM (short for *random access memory*) is used for temporary data storage, as already mentioned. The data stored there is lost when the power is switched off.

ROM (short for *read-only memory*) is more-or-less permanent. It is intended to be read from often but not often written to. It stores instructions as binary code that the controller can 'understand' and act on.

In some controllers, the data in the memory is programmed into it when the chip is made. It can not be altered. This type of ROM might be found in a controller for a washing machine. Thousands of ready-programmed ROMs are mass-produced for a particular model of machine.

Many controllers, such as the one in Figure 43.1, have ROM in which the program can be erased and then rewritten as many times as you want. Figure 43.2 shows a device used for programming the ROM of this kind of controller.

The programmer has circuits concerned with providing the voltages necessary for storing data in the ROM of the controller. Programming is controlled by a computer, on which the user creates and tests the

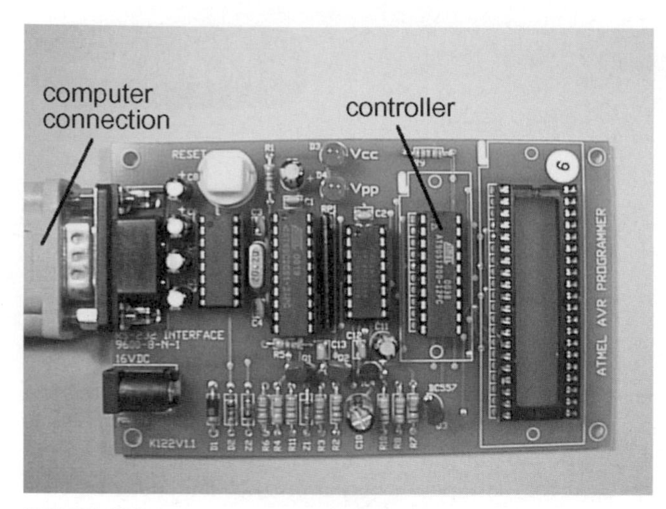

FIGURE 43.2

program. The board has a special 20-pin socket into which a controller is plugged for programming.

When the program is complete and tested, the controller is removed from the socket. It is then put in a socket in the device (such as a home security system or an electronic kitchen scales) that it is intended to control. The 40-pin socket on the right of the controller is for programming the more complicated controllers that have more input and output pins.

Writing the binary code of the program into the ROM of the controller is under the control of software running on a computer. The person creating the program types in instructions telling the controller which of its inputs to check on and what to do if the inputs are at logic high or low (depending on the sensors attached to the inputs). The program tells the controller what to do (that is, what outputs to activate) when it detects certain combinations of inputs.

When programming software is run, the computer screen looks something like that shown in Figure 43.3.

The operator is typing the program into the window at top left. The operator uses a special 'language' to tell the controller what to do (more about that in Topic 44). The other windows tell the operator about the state of various parts of the controller. This includes a list of all the input and output pins and the present state of each of these. Using software such as this, the operator creates a program and tests it before actually writing it into the controller. The program can be run at its usual speed or just one step at a time. Stepping through gives the operator time to check that the program is working correctly at each step.

The operator can also decide on key stages of the program at which the run should stop. When the program stops at the breakpoint, the operator can study the contents of memory and registers, and the state of the outputs at the end of each stage.

Up to this stage, the controller itself has not been programmed. Everything is being done in the operator's computer. When the operator is satisfied that the program works correctly, it is downloaded into the ROM. The software in the computer converts the instructions keyed in by the operator into a binary code that the processor can 'understand'.

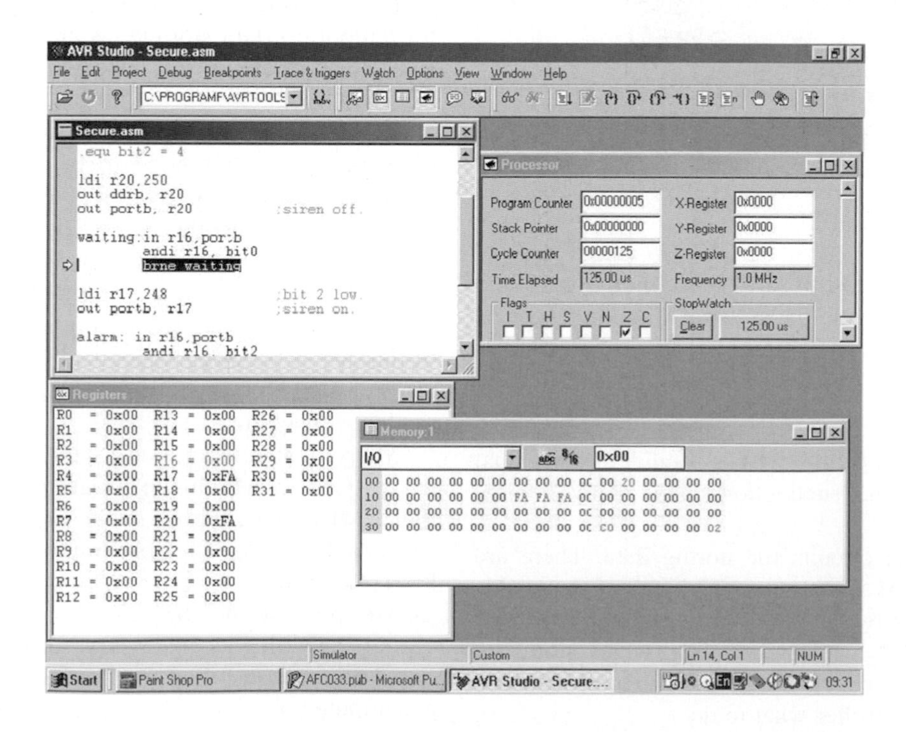

FIGURE 43.3

If the program is still not quite right, it is corrected on the computer and the new version downloaded into ROM, to replace the original version.

INPUT INTERFACING

An input pin of a typical controller can accept up to 25 mA. This does not mean that we can feed 25 mA into all 20 pins at once. That would be a total of 0.5 A, which would be too much for the chip to handle. The data sheets state maximum limits for input currents. Normally, the input current from a sensor may be only a few microamps, so the problem does not arise.

The other point about input to a microcontroller is that the voltage of the signal must be within a certain range. Normally the minimum voltage is 0 V, or close to 0 V, and this is accepted as a logic low. The highest voltage is normally the supply voltage, or close to it. This is accepted as a logic high.

Most microcontrollers run on a range of supply voltages. The PIC microcontroller illustrated in Figure 43.1 runs on any voltage between 2 V and 6.25 V. The Atmel controller Figure 43.2 runs on 2.7 V to 6 V. These ranges allow the chip to be conveniently powered from batteries in portable equipment.

Inputs to the controller usually come from switches or sensors. Here are some examples:

Switches: This category includes pressure mats, microswitches, push-buttons and keyboard switches. Typical circuits are shown below.

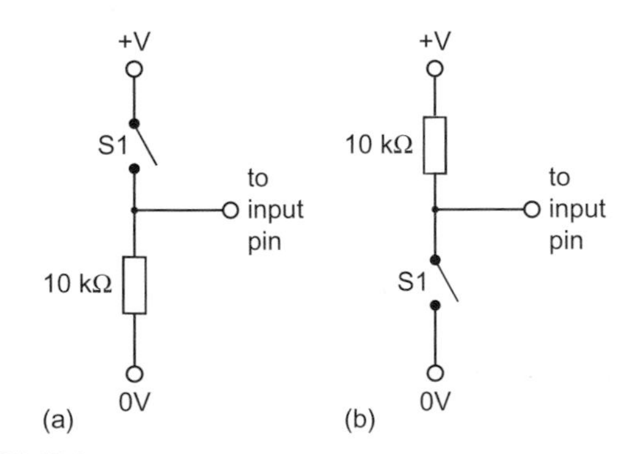

FIGURE 43.4

Circuit (a) normally gives a low input, but gives a high input when the switch is closed. Circuit (b) normally gives a high input, but gives low when closed.

Voltage dividers: Sensors such as thermistors and LDRs are usually a part of a voltage divider network. We use a version of the circuit on p. 95 to obtain a binary input .

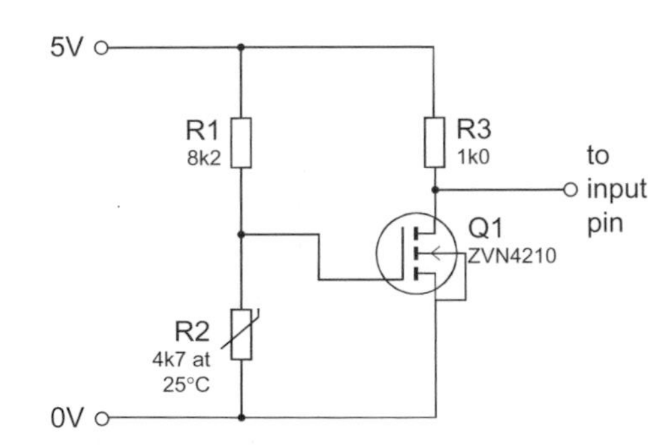

FIGURE 43.5

When the temperature is above a certain level, Q1 is off. There is no voltage drop across R3, so the input to the pin of the microcontroller is logic high.

When the temperature is below that level the input is logic low. Switching occurs fairly sharply as temperature crosses the set point. If a more precise action is required, use a Schmitt trigger, as on p. 130.

Another way of obtaining the same action is to use a microcontroller that has a comparator on the chip. The circuit is the same as that on p. 131, except that the comparator is included in the microcontroller. The output of the comparator sets or resets a bit in the memory, so the ALU can read '0' or '1' from this, depending on the temperature, and act accordingly.

Analogue input: For some projects we need to input the actual voltage across the sensor, instead of simply '0' or '1'. For example, we might need to record air temperature once an hour throughout the day, and also find minimum, maximum and average values. Although the input will actually be a voltage in the range 0 V to the supply voltage, there is no problem in converting this to a Celsius temperature. For analogue quantities we need an **analogue-to-digital converter**, or **ADC**. If you need this feature, check through the data sheets to find a controller that has one (or more) ADCs on board. The data sheet will tell you how to use it. Alternatively, use an external ADC such as the ADC0804LCN. This has an 8-bit output that can be fed directly to a microcontroller.

OUTPUT INTERFACING

The output pins of a controller can supply only a limited current. Typically, this can not be more than 20 mA. Outputs can burn out in a second or two if this limit is exceeded. As with inputs, there is an overall limit to the total output current that can be supplied at any one time.

Light emitting diodes: These can be driven directly, provided that there is a current-limiting resistor in series with the LED. To be on the safe side, limit the current to 10 mA. Given the supply voltage is 6 V, a 390 Ω resistor is suitable.

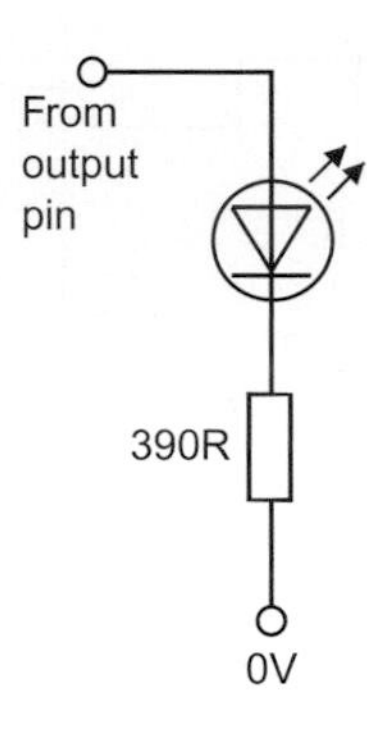

FIGURE 43.6

LEDs are useful for displaying output logic levels when testing a new project.

Transistor switches: Most output devices, such as motors, solenoids and loudspeakers, take such large currents that they are best driven by a transistor switch. A MOSFET switch is ideal (p. 95) as it takes virtually no current. It can not overload the output circuit of the controller.

The circuit on p. 108 shows how to interface a relay using a MOSFET switch. This is useful for controlling heavy-current devices. Smaller loads can be switched by reed relays, directly from the output pins.

Pulsed switching: On pp. 141–2, a method is described of varying the duty cycle of a 555 astable. This is used for lamp dimming and for controlling the speed of a motor. The same effect can be produced by a controller that has a built-in pulse generator. Pulse width is set by programming the controller.

QUESTIONS ON MICROCONTROLLERS

1 Name three appliances or pieces of equipment that contain a microcontroller.
2 What are the main units present in a microcontroller, and what does each do?
3 What are RAM and ROM? Describe the main uses of each of these.
4 Using a data sheet, list the amounts of RAM and ROM present in a named microcontroller.
5 In what type of applicance or equipment would you expect to find permanently programmed ROM.
6 Outline the process of programming a controller from a computer.
7 Describe how you would interface (a) an LDR and (b) a tilt switch to a controller.
8 Why is it essential to know the current needed to drive an output device?
9 Describe how you would interface (a) a 6 V 360 mW filament lamp and (b) a mains-voltage electric fan to a controller.

Programs

Programming is not included in some Electronics specifications.

The **program** stored in the ROM of a microcontroller tells it what to do. It consists of a series of **instructions**. An instruction might *mean* 'Find out the logic level at input pin 5'. But, although this is what the controller will do when it comes to that instruction in its ROM, the instruction is not there as a sentence made up of words. It is coded, often as a single byte of a binary code.

To help you understand what a program really is, we will look more closely at a very short section of an actual program. It runs on the controller pictured in Figure 43.2. It is part of the security program featured in the screen shot in Figure 43.3. Follow the descriptions, noting the main features that are printed in **bold type**.

If we were to check the contents of the ROM in which the program is stored, we would find this:

Address	Byte
0	11101111
1	01001010
2	10111011
3	01000111
4	10111011
5	01001000
and so on ...	

This is what the controller has to read, byte by byte. The '0's and '1's represent bistables that are either reset or set. This is what the programmer has to put into the ROM in order to make the controller do anything useful. This array of '0's and '1's is called **machine code**. Few operators try to program directly in machine code.

Summing up:

A program contains a series of instructions, coded by the states of bistables in ROM.

CODES FOR PROGRAMMING

The screen shot in Figure 43.3 shows the program being written by using a special kind of software, called an **assembler**. The program begins with the line:

```
ldi r20, 250
```

This too is a code but a little easier to understand than machine code. The first part of the code 'ldi' means 'Take the value that comes next and *load immediately* into a specified register'. The byte stored at address 0 is the machine code version of this instruction. This means something to the controller but not to us!

The second part of the code 'r20, 250' means 'The value is 250 and the place to load it is register 20'. This data is coded in the byte at address 1. Register 20 is one of 32 sets of 8 bistables that form part of the RAM. So the first two bytes of the program tell the controller to put the value 250 in register 20. Register 20 will then look like this:

11111010

This binary number is the equivalent of decimal 250. Note that byte 1 is not an instruction. It is coded data: what to put and where to put it.

As well as instructions, a program contains coded data.

The next part of the program sets up the controller to operate the security system. We will not go into the details of how bytes 2 to 5 code this. After that, we have the instructions for waiting for input and producing output.

A program written in assembler lists an instruction, and often a byte of data, for every operation that the controller does. There is no way of combining the two into one statement. The instructions all specify very simple, short processes. For example, when two numbers are to be be added together, there must be instructions to place the numbers in different registers, then add them, then transfer the result to some other register.

Electronics: A First Course.

Processing is broken down into very elementary steps.

As a result of this, a program has many instructions, and takes a long time to write.

Other codes, or languages, are often used for programming, using different software. Some use BASIC, which is easy to learn. The instructions are written in statements that are very like ordinary English sentences. For example, the instruction on the previous page would be:

LET r20=250

This is equivalent to several lines of machine code or assembler code. Also it is much easier to understand.

Another language that is often used for programming controllers is called *Java*. This has many useful features though it takes longer to learn. Summing up:

Programming is done in assembler or a high-level language, such as BASIC or Java.

INPUTS AND OUTPUTS

The controller shown in Figure 43.2 has a set of 8 pins for input or output. These can be programmed singly or as a byte. A set of pins such as this is often called a **port**.

In this very simple programming example, we need only three bits. Two are programmed to be inputs (from the sensor and from the reset button). One is programmed as an output (to switch on the siren).

The sensor circuit includes a NAND gate with its inputs connected together. It acts as a NOT gate. The output goes low when the beam is broken by the intruder, making bit 0 low.

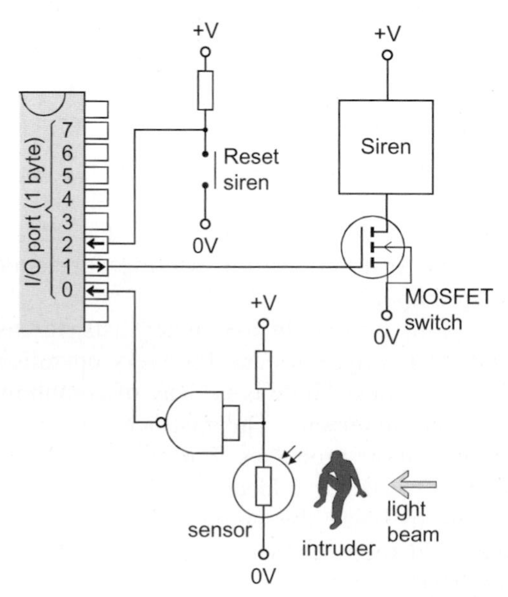

FIGURE 44.1

The output bit 1 is normally low. When it is made high it switches on the MOSFET and the siren sounds. Bit 2 is an input that reads the state of the reset button. Pressing the button makes bit 2 go low.

There are 5 unused bits on the port and the controller also has another port with 7 bits. We could use these to control a more complicated security system. In fact, it is barely worthwhile to use a controller on such a simple system as our example. The same action can be obtained with a few logic gates. However, the point of this example is to show that a system built up from **hardware** (logic gates, op amps and other electronic components) can be mainly replaced by **software** (programs run by a controller).

The advantage of software over hardware is more easily seen when the system is a complicated one. Designing, writing and possibly amending software is much easier than wiring hardware, especially if the hardware needs to be rewired to correct it or improve it.

PROGRAM FLOWCHARTS

To program a controller, you will need to learn more about the assembler or high-level language of the programming software. The manual or help screens of the software will guide you in this. Usually, the first step in writing the program is to prepare a flowchart. This outlines the main sections of the program without going into the step-by step details of the instructions to the controller.

Flowcharts are drawn as a number of linked boxes. There is a box for each major step of the program. The shape of the box depends on what kind of operation is involved:

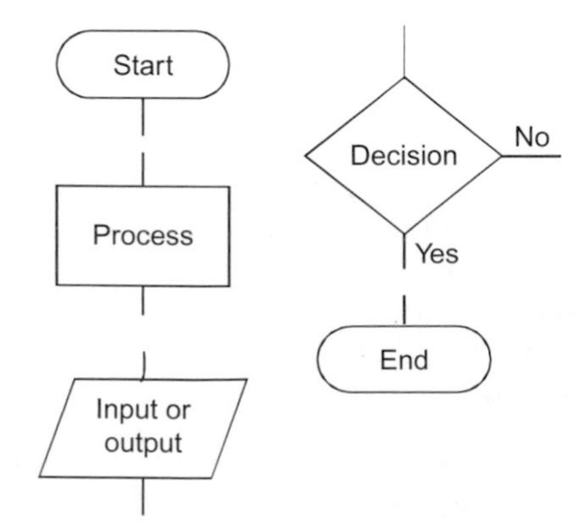

FIGURE 44.2

To illustrate the way these are used, on the right is the flowchart of the security system referred to earlier.

The program begins by setting up the three pins as two inputs and an output. This is a *process* within the controller. Next, the controller inputs the state of bit 1, the output from the sensor. If it is not '0', the light beam is unbroken. This is a decision box, and the answer is NO. The program loops back to read the input again. It does this repeatedly, looping round and round, waiting for an intruder to break the beam. This is a **wait loop**.

If the beam is broken, the answer to the question in the decision box is YES. Then the controller continues to the next step of the program. Output at bit 1 is made high, turning on the siren.

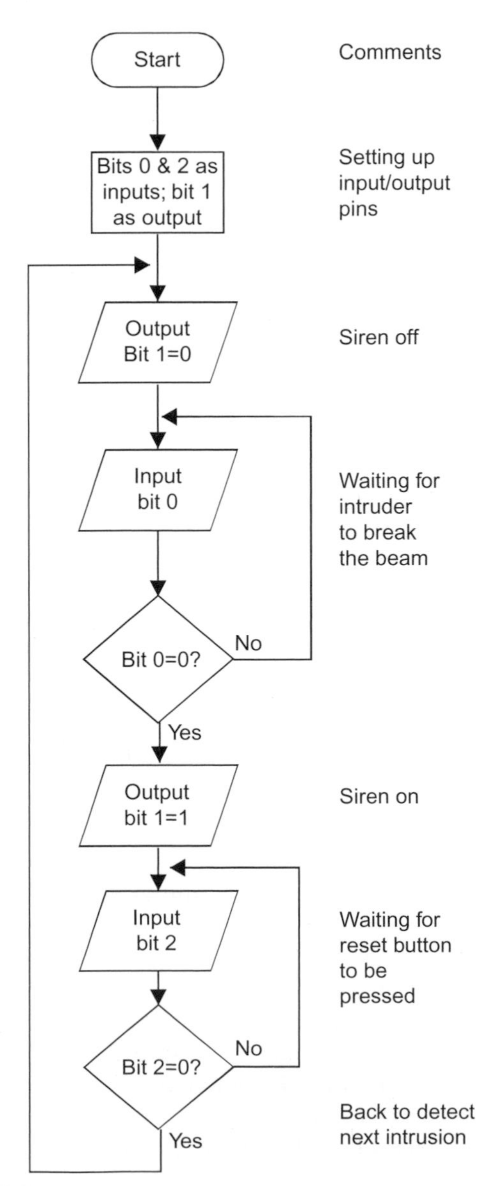

FIGURE 44.3

The siren continues to sound while the controller waits in the next loop. This time it is inputting data at bit 2, the state of the reset button. It waits until the button is pressed and bit 2 becomes '1'. Then the answer to the lower decision box is YES. The controller loops back almost to the beginning of the program. There it turns off the siren and waits in the first wait loop for the next intruder to break the beam.

Note that this program runs continuously in loops and does not have an END box.

AUTOMATIC DOORS

The flowchart on the right is the first step at writing the program. Note that the flowchart shows the software.

The flowchart begins with the usual setting-up of input and output pins. Then the doors are closed, if they are not already closed. It is usually safer to begin a program with an action such as this, to ensure that the system is in a specified state to start with.

Next, the input from the microwave sensor is read. If a person is detected approaching the doors, the program runs on to open the doors. If not, the program loops around, reading the sensor repeatedly until someone approaches. After the doors are opened, there is a delay to give the person a chance to pass through. After that, the program jumps back to read the sensor again. The person may have been walking too slowly or a second person may be approaching. In these cases the doors remain open. However, if no one is approaching, the doors are closed.

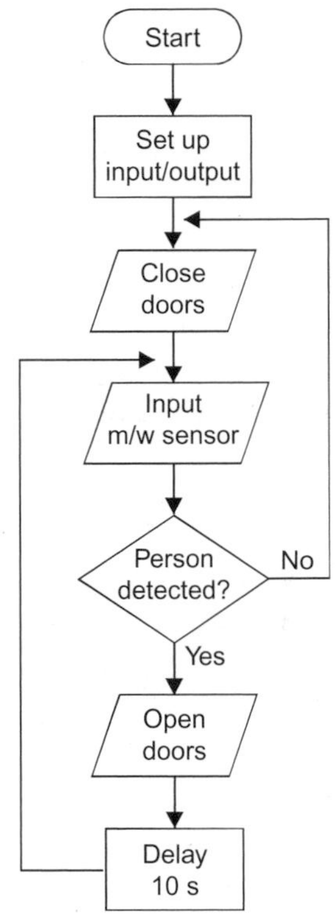

FIGURE 44.4

It is a good approach to a program to map out the main stages first. Details follow later. In this example, it is important to program the opening and closing of the doors more carefully. A practical point is that the doors do not need opening if they are already open. Another point is that just switching on a motor for a given time does not guarantee that the doors are fully open. We need positive input about whether the doors are fully open or not. This data about what has (or has not) happened is known as **feedback**.

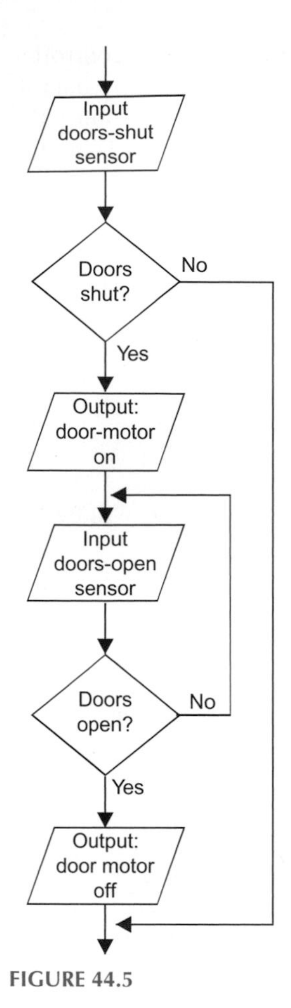

FIGURE 44.5

The system needs two more inputs, perhaps from two microswitches. One closes when the door is fully open and the other closes when the door is fully shut. Using these sensors, we expand the 'Open doors' stage of the previous program.

The first step is to check if the doors are already open. If so, the program skips to the end of the open doors routine. The motor is turned on to open the doors. Then the sensor is read to find out if the doors have opened as wide as possible. If not, the routine loops around, with the doors motor still running, to check again. When the 'doors-open' sensor confirms that the doors are open, the motor is switched off. This is an example of feedback.

A matching routine should be added to the main program to supervise the closing of the doors.

Things to do

Even this is not the end of the programming. What happens if someone leaves a supermarket trolley in the doorway as the doors close? How can we provide a manual over-ride button to open or close the doors at the beginning and end of the day, or when the window cleaner needs to work on the doors? It would also be a useful feature to have automatic locking, so that the doors can be made secure when the supermarket is closed. There are lots of refinements possible. Can you think of more? Write the programs.

TWISTY WIRE GAME

The microcontroller version of this game has one input. This goes high (=1) when the loop contacts the twisty wire. It has one output, to a MOSFET switch that turns on a buzzer.

The program begins by setting up the input and output pins. Then a register, called *count*, has zero loaded into it. *Count* has 1 added to it every time contact is made between the loop and the wire.

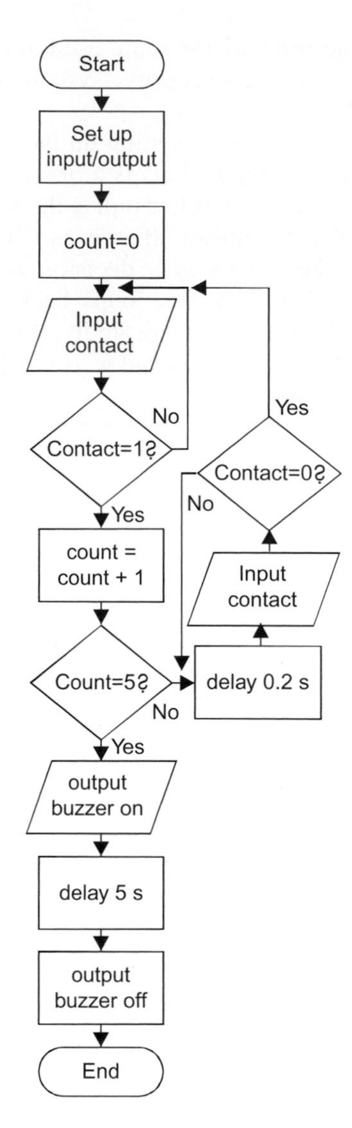

FIGURE 44.6

The player is allowed five 'contacts'. When *count* reaches five, the buzzer is switched on for 5 seconds and the game ends.

Note the 0.2 s delay. This allows for the **contact bounce**. Multiple makes and breaks of contact are registered as only one contact on each occasion. It gives time for the bounces to settle.

Contact debouncing can be done by using a capacitor or a Schmitt trigger circuit (hardware debouncing) but here we use **software debouncing**. The routine used here also checks the input until it has returned to '0' before returning to wait for the next contact.

QUESTIONS ON PROGRAMS

1 Where and in what form is the program stored in a microcontroller?
2 What are the two main kinds of information in a program?
3 Name one of the programming languages that are used by programming software. Give an example of an instruction written in that language.
4 Describe a circuit suitable for interfacing a named sensor to an input pin of a microcontroller.
5 Explain how you would instruct the controller to read the input from the sensor circuit of Q. 4.
6 Describe a circuit suitable for interfacing a named output device to an output pin of a microcontroller.
7 Explain how you would instruct the controller to activate the output device.
8 When we use a microcontroller, much of the hardware of a circuit is replaced by software. What are the advantages of this?
9 Describe the meanings of the shapes of the two flowchart boxes drawn below:

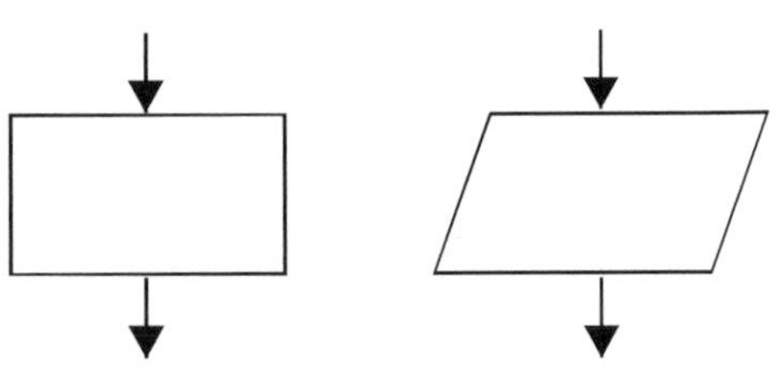

FIGURE 44.7

10 Why is a flowchart the best first step toward writing a program?
11 What is the difference between a program flowchart and a system diagram?
12 What are normally the first processes at the start of a flowchart?
13 Draw a flowchart for a panic button program, operating as described on p. 149.
14 Draw a flowchart for the operation of a washing machine (with only one washing program).

DESIGN TIME

Design programs for the systems described on this page. Draw the flowchart and, if your course includes this, write and debug the program. If possible, program a microcontroller and attach the hardware needed to complete a fully operational system.

Make the twisty wire game more competitive. Extend the program to put a time limit on the player's turn.

The twin-loop model railway system on p. 164 can be made more realistic by putting it under the control of a microcontroller. It could route the train in several different ways and perhaps use a more complicated layout. You may want to add extra sensors. If you have the model equipment, you could add electrically operated signals. This makes an effective Open Day exhibit.

Make the twisty wire game more rewarding. Add a special contact sensor at the far end of the wire. Then extend the program to flash LEDs when the player makes a clear run.

Add to the realism of the model railway by designing a circuit sensitive to the sound of a whistle. Program the controller to start the train when the whistle is blown once and to stop it when blown twice.

Microcontrollers are ideal for programming lighting displays. Connect MOSFET switches (or possibly relays) to half-a-dozen or more output pins. Then wire up panels or strings of lamps. Program the controller to produce a scintillating display. Fun at Christmas time.

Devise a reaction tester program for a microcontroller that has the same action as the 4017 logic system described on p. 171. Then extend it to include other features that you would like it to have.

The intruder system on p. 143 can be implemented and extended by programming a microcontroller to drive it. Most controllers can perfom the AND operation so the hardware AND gate can be replaced by its software equivalent.

Design and program a weather station system. This could be done by several students, each working on a different section. It could show present temperature, and 24-hour maximum and minimum. You might design simple sensors for wind direction, wind strength, pressure and rainfall.

Design a metronome system based on a microcontroller. Devise a system of lamps or sounds ('clicks') to emphasise the beat at the beginning of each bar. Try programming Latino rhythms.

Build and program a working model lift or car-park entrance gate.

Visual Output

Most important visual outputs are based on lamps, LEDs, and liquid crystal displays.

LAMPS AND LEDS

LEDs are probably used more widely than any other device for indicating the output of a system. They are very efficient, use little current, last much longer than filament lamps, and are less easily broken. They are made in a wide range of colours, shapes and sizes. They are also available with a built-in flashing circuit. Because they take little current, they can be driven directly from the output of a 7555 timer IC, TTL, CMOS buffer gates (4049 and 4050), microcontrollers and many other devices. This simplifies circuit design and construction. Super bright LEDs, requiring larger currents, are used in applications such as car brake and indicator lamps and in torches.

Filament lamps are gradually being replaced by LEDs, especially as ultra-bright LEDs are becoming available.

7-SEGMENT DISPLAYS

These are described on p. 72. The 4511 CMOS decoder described there has outputs capable of driving the segments of an LED display. The outputs are active-high. This means that when a segment is to be switched on the corresponding output goes high. The display used with this IC must be the common cathode type.

A decoder-driver from the TTL family is the 74LS47 IC. This has four BCD inputs and seven outputs to the segments. The outputs are active-low and need a common-anode display.

The anodes are connected inside the display and to a terminal pin that is connected to the positive supply line. The individual cathodes are connected through series resistors to the outputs of the 74LS47.

LIQUID CRYSTAL DISPLAYS

An **LCD** consists of a special fluid sandwiched between two sheets of glass. There is a **backplane**, which is a transparent electrode coating the inner surface of the rear glass. Normally the liquid is clear. Regions of it appear black when an alternating voltage is applied between the backplane and the pattern of transparent electrodes that is plated on to the inner surfaces of the front glass. LCDs are usually made with several digits and decimal points. Often there are extra symbols and words to suit the LCD for special applications. The digital multimeter shown on p. 16 has an LCD. There are hundreds of other applications for LCDs, including pocket calculators, kitchen scales, microwave ovens and CD players. Colour LCDs are used in mobile phones, digital cameras, and computer monitors.

LCDs can be driven by the BCD-to-7-segment ICs already described. However, it is simpler to use the 4543 LCD driver. Like the other ICs, it includes a BCD decoder with 7 outputs. There is an input for the **display frequency**. This is a square wave at about 200 Hz, generated by an astable. When a segment is to appear black, the IC generates an alternating field between the appropriate electrodes.

LCDs have the advantage that they are easy to read in bright daylight. In dull light or darkness, there must be a lamp mounted behind the display. A big advantage of LCDs for portable equipment is that the LCD takes only microamps of current (compared with about 20 mA *per segment* for an LED display). Clocks, watches and thermometers with LCDs run for months on a small button cell.

Audible Output

ALERTS

A piezo-electric **buzzer** produces a low-pitched sound suitable for a door alert.

FIGURE 46.1

The buzzer shown above operates on 6 V and requires 20 mA. Its power leads are coloured red and black. Connect red to positive and black to negative. There are two fixing lugs. Bolt the buzzer firmly to the case or circuit board to obtain maximum sound.

A small piezo-electric **siren** produces a penetrating high-pitched note. The one below operates on 3–16 V and requires only about 5–7 mA.

The note is continuous but is made much more noticeable by sounding it in short bursts. Drive the siren from an astable running at about 1 Hz.

FIGURE 46.2

The sirens can also be used as alerts or alarms. Their sound level is between 100 dB and 110 dB. For maximum loudness, the siren must be firmly attached to the enclosure or circuit board.

A **sounder** is a brass disc coated with piezoelectric material. It may be driven directly by a 3 V peak-to-peak square wave generated by CMOS logic. It is useful for producing a variety of 'alert' sounds. Sounders are made in different diameters with different resonant frequencies. For maximum effect, the sounder should be mounted firmly on the enclosure or circuit board and be driven by a signal close to its resonance frequency.

ALARMS

These produce much louder sounds than alerts. Also the sound itself is usually an intermittent bleeping or warble to catch attention.

FIGURE 46.3

A wide range of alarm sirens is available. The multiple 'screecher' (above) produces a warbling tone. It requires 50 mA at 6 V or (for a louder sound) up to 200 mA at 15 V. This siren is for indoor use but weatherproof sirens are made for use outdoors. Alarm

Electronics: A First Course.

FIGURE 46.4

sirens produce about 130 dB of sound. Very loud sirens such as these can damage your hearing if you are close to them. Because of their loudness, they are effective at forcing an intruder to leave the premises as quickly as possible.

LOUDSPEAKERS

These can be used as alerts and alarms. An example of such use is a 'musical chimes' door alert. The photo shows a suitable miniature speaker, 26 mm in diameter.

The mylar cone of this speaker makes it weatherproof. The coil resistance of speakers is typically 8 Ω, so a power transistor is needed in the output circuit. For maximum sound, mount the speaker on a baffle, that is, a panel with a circular aperture cut in it to accept the speaker.

Mechanical Output

MOTORS

Small DC motors are used for driving many mechanical projects. They normally run on low voltages, such as 6 V or even as little as 1.5 V. They usually require a few hundred milliamps to drive them. This means that they must be switched with either a power transistor switch or a relay.

This photo shows a reversing switch, based on a DPDT switch.

FIGURE 47.1

The DPDT switch can be replaced by a relay that has two changeover contacts (p. 97). This allows the flow of current through the motor to be reversed. A second relay with single normally-open contacts is used to switch the motor on and off.

Small motors usually run at several thousands of revolutions per minute. Gearing can be used to reduce the rate of revolution and to increase the turning force. As an example, the motor in the photo has a worm gear on its shaft. This can engage with a large cog-wheel to act as a reduction gear.

Another way to obtain reduced speed and also to control the speed is to drive the motor with pulses of variable duty cycle.

SOLENOIDS

A solenoid consists of a **coil** in which slides a soft-iron **plunger**. The plunger normally has just one end in the coil. When a current is passed through the coil, the plunger is pulled strongly into the coil.

The arrow in Figure 47.2 shows the direction of forceful motion. The outer end of the plunger may be coupled to parts of a mechanism.

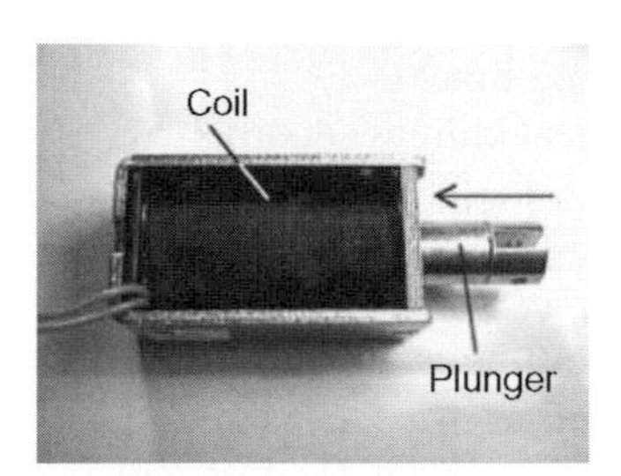

FIGURE 47.2

Note that this solenoid can *pull* but it can not *push*. Most mechanisms need to be able to return to their initial position. This can be done by using a spring or a rubber band. In some cases we can rely on gravity to pull the plunger back out of the coil. A third solution is to have two solenoids, one to pull the mechanism one way and the other to pull it back again.

In the image below, the plunger is on the left. There is a spring to force the plunger back out of the coil when the current is turned off.

FIGURE 47.3

Like the previous solenoid, this solenoid can be used to pull. However, there is a rod attached to the plunger. This projects on the right. When the current is switched on, the plunger is drawn into the coil and the rod *pushes* forcibly toward the right.

Solenoids require several hundred milliamps to power them so are best controlled by power transistor switches or by relays.

OTHER DEVICES

These are often driven by solenoids. They include valves, taps, and door latches. A stepper motor is a useful way of obtaining controlled rotary motion.

QUESTIONS ON OUTPUT DEVICES

1 List three visual output devices and suggest one use for each.
2 What is a seven-segment display? Mention one type of IC that can be used to drive such a display from a BCD input.
3 What are the advantages and disdvantages of an LCD compared with an LED display?
4 What audible output device is most suitable for use with:
  (a) a cooking timer?
  (b) a 'musical chimes' doorbell?
  (c) a smoke alarm?
  (d) a circuit that detects when a customer has entered a shop?
5 Describe the action of a solenoid. What type of circuit is needed to interface it to CMOS logic?
6 Design a relay circuit to start and stop a small DC motor and change its direction.

MULTIPLE CHOICE QUESTIONS

on storing data, microcontrollers, programs, and output devices.

1 A bit is:
  A a binary number.
  B 0.
  C high logic level.
  D a binary digit.
2 A megabyte is
  A about 8 million bits.
  B 1000 Kb.
  C 1024 gigabytes.
  D about 8 million bytes.
3 To read data stored in a memory chip we make the:
  A $\overline{CS}$ and $\overline{WE}$ inputs low.
  B $\overline{OE}$ input high.
  C $\overline{OE}$ and $\overline{WE}$ inputs low.
  D $\overline{CS}$ and $\overline{OE}$ inputs low.
4 The number of address inputs for a 4 Kb memory chip is:
  A 1024.
  B 4096.
  C 2047.
  D 8192.
5 The highest address in an 8 Kb memory is:
  A 8192.
  B 4096.
  C 4095.
  D 8191.
6 A data bit is stored in flash memory by:
  A Setting a transistor flip-flop.
  B Charging a gate negatively.
  C Resetting a flip-flop.
  D Turning on a MOS transistor.
7 A disadvantage of flash memory is that:
  A it eventually wears out.
  B it is removable.
  C it is easily damaged by mechanical shock.
  D data can only be written and erased in large blocks.
8 A microcontroller chip has:
  A a memory.
  B a unit for processing data.
  C a clock.
  D all of the above.
9 In a microcontroller, RAM is used for:
  A permanent data storage.
  B temporary data storage.
  C data processing.
  D storing the program of the operating system.
10 The program is stored in a microcontroller's memory as:
  A binary code.
  B assembler.
  C BASIC code
  D a flowchart.
11 Most of the pins of a microcontroller IC are used for:
  A data input.
  B data input or output.
  C the power supply and the clock.
  D data output.

DESIGN TIME

Design and build a minute timer. Use a 7555 astable to time the minutes and a CMOS counter to count the minutes from 0 to 9. Add a decoder and single-digit LED display to show the time elapsed.

An automatic plant-pot watering circuit turns on a water tap for 1 minute twice in every 24 hours. Design and build the circuit, including the watering mechanism. Use a ready-made valve or devise a water tap actuated by a solenoid.

Design and build a two-tone alert circuit using a 7555 astable and a frequency divider based on CMOS ICs. The 7555 would run at about 2 kHz. The divider produces a range of lower frequencies. Use two of these frequencies and design CMOS gating circuits to send these to a suitable audible output device.

Devise a solenoid actuated mechanism to open and close a greenhouse window. Add a circuit to open the window when the temperature inside the greenhouse exceeds, say, 25°C. The circuit is to close the window when the temperature falls below, say, 20°C.

Design a garage lamp to be switched on when a sensor detects the headlights of an approaching car at night. There is to be no response during the day. The light is to be switched off 5 minutes later, unless a push-button is pressed. A warning bleep is to sound 20 seconds before the lamp goes out. Another bleep is heard if the button is pressed. This circuit might need a microcontroller.

Design and build a one-way intercom system, with an alert to call someone to the receiving station.

Search the data sheets to find out about stepper motors and how to control them with a microcontroller. Design a project that is based on a stepper motor.

Test the properties of Flexinol wire. This is sometimes known as 'muscle wire'. It contracts forcibly when it is heated by a current passing through it. About 180 mA is needed. Measure response and recovery times.

Design a mobile robot, but read Topic 53 first. Decide what you want it to be able to do before you start on the design.

If you decide to build it, you will need to learn about programming software.

Build a simple mechanical model to illustrate the use of Flexinol wire. Design and build a circuit to switch the current on and off. It could be a slow astable and might include a sensor or two. It might even be part of your robot.

Part 4

Electronic Systems in Action

Audio Systems

The term 'audio system' could describe any system operating at audio frequencies, from a basic intercom to the sound reproduction system of a super-cinema. In this book, we look only at a typical stereo system for use in the home.

A system diagram (right) shows the main features of such a system. Typical inputs to the system are on the left of the figure.

Radio tuner: This receives radio signals carrying audio signals and converts them into electrical signals. The way it does this is described in more detail in Topic 50.

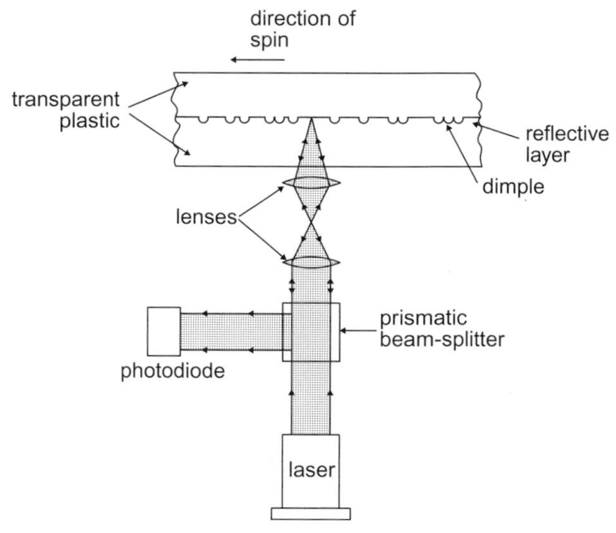

FIGURE 48.1

CD player: Music and sound, recorded digitally on a compact disc, are converted to analogue electrical signals by this unit. The disc consists of two plastic layers sandwiched together with a silvered coating between them (above).

The data is recorded as a series of dimples with spaces between them. The dimples represent logic '1' and the spaces between them represent logic '0'. As the disc spins, a beam of light from a low-power laser is focused on its underside. If there is no dimple, the beam is reflected back and detected by a photodiode.

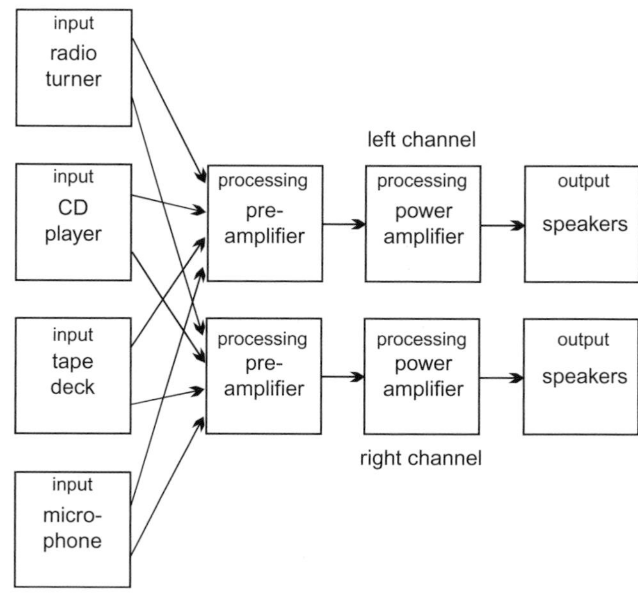

FIGURE 48.2

Where there is a dimple, the beam is scattered sideways and is not detected. This provides a stream of bits, either '0' or '1'. This is processed by complex logic circuits, eventually producing two analogue signals for the left and right stereo channels.

Tape deck: Many of the newer audio systems do not have a tape deck, but there are still plenty of older sytems in use. Also the recording technique is still employed in computer hard disc drives (see Topic 52).

The plastic tape is coated with a layer containing a magnetic substance such as chromium dioxide. This becomes organised into microscopic regions known as **domains**. Each domain is equivalent to a very small magnet. In an unrecorded tape, the domains are arranged irregularly, so there is no overall magnetisation.

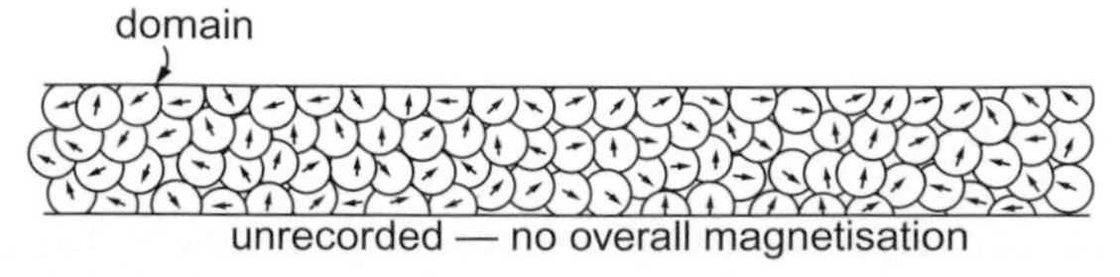

FIGURE 48.3

When sound is being recorded on a tape deck the tape passes a gap in a magnet in the recording head. A signal from an amplifier causes an alternating magnetic field in the gap, and this causes the domains to change direction.

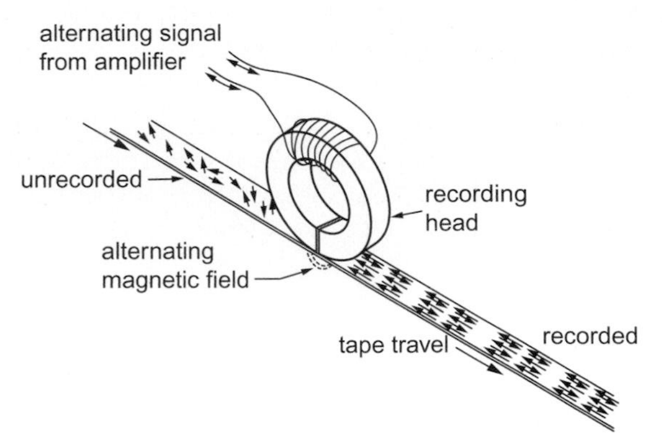

FIGURE 48.4

In some regions they are mainly pointing one way (→ below). In other regions they are mainly pointing the other way (←). The directions and the proportions of domains affected correspond to the waveform that is being recorded. The diagram below shows the original analogue audio signal and the corresponding arrangement of the domains.

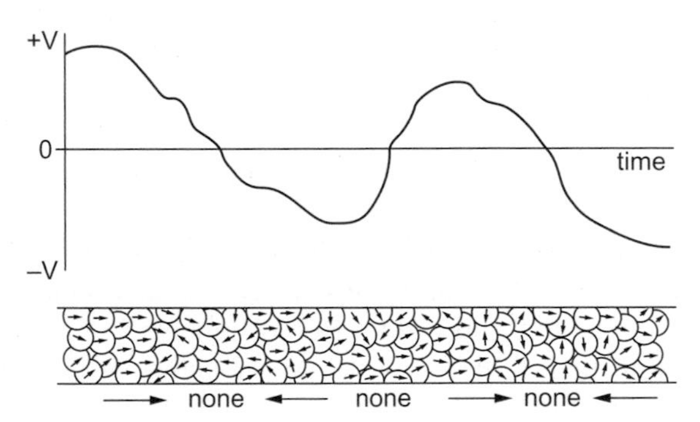

FIGURE 48.5

When the tape is played back, it passes under the playback head. There the magnetic fields produced by the domains on the tape induce alternating currents in the coil. These currents are a reproduction of the original signal current.

Microphone: These are described on p. 119. Two microphones are needed to produce the left and right stereo signals.

PROCESSING

There are two identical processing channels for the left and right stereo signals. The **preamplifiers**

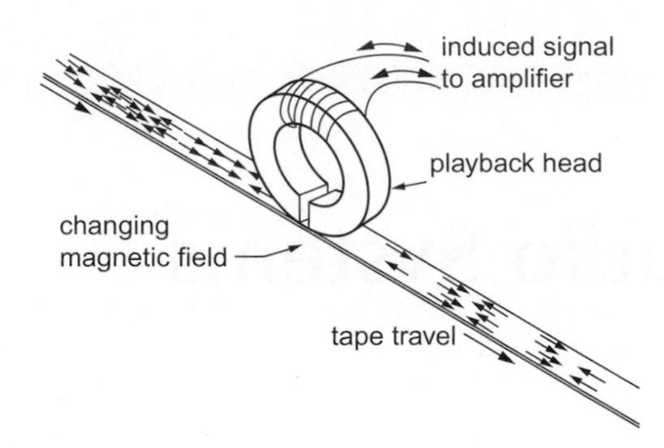

FIGURE 48.6

accept the signals from the input stage that is switched through to them. These are **voltage amplifiers**, as described in Topic 33. The amplitude of signals from the input stages is usually no more than a few milliamps. Typical supply voltages for amplifier circuits are 15 to 30 V, and amplification increases the signal amplitude in this range. Currents in these amplifiers are small, to avoid generating noise during the amplification of the low-voltage input. It is essential to avoid introducing noise at the early stage, when the signal is small, for the later stages will amplify the noise along with the signal. There may be several stages of voltage amplification.

The final stage of amplification is performed by **current amplifiers**. Current amplifiers often have no voltage gain, but the current gain may be 100 or more. The combined effect of amplifying voltage and then current is a high amplification of power. ($P = IV$, p. 32). The loudness of the sound produced is proportional to the power.

OUTPUT

The output from the power amplifiers goes to two sets of speakers. A single speaker for each channel is all that is essential, but quality systems have twin arrays of speakers.

Typically, an array consists of three speakers mounted in one enclosure. Between them they cover the range of frequencies perceptible to the human ear. This is the range from 30 Hz to 20 kHz. Below 30 Hz the sensation is one of vibration rather than sound. Above 20 kHz the ear hears nothing.

The smallest speaker in a typical array is a tweeter, for sounds in the highest frequency range, 2 kHz to 20 kHz. The mid-range speaker has a greater diameter and covers the range 50 Hz to 5 kHz. The bass speaker, or woofer, deals with the low frequency range 30 Hz to 800 Hz. There may also be a fourth speaker, a sub-woofer, that handles 20 Hz to 200 Hz.

The signal from the power output amplifier is fed to these speakers through a cross-over network. This feeds each speaker with signals in its frequency range.

OTHER CONNECTIONS

A home audio system is most often used for listening to sound from one of the four input units shown in the diagram on p. 195. However, other connections may sometimes be used. For example, the signal from the radio tuner may be sent direct to the tape deck. This allows a radio programme to be recorded on tape to be listened to later.

Other inputs devices may also be attached to the system. For example, an MP3 player can be used as a source of recorded music. The term MP3 refers to a technique for compressing digital music files so that they become small enough to be stored in a reasonably small memory chip. MP3 files may be downloaded from sites on the World Wide Web.

The development of cheap 'flash' memory chips has made MP3 players affordable. This has made MP3 and more recent systems very popular.

Digital versatile discs (DVDs) are another possible audio source, although they are more frequently used for recording films with multiple sound tracks. They may include special multimedia features, such as subtitles in several languages. DVDs can be used for audio recordings of high quality. They are very similar to compact discs but have smaller dimples, which are more densely packed, and so store a much larger amount of data (up to 17 Gb compared with 700 Mb on a compact disc).

MUSICAL SOUNDS

When a musical instrument emits a note, the note is not just a pure sine wave. It is a mixture of sine waves, all sounding at the same time, but not all equally loud. Consider a violin or guitar string fixed at both ends, then plucked. The string vibrates as in the top drawing below. The frequency of its vibration depends on the tension in the string and on its mass per metre.

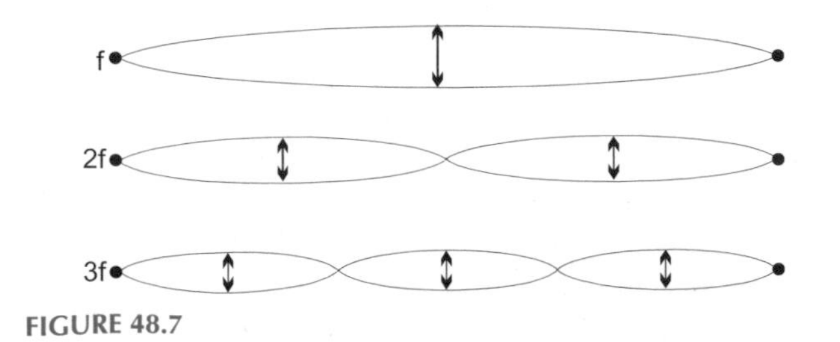

FIGURE 48.7

The frequency when it is vibrating as in the top drawing is f. It is also vibrating as in the middle and lower drawings, but not as strongly. The frequencies of these two vibrations are $2f$ and $3f$ respectively. It may also be vibrating in four or more sections, giving frequencies of 4f, 5f, and so on.

Frequency f is called the **fundamental** frequency. The higher frequencies are called the **overtones**, or **harmonics**.

If we use a microphone to pick up the sound made by the vibrating string and display it on an oscilloscope, we see the waveform labelled '$f + 2f + 3f$ ' in the figure below.

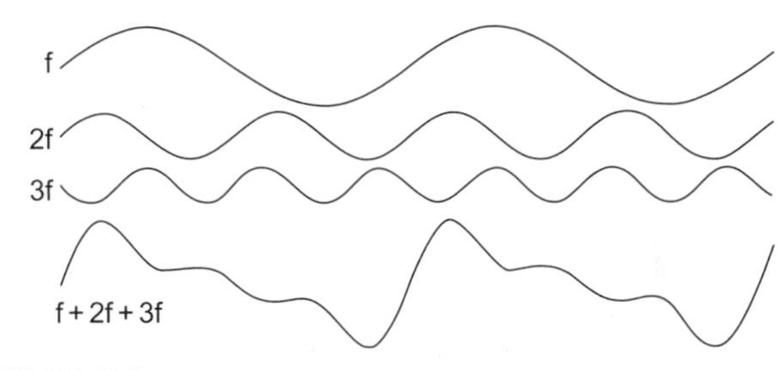

FIGURE 48.8

This is the sum of the fundamental note of the string plus the first two harmonics, drawn separately. The fundamental and harmonics are all sine waves of different amplitudes. But the frequencies of the harmonics are multiples of the fundamental.

If we alter the length of the string to get another note, we get a different frequency and a different but related set of harmonics. In all cases, the combination of fundamental plus harmonics has the characteristic sound of a violin or guitar.

Similarly, the air in a flute vibrates in several different ways, producing a fundamental and a different set of harmonics. Their relative amplitudes may differ and some may be missing. They sound like a flute when added together. Differences between the sets of harmonics are not the only features that make a flute sound different fron a guitar but they are an important difference.

The **bandwidth** of an audio system has a big influence on the quality of the sound it produces. The amplifier that gave the graph in Fig. 37.2 has a bandwidth from 800 Hz to 5 MHz. This might be suitable as a radio-frequency amplifier, but it is not suitable for audio frequencies. It does not amplify the lower frequencies. The graph shows that the signal is much reduced in amplitude between 30 Hz and 800 Hz. The bass response of the system would be very faint indeed. Ideally, the bandwidth of an audio system should extend from 30 Hz (or a little lower) up to 20 kHz.

Within the bandwidth, the amplifier should amplify all signals equally. In other words, the top of the curve should be flat, as in Fig. 37.2. Then the system will reproduce the original sound with the loudness of all its harmonics in the correct proportions. Even though a musical instrument does not play fundamental notes as high as 10 kHz, some of its harmonics may be over 10 kHz. If these are not reproduced at their correct relative volumes, the 'mix' of harmonics in the sound is wrong, and the quality of reproduction is poor.

TONE CONTROL

Ideally, the output from the speakers is an exact replica of the original sound. The amplifier and all other stages in the system must have a bandwidth extending over the complete audio range. Unfortunately, there may be stages in the system in which the bandwidth is too narrow, or the frequency response is not flat-topped. For instance, the recording microphone may emphasise some frequencies and reduce others. The amplifier, the speakers and their enclosures, and even the furnishings of the room exert their effects on the final sound as heard by the listener.

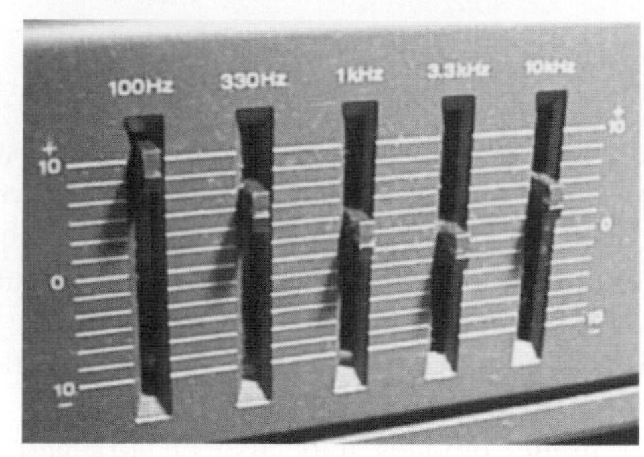

FIGURE 48.9

To compensate for these effects, most audio systems have **tone control** circuits (above). These consist of capacitors, resistors and op amps. They allow bands of frequencies to be boosted or cut. The frequency response can thus be adjusted to give high-fidelity sound reproduction.

Radio Transmission

If we apply the output from an oscillator to a pair of metal rods, electrons rush to and fro along the rods. Their rapid motion generates an **electromagnetic field**. Electromagnetic waves spread outward from the rods, like ripples on a pond, only in three dimensions. If the frequency of the oscillator is between 30 Hz and 30 GHz, the waves are **radio waves**.

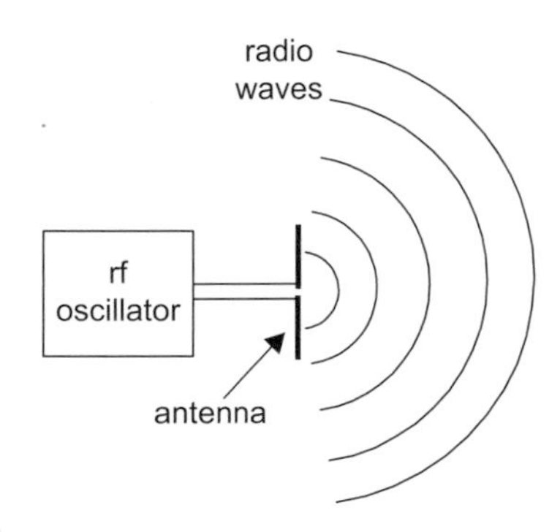

FIGURE 49.1

The paragraph above describes a basic **radio transmitter**. In the case of the transmitter we have described, its **antenna** or **aerial** is the pair of rods (or **dipole**), but it may be a piece of wire several metres long, suspended from masts or a similar structure. It may have a reflector of metal rods or wire mesh to concentrate the radio waves into a beam.

Radio waves travel from the transmitting antenna. They may travel for millions of miles through Space, as when we communicate with a Space probe approaching the planet Pluto. However, not all transmissions travel so far. Depending on their frequency, radio waves may be reflected back toward the Earth's surface from layers high in the atmosphere. Others stay fairly close to the Earth's surface.

Focused beams of radio waves are usually directed at a receiver that is a few tens of kilometres away. The receiver is in line of sight of the transmitter. Radio waves that are reflected in the Earth's upper atmosphere may be detected at distances of many hundreds of kilometres from the transmitter.

INFORMATION CARRIER

If we receive radio waves from a transmitter that is steadily operating at, say, 10 MHz, this tells us only that the transmitter is switched on. To be useful, the radio waves must be made to *carry information*. We use the basic radio-frequency as a **carrier wave**.

One way to carry information is to switch the output stage of the transmitter on and off. The transmission is a series of short bursts or **pulses** of the carrier.

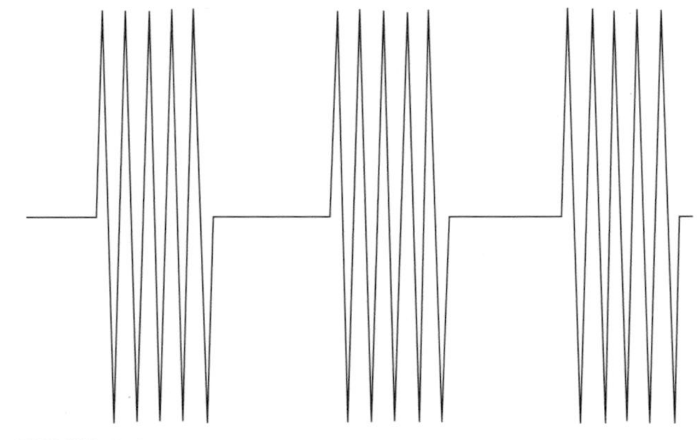

FIGURE 49.2

We say that the carrier is **modulated** into pulses. The pulses are all the same amplitude but the pulses and the gaps between them may vary in length. The pulses are a code, which represents the information that is being carried. This type of transmission is known as **pulse code modulation** (PCM). It is the most widely used technique for carrying digital information, and there is more about this in Topic 51.

When a PCM transmission is received, it is first **demodulated**. The radio-frequency waves are removed, leaving the pulses. These are sent to a decoder, to convert them to understandable data. As well as being widely used for data transmission, PCM is used in remote control systems, such as car immobilisers and domestic security systems.

AMPLITUDE MODULATION

AM is used for transmission of analogue signals. Before transmission, the carrier wave is passed through a modulator. This modulates the *amplitude* of the analogue signal on to the carrier wave.

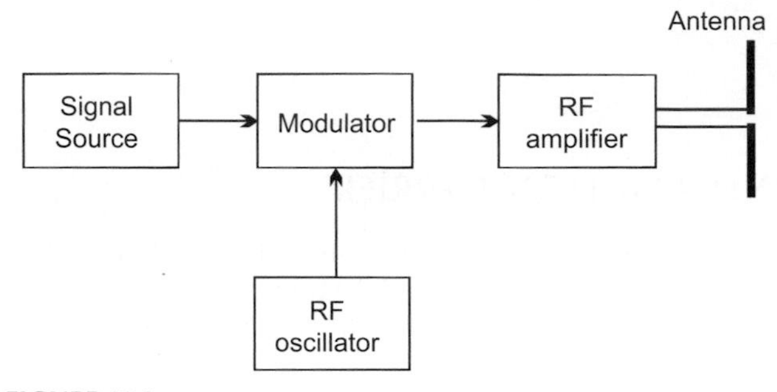

FIGURE 49.3

The result is shown below:

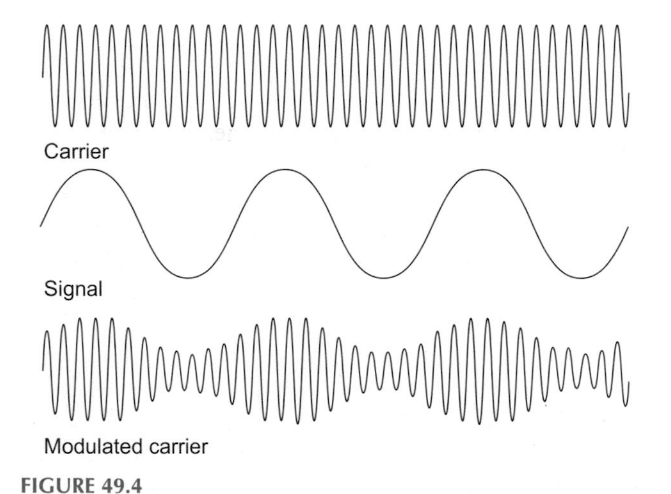

FIGURE 49.4

The modulated carrier is transmitted. It is demodulated at the receiver.

FREQUENCY MODULATION

FM is an alternative technique for placing an analogue signal on a carrier wave. We modulate the *frequency* of the carrier according to the analogue signal. The amplitude of the carrier is constant. The effect of FM on the carrier is shown below:

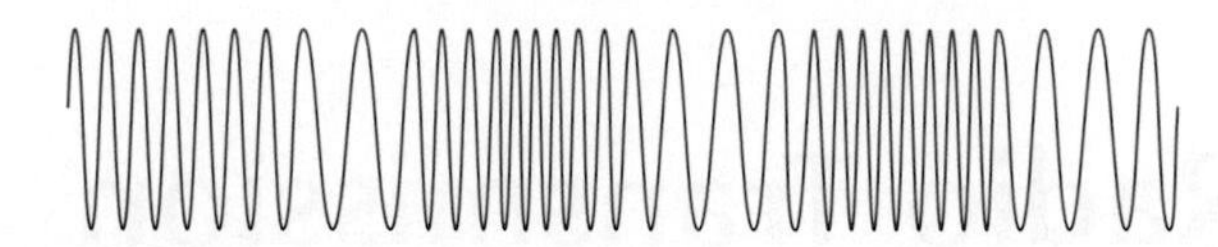

FIGURE 49.5

The frequency increases as the signal voltage goes more positive. It decreases as it goes more negative. To make the effect clear, the diagram shows a large increase and decrease in frequency. In practice, the variation in frequency is much less than this.

FM is not affected by changes in the amplitude of the modulated signal. This means that if reception is poor and the amplitude varies, the strength of the demodulated signal remains unaltered. Another related advantage is that sudden spikes in the amplitude of the signal, caused perhaps by lightning discharges or local electrical disturbances, have no effect on the frequency. There are no crackling sounds of interference such as are common with AM transmission.

QUESTIONS ON AUDIO SYSTEMS AND RADIO TRANSMISSION

1 Draw a system diagram of a domestic audio stereo system. Explain the function of three of its parts.
2 What is the principle of a CD player?
3 Outline the way in which a tape recorder and player work.
4 Why does the pre-amplifier amplify voltage and not current?
5 What are the lowest and highest audio frequencies?
6 Explain the meanings of the terms 'fundamental' and 'harmonic'.
7 What is meant by the term 'bandwidth'? With the help of a diagram describe the frequency response of an ideal audio amplifier.
8 Why is tone control needed in an audio system?
9 Explain how an oscillator is made to produce radio waves.
10 What is meant by the terms 'carrier wave' and 'PCM'.
11 Describe one way of transmitting analogue information by radio.

Radio Reception

SENSITIVITY AND SELECTIVITY

These are two important features of a radio receiver.

A *sensitive* radio receiver is able to pick up signals from weak or distant transmitters.

A *selective* radio receiver is able to pick out the signal of a particular transmitter from the signals on close-by frequencies that are arriving from other transmitters.

RADIO RECEIVER

The diagram shows the main stages in a tuned radio frequency receiver.

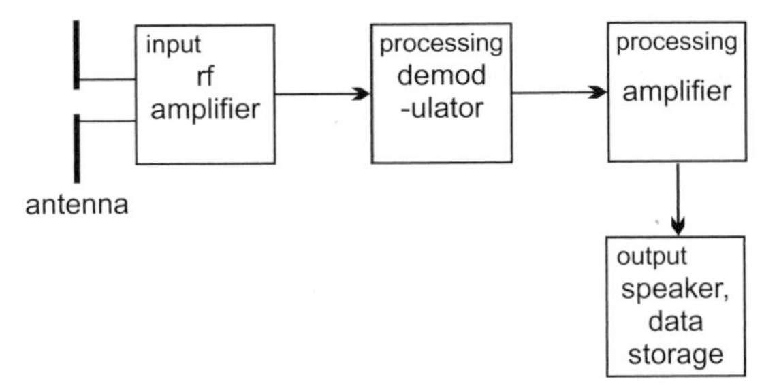

FIGURE 50.1

Antenna: This may be a **dipole**, similar to the antenna of the transmitter. Other types of antenna may be used, for example, a **long wire** suspended well above the ground on two or more masts. When radio waves strike the antenna, their electromagnetic field makes the electrons in the antenna oscillate.

The electrons in a dipole antenna oscillate most strongly when receiving waves of a particular frequency. This makes the antenna more sensitive to transmissions of a given frequency. It is more *selective*. It helps to tune the system to receive a particular station.

Dipoles often have extra elements to make an array that is more *sensitive*. They also make it more directional, which is another way of making it more *selective* (Figure 50.2).

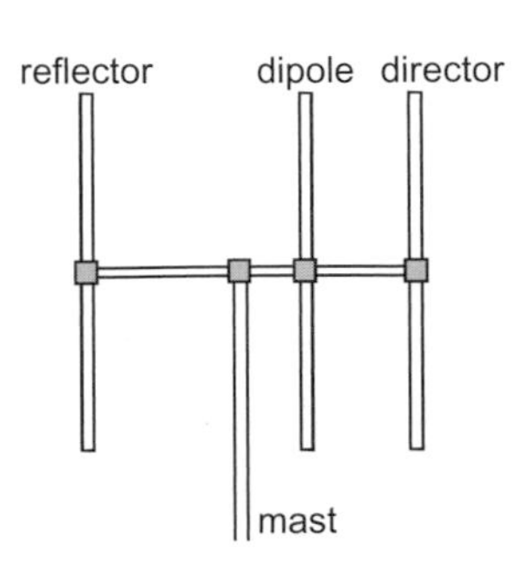

FIGURE 50.2

A **ferrite rod** is a compact type of antenna, which is often used in portable radio receivers. Ferrite is a magnetic material. It makes the lines of force of the received electromagnetic field tend to bunch together through the rod. This increases the apparent strength of the signal and makes the receiver more sensitive.

One or more coils of fine wire are wound on the rod and the magnetic fields induce currents in these coils. The coils are connected to variable capacitors that, together with the coils, form tuned circuits. These tuned circuits make the receiver more selective.

RF amplifier: This includes a resonant circuit (often a combination of a capacitor and an inductor) that makes the receiver amplify one particular frequency very strongly. In other words, the tuned receiver has very *narrow* bandwidth in the radio-frequency band.

The tuned circuit is selective and does not amplify the signals from transmitters with slightly lower or higher frequencies. Usually there are two or three stages of tuned amplification, so increasing selectivity and, at the same time, increasing sensitivity.

Demodulator: In an AM system, there are two stages in demodulation. The first stage is to rectify (p. 69) the output from the rf amplifier. The way this is done is shown in Figure 50.3. The effect of demodulating an audio transmission is shown in Figure 50.4.

The second stage is to 'average out' or smooth the amplitude of the rectified signal. The capacitor does

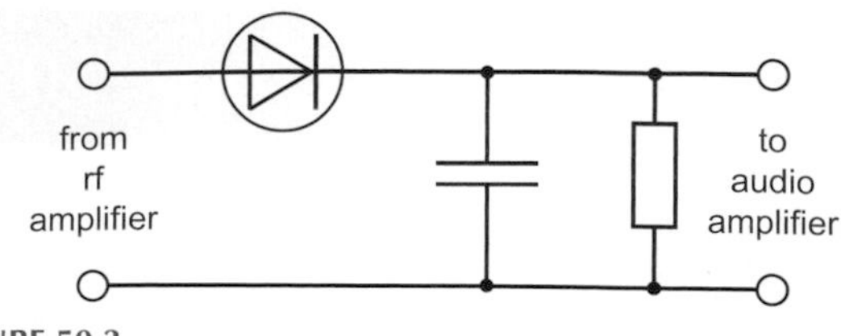

FIGURE 50.3

this. The result is an audio signal or data signal across the resistor. This is then sent to an amplifier.

Another way of thinking of this is to say that the capacitor acts as a low-pass filter, passing the low-frequency audio or data signal, but conducting the high-frequency carrier signal to the 0 V line.

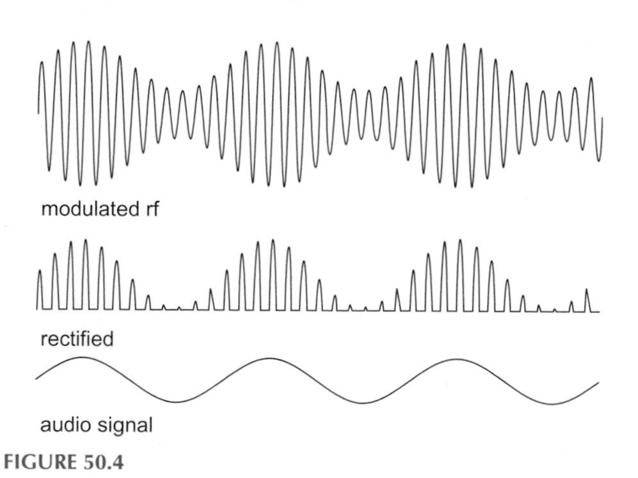

FIGURE 50.4

Demodulation is different in an FM receiver. The receiver includes an rf oscillator that is tuned to the same frequency as the transmitter. A comparator circuit compares the frequency of the built-in oscillator with the frequency of the received radio signal. It produces a voltage that is proportional to the difference between the frequencies. The voltage rises and falls as the frequency of the received signal increases or decreases relative to the constant frequency of the oscillator. This voltage signal is a replica of the original audio signal.

Audio amplifier: The output from the demodulator may need amplification. A wide band amplifier is needed for an audio signal.

Output: A speaker is required for speech and music. A data transmission signal may first go to a decoder. After that it may go to various kinds of control circuit, to a computer screen, or to a memory for storage.

RADIO ASTRONOMY

Events in distant galaxies generate electromagnetic waves at radio frequencies.

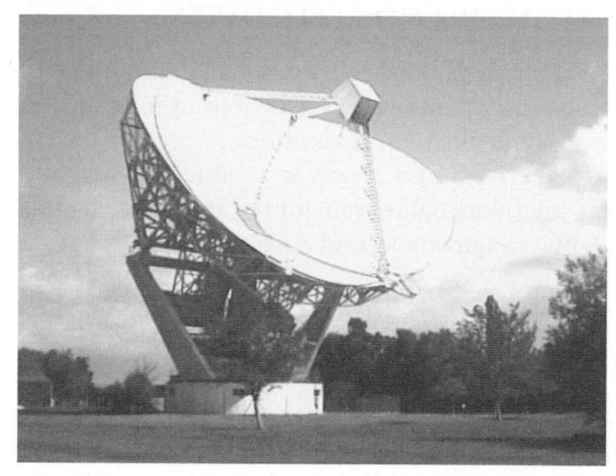
FIGURE 50.5

By receiving and analysing these signals, we can learn much about the Universe. Above is one of the radio telescopes at Jodrell Bank, Cheshire. The parabolic reflector concentrates the received radiation on to an antenna, which is mounted at its focal point. The rf amplifiers are also at this point. Signals from these are passed to computers for analysis.

QUESTIONS ON RADIO RECEPTION

1 Draw a system diagram of an FM radio receiver and explain what each stage does.
2 Explain how an AM signal is demodulated.
3 At what stages in a radio receiver is there (a) a wideband amplifier, and (b) a narrowband amplifier?
4 What features of a radio receiver make it more selective? Why is selectivity important?

Topic 51

Digital Communications

SERIAL TRANSMISSION

In computers and other data processing devices, data is usually handled in bytes. Eight parallel lines carry one bit each. However, setting up eight separate transmission channels between a transmitter and a distant receiver is not practicable.

Normally, data is transmitted one bit at a time, or *serially*. The diagram shows exchange of data between two computers. A special logic interface IC receives bytes of data from the computer.

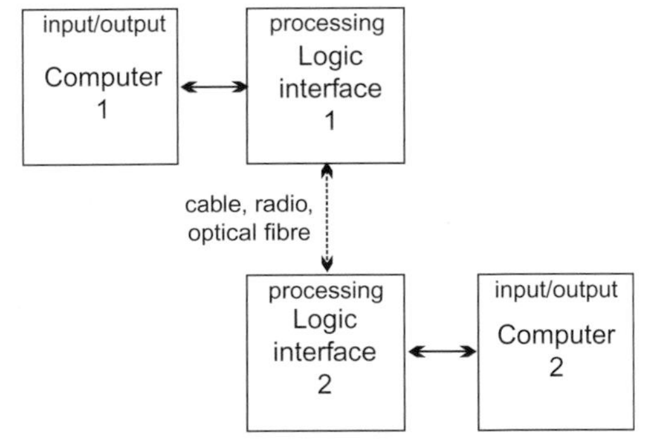

FIGURE 51.1

Usually the IC adds additional bits to this to prepare it for transmission. These include a start bit, a parity bit (see 'Parity', on next page) and a stop bit. The receiver is waiting (or 'marking') while the received voltage is low. The arrival of the start bit tells it that there are 9 more bits to follow, equal in length to the start bit. Reception begins when the start pulse arrives. From then on, the system clock in the receiver has to time only 9 bits.

Even if the receiver clock does not run at precisely the same speed as the transmitter clock, it is unlikely to get out of step during such a short transmission. Because the clocks do not need to be synchronised, this is known as an **asynchronous system**.

ANALOGUE DATA

Analogue data includes speech, music, and video. It may also include data from sensors, such as when a satellite sends temperature readings back to Earth.

Analogue data can be transmitted as an AM or FM signal. However, because of the advantages of digital communication it is better to convert it to digital form before transmission (using an ADC) and then convert it back again (Using a DAC) after reception.

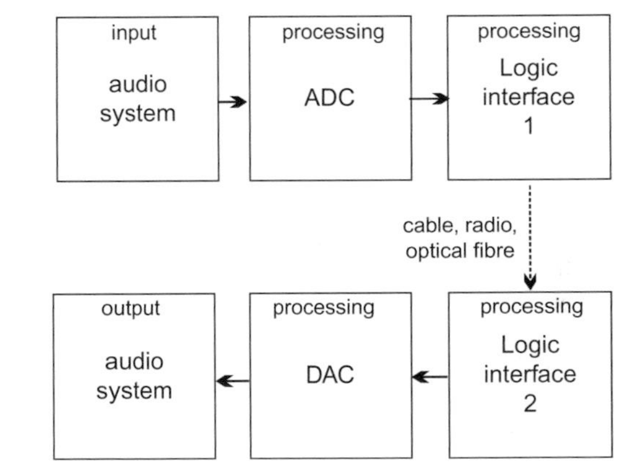

FIGURE 51.2

The first stage in analogue to digital conversion (ADC) is sampling. The signal is sampled at regular intervals and each sample is converted into its digital equivalent.

Figure 51.3 shows the signal being converted into 4-bit samples but this gives only 15 possible values. For greater precision, most ADC converter ICs produce at least a 12-bit sample, which gives 4096 possible values. If the sampling is to keep track of the varying signal voltage, it is essential to take samples at short intervals. It can be shown that the sampling frequency should be at least double the highest

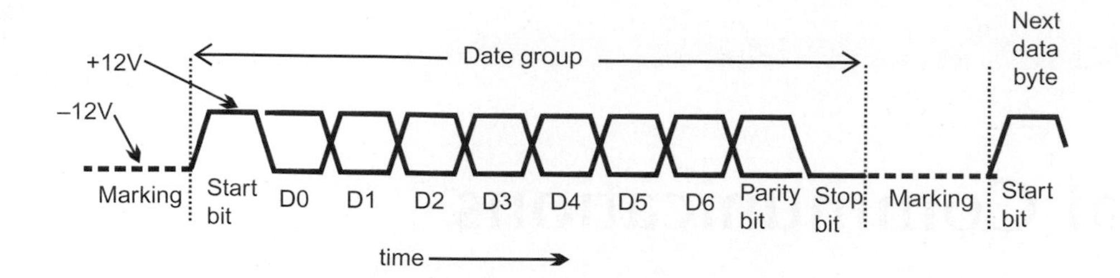

FIGURE 51.3

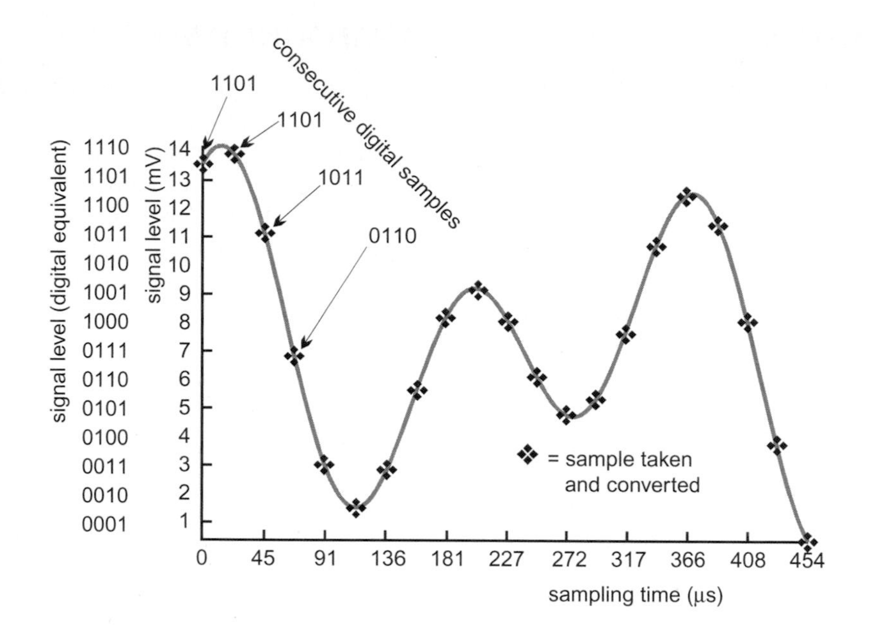

FIGURE 51.4

frequency present in the signal. In the example in the diagram, the signal is sampled at 44 kHz. This just covers the top end of audio frequencies, at 20 kHz.

The logic at the transmitter receives the output from the ADC. It then transmits it serially. At the other end of the channel, a similar IC in the receiver is able to collect the bits as they arrive and assemble them into bytes again.

It is important to check that there have been no errors in transmission. The only error that can occur is that a '0' may become a '1', or the other way about. The simplest way to check that this has not happened is to check the **parity** of the bytes as they arrive at the receiver.

PARITY

The data byte comprises seven data bits, plus one bit that is known as the **parity bit**. The parity bit is either '0' or '1', and is selected automatically so as to give an even number of '1's in the byte. For example, suppose the data bits are:

1101001

This has an even number of '1's so the parity bit is '0' and the complete byte is:

11010010

The byte is checked for parity by the logic interface at the receiver. If the byte arrives with an odd number of '1's, it shows that something is wrong.

As another example, take these 7 data bits:

0110010

They have an odd number of '1's, so the parity bit must be '1' to make the number even. The completed byte is:

01100101

The parity described above is called **even parity**. Some systems use **odd parity**, in which the number of '1's is made odd. There are ICs that automatically check parity and indicate when there is an error.

Parity checking is a simple operation but is itself subject to error. For example, if *two* bits are wrong, the errors may cancel out. Other similar but more complicated techniques are used for error checking when accuracy is essential.

Self Test

If parity is even, what parity bit is added to these seven bits:

0101100

PULSE CODE MODULATION

Pulses may be sent along a cable as a sequence of high and low voltage levels but, for transmission by radio, optical fibre and also by cable, they are more often modulated on to a carrier frequency. This is pulse code modulation, as described in Topic 49.

PCM is the most widely used technique for transmitting digital and (after conversion) analogue data. Bursts of the carrier frequency, representing '0's and '1's, are transmitted and then demodulated at the receiver.

The use of carriers means that many different messages may be sent along the same channel at the same time. This is a valuable feature. Communications links such as satellites and under-sea cables are expensive to build and maintain. The more signals that can be carried at any one time, the more economic the system.

The technique is to modulate the pulsed signal on to one of a number of carrier frequencies. The frequencies are spaced sufficiently widely apart that they do not interfere with each other. Yet they are close enough together to allow many carrier frequencies to be fitted into the bandwidth of the channel. This is known as **frequency division multiplexing**.

With FDM, the modulated carriers all pass through a single channel at the same time. The signals keep separate from each other because of their different carrier frequencies. Using only a single channel means that only one broadband amplifier is needed for transmitting and for receiving the multiplexed signals.

On reception, the individual tranmissions are separated out by using circuits tuned to each of the carrier frequencies. The separated signals are then routed to computers, and to the public telephone network.

Digital signals that were originally audio or analogue signals are restored to their original form by analogue to digital converters.

ADVANTAGES OF PCM

Pulse code modulation is the basis of today's 'information explosion'. Ever-increasing amounts of data are being transmitted across the world at ever increasing speeds. Below are some of the reasons for the success of PCM:

- Digital signals consist of '0's and '1's. There are no intermediate levels that may become difficult to recover after a signal has been distorted during transmission.
- Digital signals are highly immune from noise. The signal below, though noisy, is still recognisable as a pulsed signal.

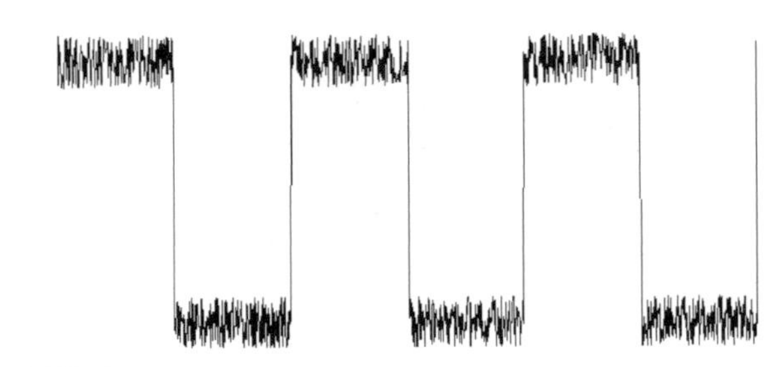

FIGURE 51.5

- Distorted or noisy signals can easily be 'cleaned up'. Passing the signal through a Schmitt trigger circuit (p. 130) restores it to its original pulsed form. In this way, noise is eliminated.
- Distortion of the shape of signals is unavoidable, but the effect is much less serious with digital signals than with analogue signals. The original pulses are easily restored.
- On a long transmission path (for example, England to Australia) the signal can be cleaned up, amplified and re-transmitted at a series of **regenerators** along the route. Digital signals are much easier to regenerate accurately than analogue signals. Fewer regenerators are required along the route.
- Digital signals are suited to processing by computer or digital logic circuits. This makes such procedures as automatic parity checking and multiplexing much easier. The channel can be completely under computer control.

TIME DIVISION MULTIPLEXING

This is another technique for transmitting several signals along the same channel. In TDM, all transmissions are on the *same* frequency. Each can use the full bandwidth of the channel.

In TDM, signals from several sources are allocated to short time slots. Usually a time slot is about 3.9 μs long. A channel is allocated a time slot every 125 μs. The signals from different sources are transmitted one after another in quick succession. Each signal is automatically placed in its allocated slot. At the receiving end the signals are sorted out automatically and routed to their correct receivers.

One of the complications of this system is that it has to be **synchronous**. The transmitter and receiver are synchronised so that time slots at the transmitting and receiving ends occur at exactly the right times. In contrast to this, frequency division multiplexing is **asynchronous**. The channels are independent and transmitters can send messages at any time.

TDM is used on the telephone network and in radio satellite communications. The equipment required for TDM is more complicated but the system is able to carry twice as much data as an equivalent FDM system.

RS-232 PORTS

The RS-232 standard defines a serial port for transmitting data over cables to a distance of up to 15 m. The system is often used for communications between computers. The standard specifies the type of connector as either a 9-pin or a 25-pin D-type connector. The logic levels are −12 V for digital '1' and +12 V for digital '0'. ICs are available for converting between TTL logic levels and RS-232 levels. The standard specifies that the connecting cable has nine lines. Three of these are concerned with sending the signal: SG, the signal ground line; TD, transmitted data; RD, received data. In this way, the system allows for two-way transmission.

Many simpler (not true RS-232) systems use just these three lines.

In RS-232, the other six lines carry **handshaking** signals. These are high or low logic levels by which the two stations coordinate their actions. For example, a high level on the RTS line is a 'request to send' telling the receiver that the transmitter is waiting to send data. If it is ready to receive data, the receiver puts a signal on the CTS line, indicating 'clear to send'. On receiving this signal, the transmitter begins sending data on the TD line.

ASCII CODE

This is a widely used code for transmitting alphabetic and numeric characters and most punctuation marks. The code consists of the 128 seven-bit binary numbers from 0000000 to 1111111. The seven bits are transmitted serially as bits D0 to D6 (see Fig. 51.3). The eighth bit of the byte is D7, which is a parity bit.

The code includes 'control characters', which do not appear in the printed text but are used to control the transfer of data between sender and receiver. For example, 0000100 indicates 'end of transmission', 0001101 is 'carriage return', and 0000111 is 'bell' which sounds a 'beep'. The names come from their use in early teletype machines.

CABLE TRANSMISSION

Telecommunications cables are of two main kinds:

- **Twisted pair:** Two insulated wires twisted around each other.
- **Coaxial cable:** This has a solid inner copper conductor, which carries the signal, and a braided copper wire screen or outer conductor.

The two types of cable are illustrated below:

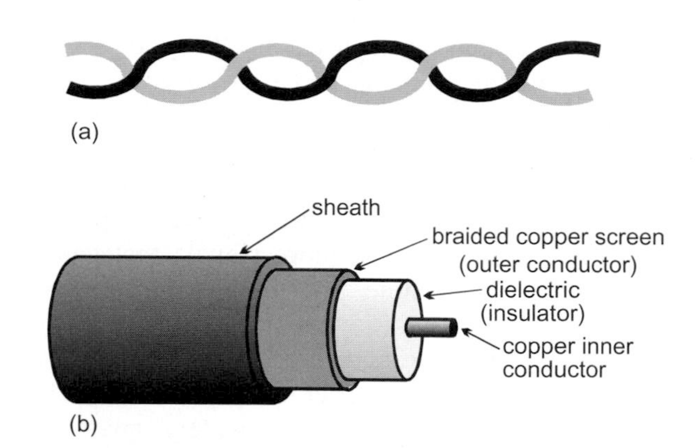

FIGURE 51.6

It is important that a cable should not be affected by external magnetic fields. These would induce currents in the wires and add noise to the signal.

The inner conductor of a coaxial cable is shielded by the outer conductor. Although the wires of a twisted pair are not shielded from external magnetic fields, the twisting of the wires reduces their effects. In Figure 51.7, the loops of the twisted pair are threaded by a magnetic field coming upward through the book. This induces a clockwise current in each loop. Because of the twisting these tend to cancel out between adjacent loops of the wires. Alternate loops are equally but oppositely affected and the magnetic field has little overall effect.

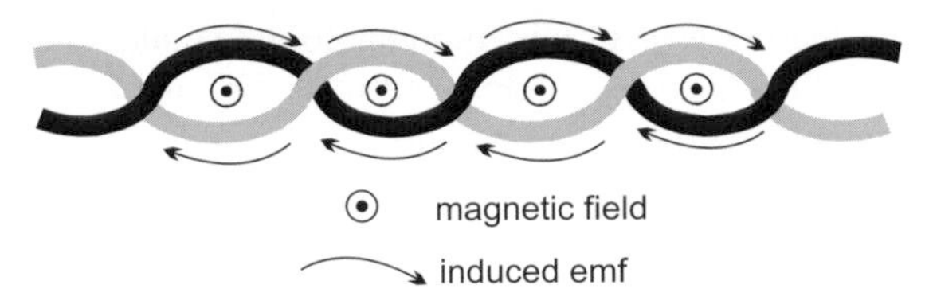

FIGURE 51.7

The diagram below shows how a signal is sent along a twisted pair, using a **line driver**.

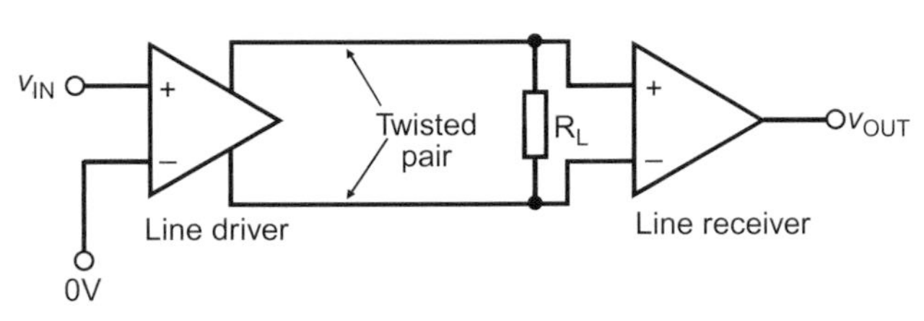

FIGURE 51.8

A line driver is an IC similar to an op amp but with two output terminals. When one output goes high the other one goes low. The driver has high bandwidth.

There is a load resistor at the far end of the line and a voltage develops across this when one line goes high and the other low. The **line receiver** is also similar to an op amp. Its output depends on the voltage across the load resistor.

Logic gates such as NOT and NAND can be used as drivers and receivers for digital signals, but special ICs with higher bandwidths are preferred.

When using logic line drivers and receivers with coaxial cable, the outer conductor (shield) is usually connected to the 0 V supply terminals of the ICs, as shown in this diagram.

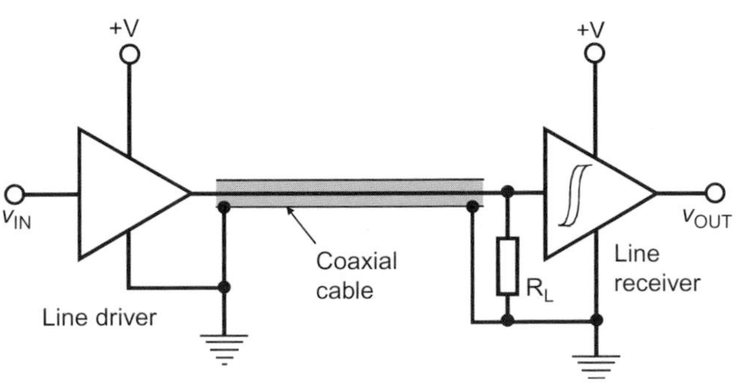

FIGURE 51.9

The receiver has a Schmitt trigger input to 'square up' the distorted incoming signal and to reduce the effects of noise on the line.

OPTICAL TRANSMISSION

As a channel of telecommunications, optical fibre has several advantages. It relies on the fact that a ray of light travelling in a glass fibre is totally reflected when it strikes the surface of the glass.

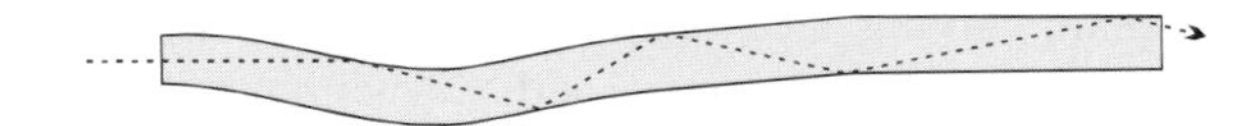

Even if the fibre is curved, the light travels from one end to the other without emerging. For best results, with cables many kilometres long, the glass is of the highest purity to minimise loss of light. Light is lost where the surface is scratched. To prevent this the fibre consists of a core of glass of standard refractive index surrounded by a cladding (surface layer) of glass of lower refractive index. Reflection takes place at the boundary between the core and the cladding.

In graded index fibre the refractive index decreases gradually from the centre of the core to its boundary with the cladding. Instead of being sharply reflected the path of the light is more smoothly curved. This reduces the distortion of high-frequency pulses.

At the transmitter, the source of light is an LED or a solid-state infra-red laser. These are mounted in a holder that clamps the source against the highly polished end of the fibre.

The light from the LED or laser is amplitude modulated with an effect similar to that of modulated radio waves.

At the receiver the light is detected by an avalanche photodiode, a PIN photodiode or a phototransistor. For the highest frequencies we use the APD, which has an extremely short response time of only 200 ps.

The advantages of optical fibre are:

- **Bandwidth:** Because of the high frequency of light, optical fibre has a greater bandwidth than any other system of telecommunication. This allows many more channels to be frequency multiplexed on a single channel.
- **Cable cost:** Dearer than cable, but the cost is falling. However, repeater stations can be further apart.
- **Electromagnetic interference:** Optical fibre is not subject to this.
- **Safety:** Faults in copper cable may generate heat and lead to fire. This is unlikely to occur with optical fibre.
- **Security:** Signals on copper cables and radio can be read by surveillance equipment. This can not be done with optical fibre.
- **Corrosion:** Unlike copper cable, optical fibre does not corrode.

QUESTIONS ON DIGITAL COMMUNICATIONS

1. What is PCM? What advantages does it have over analogue transmission?
2. What is frequency division multiplexing? What are its advantages?
3. Explain, giving an example of each, the difference between asynchronous and synchronous data transmission.
4. Outline the way in which an analogue signal, such as a musical signal, is converted to a stream of bits for transmitting by cable or radio.
5. If seven data bits are 0100100, and the parity is even, what will the eighth bit be?
6. Describe how a byte of data is transmitted asynchonously. What bits are added to the 8 data bits and what is their function?
7. What is time division multiplexing?

Computers

Topics 42 to 44 describe the operation of microcontroller systems. Computers have much in common with microcontroller systems, but are designed for a different purpose. Microcontroller systems are relatively simple, and centre mainly on a single IC, the microcontroller. They are intended for controlling equipment and appliances such as dishwashers and mobile phones. Computers too are used in control systems, but mostly in highly complex systems. For example, computers are used to control power stations, jumbo jets, and radio telescopes. On the right is the main computer console that controls the electricity generators at Ironbridge Power Station, Shropshire.

For most people, a computer is the familiar personal computer or its more portable equivalent, the laptop computer. These are intended for use in offices, laboratories and the home. They are used for handling data, for calculating, for communicating on the Internet and for playing games, among many other applications. The diagram below shows the main parts of a PC system:

FIGURE 52.2

The microprocessor is the heart of the system. It corresponds to the main processing units of a microcontroller, including the arithmetic logic unit. A microprocessor has to communicate with all other parts of the system. It needs many address lines, data

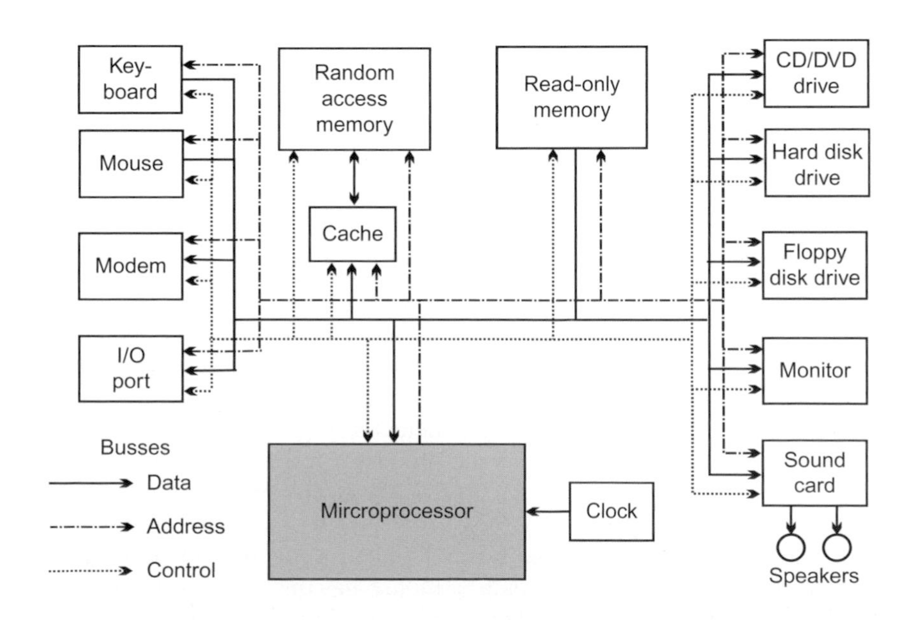

FIGURE 52.1

FIGURE 52.3

lines and control lines. The newer microprocessors, such as the Intel Pentium 4, can address several gigabytes of memory and can handle 32-bit data. Some versions can handle 64-bit data. There are also many control lines. This means that the microprocessors need several hundred pins.

The action of the microprocessor is timed by the **system clock**. This is basically an astable, but one that runs at high speed. To process such large amounts of data at high speed, the clock in a typical PC runs at several hundred megahertz.

MEMORY

There are three types of memory:

RAM: Used for temporary storage of data and programs. Typically, there is 128 Mb or 192 Mb. Memory is often expandable up to 2 Gb by plugging in extra ICs. This might be necessary if you are working with extra-large files, such as animated 3D graphics.

ROM: For storing routines used at start-up.

> **Memo**
>
> A **file** is a block of data or a program stored or transmitted in binary code. Files vary in length from one or two kilobytes to many megabytes.

Cache memory: This is a special type of RAM, which can be written to and read from very quickly. It is used for temporarily storing addresses and data that the processor might need at short notice. This helps to make processing faster. The cache memory may be on a separate chip or on the microprocesssor chip. Typically, there is 276 Kb of cache memory, but some computers have more.

INPUT AND OUTPUT

The computer needs to communicate with the outside world. It has many devices attached to it for this purpose. Some are for input and some are for output. Some are for both. Some are built into the main case of the computer while some are separate and connected to the computer by cable.

The photo above shows a typical PC system in the home. The main circuits and some of the input and output devices are housed in the tower on the left. Other input and output devices are seen in the photo as separate units.

INPUT DEVICES

A typical PC system has a number of more-or-less standard input devices:

Keyboard: This is usually the main interface between the computer operator and the system. The microprocessor scans the keyboard frequently to detect key-presses. When a key is pressed, the microprocessor responds accordingly.

Mouse: A mouse (Figure 52.4) is the most convenient way of moving the image of a small arrow (known as the **mouse pointer**) over the screen. The mouse has one or more keys on it and usually a wheel for scrolling the display. The operator uses the mouse

FIGURE 52.4

to position the pointer over images of 'buttons' (below). Clicking one of the keys on a 'button' results in an appropriate response from the computer.

FIGURE 52.5

In Figure 52.2 on p. 215, the operator is using a mouse (not a keyboard) to click on 'buttons' and 'sliders' to control the power station.

OUTPUT DEVICES

These are the means by which the computer delivers the results of its activities. They include:

Monitor: Also known as a visual display unit (VDU). This is usually a colour LCD display panel operating on digital signals from the computer. Older computers have a monitor based on a colour cathode ray tube.

Sound card: Most people use this only as an output device. It can be used for listening to CDs played on the CD-ROM drive, for playing sound files downloaded from the Web or for listening to sounds that are part of a multimedia presentation. With suitable software, the sound card can be used to generate musical and other effects. The output goes to a pair of speakers. It can also be used for recording sound signals from a microphone, a compact disc or an MP3 player.

INPUT/OUTPUT DEVICES

The remaining devices allow two-way communication.

Modem: The name is short for **mo**dulator/**dem**odulator. It is a two-way device that connects the computer to the public telephone system. It is used for accessing the Internet, to view websites and to exchange e-mail. It is also used to send or receive files from other computers.

CD/DVD drive: This drive accepts compact discs and DVDs. Provided that the computer has the necessary software, the drive can be used for a number of purposes, including:

- Playing audio CDs.
- Showing movie DVDs on the monitor.
- Downloading and running commercial software. Discs of this kind with ready-stored programs and data are often called CD-ROMs.
- Recordable CDs, or CD-Rs. These can be written to only once but may be read as often as you want. They are ideal for storing up to 700 Mb of data very quickly. They are so cheap that there is no need to fill the disc.
- Re-writeable CDs, or CD-RWs. These can be written to and read from as often as you like. So they are handy for backing up files at the end of every work session. This book was written on a CD-RW. Their capacity is 700 Mb.

Floppy disk drive: A floppy disk is a thin circular sheet of magnetic material in a plastic cover. Data is written or read using a magnetic head and circuits similar to those used in a tape recorder/player. The floppy disk can be removed from the drive, so it is useful for backing up data, and for transferring data or programs to other computers. However, the rate of transfer of data is relatively slow. A floppy disk holds up to 1.44 Mb. This is not large enough for storing the larger computer files of present-day computers, so floppy disks are less useful than they once were. CDs have mostly taken over from floppy disks, another reason being that CDs are much less destructible. In particular, they are unaffected by magnetic fields that so easily corrupt the data stored on a floppy disk.

Ports: These are ICs that provide a means of connecting the computer to a range of external devices. They may accept input or provide output or perhaps both at the same time.

Usually, a computer has at least one **parallel port**. The cable from a printer is often connected here. Typically, this has a 36-pin connector and carries 8 bits at a time along 8 data lines. Parallel data transfer is fast enough to keep the printer fully occupied. Other lines are used for handshaking p. 212 between the computer and printer.

There may also be one or two serial ports, which are often RS-232 ports p. 212. Being serial ports they transmit only 1 bit at a time and are slower than the parallel ports.

Many add-on devices make use of serial ports. Examples include joysticks for playing games, micro-controller programmers, and digital cameras.

Newer computers have sockets for the Universal Serial Bus, or USB. This is a standard bus with standard USB sockets used for connecting add-on devices to the computer. As well as the devices listed above, the USB is used for connecting flashcards, flash memory readers, printers, and modems.

HARD DISK DRIVE

This is used for fast storage of programs and data. The capacity of the drive is high, rated in gigabytes. It contains a stack of several disks turning on the same spindle. The disks are coated with a magnetic material. There is a pair of magnetic heads to read or write on the upper and lower sides of each disk. When the disks are spinning, the heads 'float' with a thin film of air between the disks and the heads. Being so close to the disk, the head is able to record a large amount of data in a small area. The disk is not removable from the drive, and is sealed in to exclude dust. This is essential, as the heads are so close to the disk that even a fine smoke particle between the head and the disk would cause the disk to crash.

The hard drive is the main storage unit of the computer. All programs currently in use are stored there, together with the files that they use or produce. When the computer is running, programs and blocks of data are continually being transferred between the hard disk and RAM.

INTERNAL BUSSES

The parts of the computer are connected together by sets of parallel lines known as busses. The **data bus** has at least 8 lines so that data can be transferred a byte at a time. The **address bus** often has 24 or more lines to address all the computer's RAM. The control bus has at least 6 lines.

Control Systems

Electronic control systems are found almost everywhere, from controlling the ignition system of a car to the control of rush-hour traffic on the motorways. Electronics controls everything from the drilling machine on the factory floor to the coffee-making machine in the boardroom. Electronics controls everything from the fan-heater in the living-room to the power station that generates the electricity it uses.

There are so many electronic control systems that we can take only a few examples to illustrate the main types of control system that have been invented.

MICROWAVE OVEN

This familiar domestic appliance has a fairly complicated system that makes it much easier to use than the ordinary gas or thermal electric oven.

FIGURE 53.1

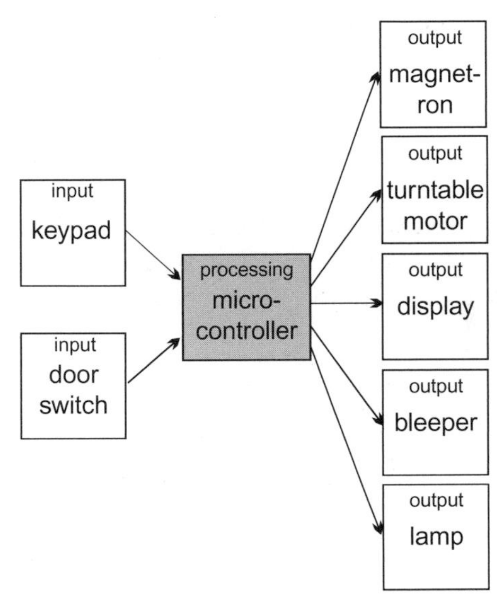

FIGURE 53.2

The centre of the system is a microcontroller or possibly a microprocessor. This is responsible for running the system. The program for this is permanently stored in its ROM. In addition, it needs a small amount of RAM and possibly three clocks. One is the system clock, another is an elapsed time clock, which provides the cooking times, and the third gives the time of day. These clocks may be on separate ICs, though they might also be on the microcontroller chip.

The system has two inputs. The keypad is for entering cooking power and times and also has keys for starting, resetting and other functions. The door switch senses whether or not the door is closed.

There are five outputs. The magnetron generates the microwaves. This is switched fully on when cooking at high power, but is switched on for intermittent periods when cooking at reduced power or when defrosting. It can not be switched on when the door is open. The turntable motor runs for the whole cooking time. The display normally shows the time of day, but displays time settings as they are entered at the keypad, and the time remaining during cooking. It may also display power levels and messages. The lamp comes on when the door is open or when cooking is in progress. The bleeper sounds when keypad entries are made and at the completion of cooking.

The microcontroller unit accepts the inputs from the keyboard and door switch. It then goes through a logically controlled series of steps to cook the food as instructed. Its clocks help it to time the processing according to the instructions that have been keyed in.

A flowchart of a possible program for the microwave oven is given in Figure 53.3. Most of the steps

Electronics: A First Course.

are self-explanatory. Some of the steps omit details. For example, the 'Keypad entry routine' would be a long one, to allow for different types of entry such as cooking instructions, setting the clock, or using the clock as a kitchen timer. In the early stages of development it is best to keep the flowchart relatively simple so that its overall action can be more easily understood. Later, the programmer can come back and fill in the details. A routine for keypad entry could be used at several places in the flowchart. We call this a **subroutine**, a program within a program.

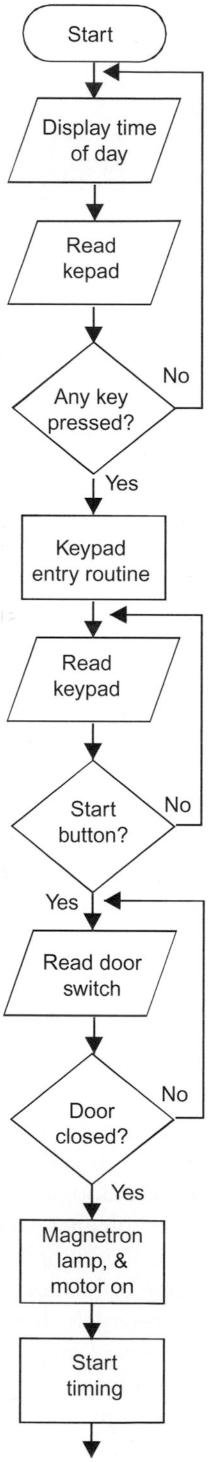

Start
Display time of day
Read kepad
Any key pressed?
No
Yes
Keypad entry routine
Read keypad
Start button?
No
Yes
Read door switch
Door closed?
No
Yes
Magnetron lamp, & motor on
Start timing

FIGURE 53.3

Note that this is a *digital* program. The door is open or closed, the lamp is on or off, the magnetron is on or off. Also there is plenty of scope for logic in the programming. A typical logical statement might be: 'IF the start button is pressed AND the door is shut, THEN start cooking now'. Microcontrollers are ideal for logical programs of this kind.

Things to do

The flowchart on the left goes only as far as the beginning of cooking. Carry it on to when the bleeper sounds at the end.

FAN HEATER

Although the *hardware* of the control system of the fan heater in the photo is mainly digital, the *action* involves an analogue quantity — the temperature of the room. A thermistor or equivalent device senses the air temperature and an analogue voltage is produced. This is turned into a digital quantity by an analogue-to-digital converter. The ADC is usually located on the chip of the microcontroller that controls the system. The digital temperature is stored in a register ready for processing.

FIGURE 53.4

The system diagram in Figure 53.5 (Note: not a flowchart) and the following description show how the system works.

There are two inputs. One is the input from the thermistor that we have already mentioned. The other is from the keypad that can be seen on the top of the heater.

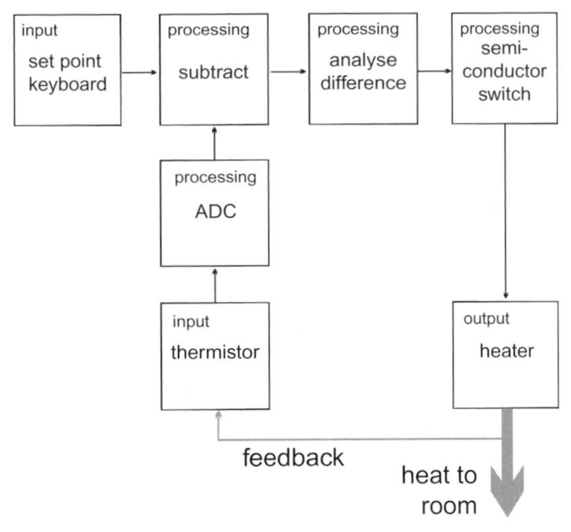

FIGURE 53.5

The keyboard is used by the operator to set the air temperature required. This is called the **set point**. The keys are frequently scanned by the microcontroller. When a new temperature is entered, the set point register in the microcontroller is updated to the new value. The value from the ADC is also updated frequently.

The processing is mainly mathematical. First, the temperature is subtracted from the set point value. If the difference is less than a given amount, no action is taken. If the difference is more than that amount a signal is sent to the semiconductor switch to increase or decrease the average current flowing to the heater element.

The microcontroller is also programmed for other more complex actions — including control of the speed of the fan — that would not be possible with an analogue circuit.

FEEDBACK

Most of the heat is carried away to warm the air in the room. A small amount of heat finds its way to the thermistor. It is *fed back* to an earlier stage in the system. We call this action **feedback**.

In this system, the action of the feedback is *negative*. If the air is too cool, it turns the heater on. If the air is too warm, it turns the heater off. This is called **negative feedback**.

Negative feedback is often used in control systems. It acts to oppose change. It gives stability. It is often used in **regulator** systems like this fan heater.

Many, but not all, control systems have sensors to provide them with feedback. Earlier we showed how the automatic door system uses sensors to confirm that the doors have actually opened or closed. This is *not* a regulator system. The sensors are simply feeding back information about the state of the doors.

When someone is flying a radio-controlled model plane, they watch it as they manipulate the controls. In this case, feedback is visual and no electronic sensors are involved.

Many examples of feedback are found in robotic systems, as the next section illustrates.

ROBOTICS

The hobby robot in Fig. 53.7 is controlled by a system based on a PIC microcontroller. We will look at a few of the things it can do.

This robot has three wheels. The two drive wheels with springy plastic tyres are each driven by an electric motor under the control of the PIC. The rear wheel is a small free-running castor, for balance.

There is no special steering mechanism. The robot turns or spins left or right by running one wheel forward and at the same time either stopping or reversing the other wheel.

The robot has several sensors, including a pair of bumpers (see system diagram below). They are each suspended on a loose hinge. When the robot runs into a solid object, such as a book or a wall, one or both of the bumpers is pushed backward. The bumper presses against a microswitch, which closes. The bumpers are interfaced to the PIC. The PIC receives a logic high input when the bumper comes into contact with an object.

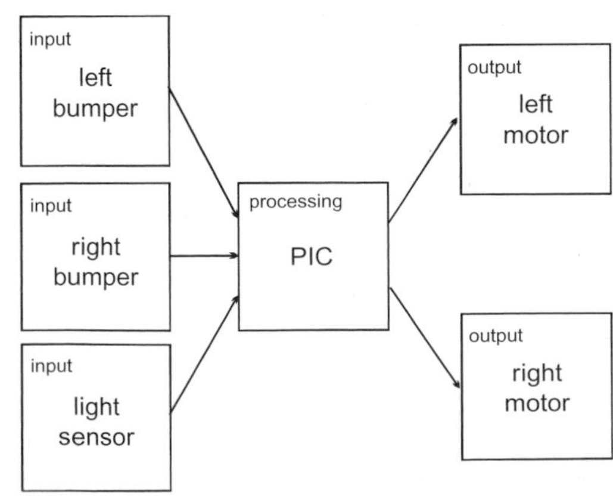

FIGURE 53.6

The third sensor is a photodiode connected in a voltage divider as shown there. The output of the divider goes to an input pin of the PIC.

The photodiode is mounted on the front of the robot and shielded by a forwardly directed open tube. When there is a bright light ahead, the voltage is high enough to count as logic high. Otherwise, the voltage counts as logic low.

Now to tell the robot how to behave! Its main task is to seek and approach a bright lamp placed on the opposite side of the room. To complicate matters, the floor is littered with books, food cans, and similar massive objects. When the robot encounters any of these, it is to forget the lamp for the present and concentrate on finding its way around the object.

FIGURE 53.7

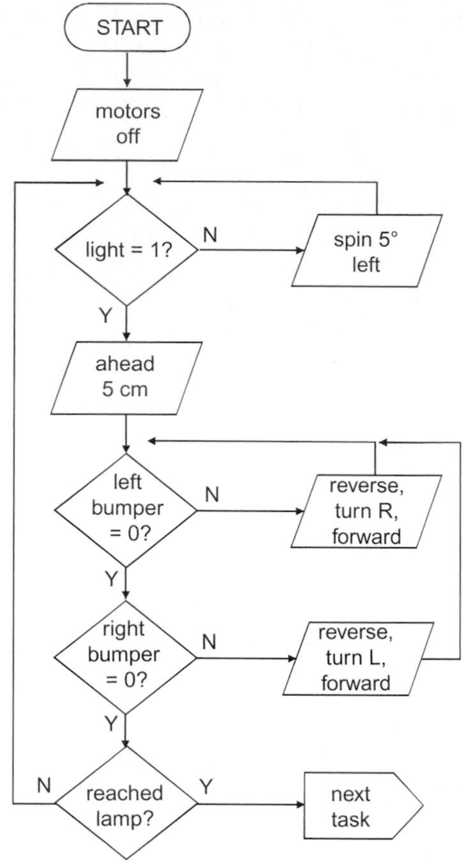

FIGURE 53.8

The flowchart on the upper right presents a simple, though not perfect, solution. It starts by turning off both motors, which is a sensible precaution to prevent the robot from running off out of control when the power is switched on. Next, it reads the input from the light sensor. If it can 'see' the lamp (light = 1), it moves forward 5 cm. If it can not see the lamp, its spins 5° to the left and checks the sensor again. It continues spinning until it does see the lamp and then runs forward 5 cm in that direction.

If there are no objects in its path, it will run forward 5 cm at a time, occasionally adjusting its direction slightly, until it reaches the lamp. However, if its bumpers detect an object, it adopts a different routine. It reverses a distance of 5 cm, turns 10° left or right (depending on which bumper made contact) and then runs forward again. This does not necessarily take it toward the lamp, particularly if it had to reverse and turn several times to avoid a large object.

When it is clear of the object (neither bumper contacted during the most recent forward run), it returns to the start of the program to scan for the lamp and run toward it. In this simple description, we have not said how it knows when it has reached the lamp. Perhaps a heat sensor could be used.

VOLTAGE REGULATORS

A circuit in Fig. 21.11 illustrates how to use a Zener diode as a voltage stabiliser. For more precise control of the voltage from a PSU we use a **voltage**

regulator. This is an IC often contained in a 3-terminal package and looking like a power transistor.

The most popular 3-terminal regulators belong to the 78XX series. The last two digits of the number indicate the regulated voltage. For example, a 7805 regulator produces +5 V, and a 7812 regulator produces +12 V. The type number may also include a letter to indicate the maximum output current. For example, a 78L09 regulator is a low-power device producing 9 V at up to 100 mA. Other types are available that produce larger currents; up to 1 A is typical.

The connections for a 78XX regulator are:

- **Input:** This usually comes from a transformer, followed by a rectifier and smoothing capacitor, as on p. 70. The voltage must be 2.5 V to 6 V higher than the required regulated voltage. Special low dropout regulators operate with a supply voltage only 100 mV above the regulated output.
- **Common:** The 0 V line for both input and output.
- **Output:** Regulated output.

The standard circuit connections are shown below.

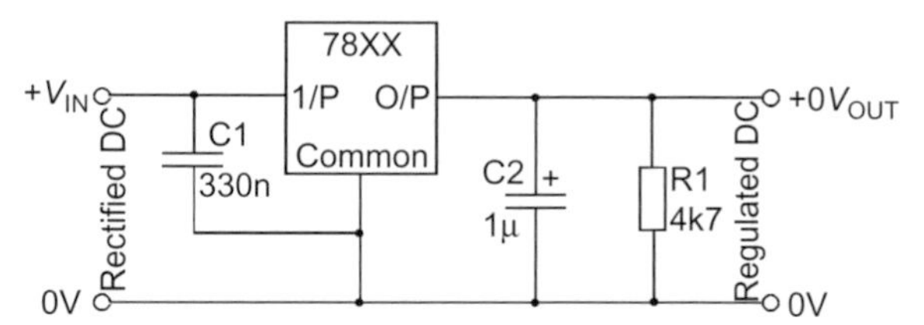

FIGURE 53.9

78XX regulators have an output precision of ±4%, which is better than that provided by a Zener regulator. The output voltage varies no more than 1% when the input voltage varies, provided it is 2.5 V or more higher. Similarly, the output voltage varies no more than 2% when variable amounts of current are drawn from the regulator. The regulator almost completely removes any ripple from the supply.

The regulator has other important features, such as **current limiting**, which drops the output voltage if excessive current is drawn from the regulator. There is also **thermal shutdown**, which cuts off the current if the regulator becomes overheated.

An equivalent series of regulators, the 79XX series, is used to regulate the negative supply to op amps and other devices.

QUESTIONS ON COMPUTERS AND CONTROL SYSTEMS

1. Describe three input devices that are part of a typical computer system, and what they do.
2. In what ways does a microprocessor differ from a microcontroller?
3. Describe three types of memory found in a computer system, and explain the uses of each type.
4. Name three different devices (other than memory) used for storing data. What are their features and for what are they used?
5. What is a bus in a computer system? What are the functions of the different kinds of bus?
6. Draw a flowchart of a computer program for controlling (a) a dishwasher, (b) a washing machine (only one wash program), (c) automatic sliding doors, (d) the entry barrier of an automatic car park, (e) a supermarket checkout, (f) disco lights or chaser lights, (g) a fan heater (as in Fig. 53.4, including an action like that of the Schmitt trigger).
7. Draw a flowchart program to replace the logic-gate system used to control the mixing tank in Fig. 40.7.

On the Web

Extend your knowledge of the topics in Part 4 by browsing the Web. Use *Google, Yahoo*! or any other browsing software. Remember to search for images too, when using Google. Often the reference given with an image will lead you to an interesting website.

Keywords for electronic systems include: CD-R, CD-RW, mouse, USB (for the Universal Serial Bus), iPod, RFID (radio-frequency identity tags), Bluetooth, Pentium, Intel, AMD (for AMD microprocessors), pcb (some of the images show factory production).

The keyword 'control systems' will lead you to hundreds of examples of practical control systems. Keywords 'robotics' or 'robots' will take you into the worlds of professional and hobby robotics.

Websites that are of interest include:
www.cdrfaq.org
www.mricher.com (info on CD-Rs)
www.roxio.com/en/support
www.wikipedia.org.wiki/CD-RW
www.intel.com (for the Pentium story)
www.usb.org
arstechnica.com (Note: no 'www')
Happy browsing!

DESIGN TIME

Design programs by drawing them out as flowcharts. Here are some suggestions for topics taken from previous 'Design time' pages.

A microcontroller version of the twisty wire game, with a few novel features (pp. 149, 190).

A rain detector that sounds an alarm (p. 132).

An egg timer (p. 149) that is based on the internal timer of a microcontroller.

A sprinkler or pot-watering system based on a moisture sensor and solenoid operated valve. The system must not overwater the soil (p. 132).

A wind detector that records the number of gusts in 15 minutes (p. 132).

A model railway controller that does what is described on p. 164, and more. There could be three or four alternative routes, selected by keypad (pp. 164, 190).

A reaction tester (p. 190).

A metronome circuit with LED and/or audio-'click' outputs (pp. 149, 175, 190).

Amplify your design by writing it as a program for a microcontroller. Try one of the simpler programs first. To complete the control system, build the input and output circuits it needs and program a microcontroller to run it.

A 'people-counter' (p. 175).

Topic 54

Robotic Systems

The popular image of a robot is illustrated by the rather chunky R2D2 in the film *Star Wars* or the robot featured in the film *I, robot*, which is even more human in appearance and behaviour than R2D2. But robots do not necessarily resemble humans, even though they are designed to do the same things as humans, relieving humans of the burden or dangers of doing them, and doing them faster, more reliably and with greater precision. In fact, a robot is more often a wheeled vehicle, such as those that wander the factory floor carrying a supply of machine parts to the work stations, where stationary robot arms assemble the parts into a finished product. Automatic robot vehicles have been widely used on the Moon and the planet Mars to take soil samples, analyse them, and transmit the results back to Earth. Other robots, include the underwater vessels used to maintain and repair offshore oil rigs.

SYSTEMS

It has been explained earlier in the book that a typical electronic system has three parts. We can represent it symbolically by a 3-stage flowchart or in words:

Input → processing → output

Some systems have more than one input or more than one output. Robots are distinguished by having several inputs, a very complicated processing stage (usually a microcontroller) which is programmable and interfaced to several outputs. We will look at each of these three stages in turn.

INPUT STAGES

If a robot is to function properly, it needs to know what is happening in the world around it. It needs sensors. These are of six main types:

1 **Light sensors:** Usually consist of a photodiode or phototransistor, with a transistor-based amplifier to interface it to the processing stage. Its output is often on a scale from 0 (dark) to 255 (maximum light). The light sensor in Figure 54.1 (centre of photo) is pointing upward to monitor the amount of light falling on to the robot from above. More elaborate robots have a video-camera to capture an image of what the robot is looking at. In such a case the image is processed by complex logic which identifies objects within the camera's field of view. They might, for example, be programmed to recognize a finger-print or (in a robotic weapon system) a military target.

FIGURE 54.1 This mobile robot has a bumper to detect when it runs into a fixed object. Like all three-wheeled robots, it has two motors and two drive wheels. Each drive wheel is driven by its own electric motor, independently controlled. The robot is steered by making one of the drive wheels turn faster than the other. Some robots have four wheels operating as in an automobile. One motor turns both rear wheels through a differential gearbox. Steering is the same as in a car, using the second motor to turn the steering mechanism.

2 **Touch sensors:** this is usually a microswitch, in contact with a bumper hanging in front of the robot (Figure 54.1). When the robot collides with a fixed object, the bumper is forced backward, closing the switch and sending a message (a rise or fall of voltage) to the processing stage.

3 **Push buttons:** used by the operator to tell the robot what to do (to start running its program, for example).
4 **Sound sensors:** Sensitive microphones are used. Figure 54.1 shows an ultrasonic sensor comprising two crystal microphones, mounted on the framework and looking like a pair of eyes. See also Figures 54.2 and 54.3.

FIGURE 54.2 In this rear view of the robot we can see the two drive wheels and the rear caster wheel. The processing circuits and the battery are contained in the rectangular box (or 'brick') at the top of the photo. The four sockets at this end of the brick are outputs to the motors. There are four input sockets at the other (front) end of the brick for connecting to sensors.

FIGURE 54.3 Getting ready to test a magnetic sensor. The direction of magnetic north is being marked on the base-board, using masking tape and taking bearings from a small pocket magnetic compass. The sensor is mounted on a mast to keep it clear of the magnetic fields of the motors. This robot has a pair of jaws for picking up and carrying objects. The jaws are opened and closed by the robot's third electric motor.

5 **Magnetic sensors:** These detect the direction of the magnetic field. They are sensitive to the Earth's field and so are used to give a robot a sense of direction.
6 **Wireless link:** The robot communicates with the computer using a *Bluetooth* radio link. Programs are written on the computer and uploaded into the Robot's memory by a *Bluetooth* transmitter in the computer and a *Bluetooth* receiver in the robot. It is also possible to use the computer to monitor the state of inputs, and the state of the battery while the robot is running.

The robot shown in the photos in this chapter were constructed and programmed using a LEGO™ Mindstorms™ NXT kit.

PROCESSING STAGE

Processing is usually performed by a microcontroller, assisted by a block of memory, and operating according to instructions in the form of a program. The program consists of a considerable number of logical and arithmetical operations, performed at high speed. The high speed is necessary because the microcontroller can work only in very short, simple steps and many such steps are needed to (for example) input two numbers, multiply them together and output the value of their product.

Because of the extreme complexity of the processing stage, a robot is capable of *learning*. For instance, *it can learn* how to perform certain operations, such as solving a maze. Having solved the maze it remembers the path it followed. Or it can be programmed to play a game, such as Chess, discover winning strategies, and win consistently against a human opponent.

The microcontroller (or maybe there are two) form the heart of the robot system. Input from several sensors can be used to generate signals that are sent to several different output stages simultaneously (not actually simultaneously, but it works so fast that its actions *seem* to be simultaneous).

OUTPUT STAGES

These are the stages which *do* something. We sometimes call them **actuators**. A robot is useless without a set of actuators. This may include:

1 **Motors:** These are for propulsion and providing motive power for mechanisms such as jaws.

FIGURE 54.4 A robot equipped with two light sensors can be programmed to follow a line. The robot runs on a dark brown hardboard base. The line is made up of strips of light-coloured masking tape. The robot runs along, straddling the line with its two sensors. As well as its photodiode, each sensor has an LED which illuminates the board below the sensor. As long as the robot has the line centrally between its sensors, very little light is reflected back up to the sensors; the robot runs straight ahead. If the line curves to one side, or the robot wanders to one side, one of the sensors receives increased light reflected from the masking tape. The situation is assessed by the processing unit and the speed of each of the motors is adjusted so that the robot follows the curving path or the robot alters course to stay on the line.

2 **Loudspeaker:** Logic circuits connected to the loudspeaker can be programmed to produce a variety of sounds such as musical tones, sound effects, and speech.
3 **Light display:** flashing an LED can convey a meaning such as 'Ready to start'. There is an LCD display screen on the front of the Mindstorms 'brick'. This displays word messages, and other symbols.

On the Web

There is much about robots on the Internet. Find out what you can about robots, robotic kits, industrial robots.

ROBOTS IN ACTION

The best and most exciting way to learn about robots is to build one from a kit and program it.

Supplements

A. SYMBOLS USED FOR CIRCUITS

Conductors

Switches

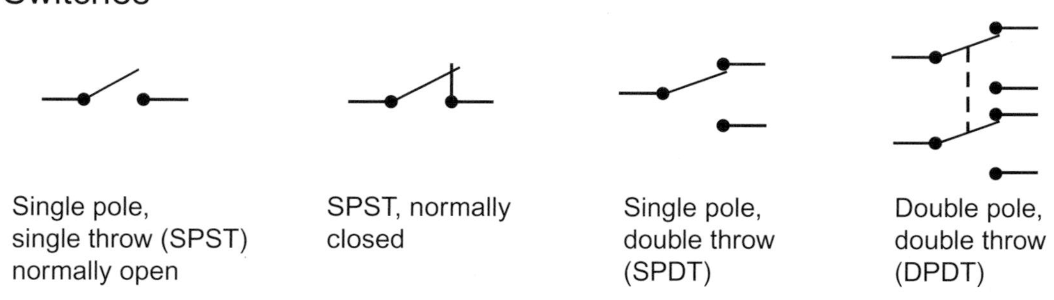

Push-buttons

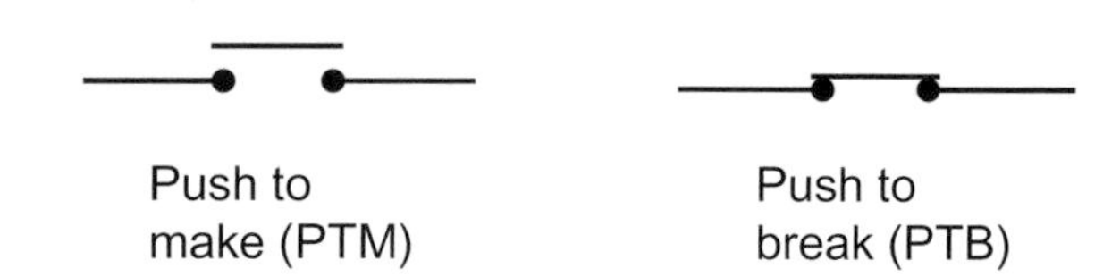

Power supply

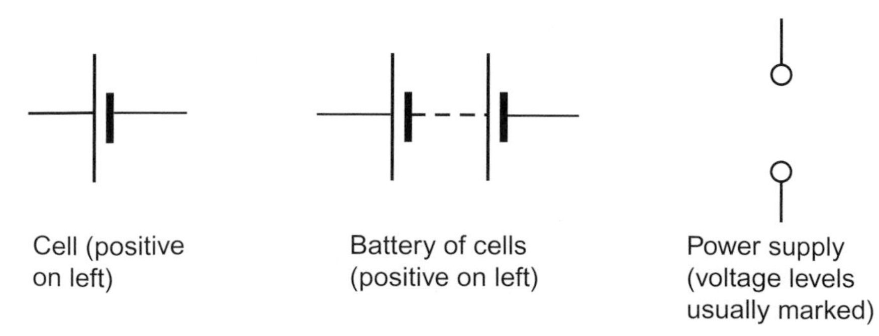

Resistors

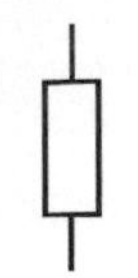

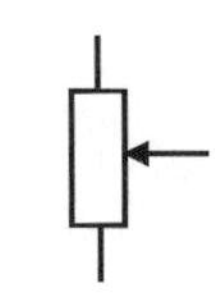

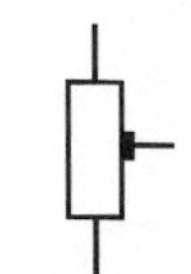

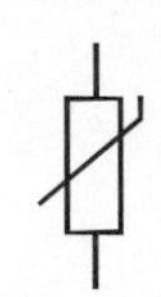

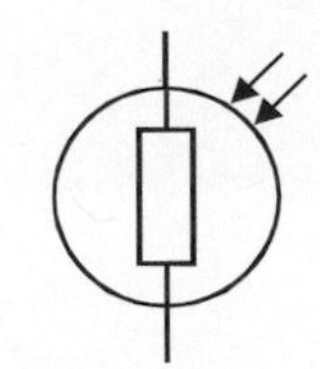

Fixed

Variable (potentiometer)

Preset variable (potentiometer)

Thermistor

Light dependent (LDR)

Capacitors

Non-polarised

Polarised

Semiconductor devices

Diode

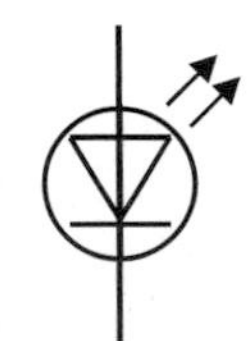

Light emitting diode (LED)

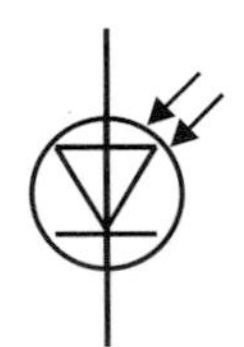

Photodiode

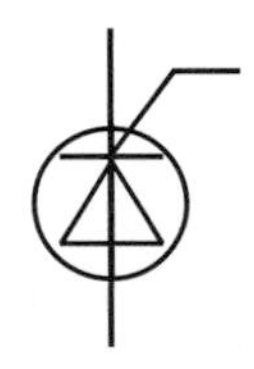

Thyristor (SCR)

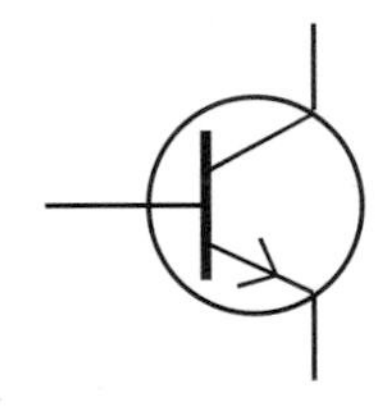

npn transistor (BJT)

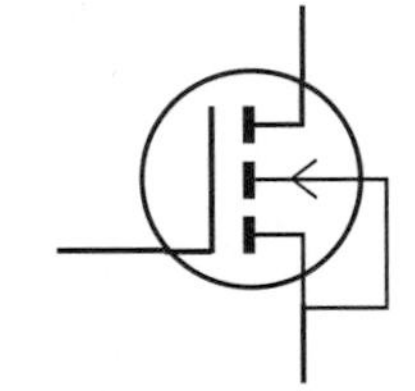

n-channel MOSFET (FET)

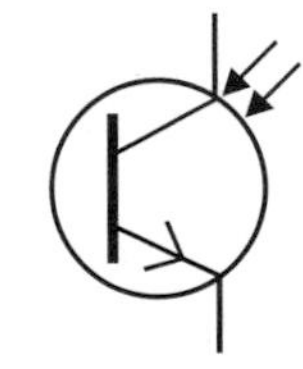

phototransistor

Logic gates

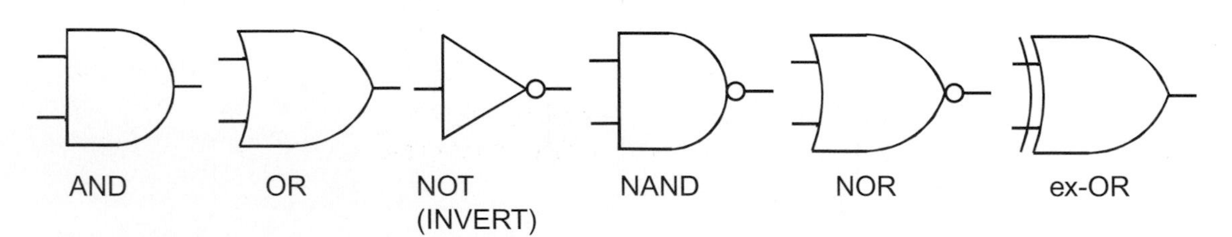

AND OR NOT (INVERT) NAND NOR ex-OR

Input/output devices

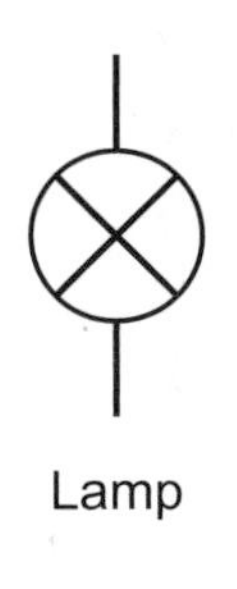

Lamp

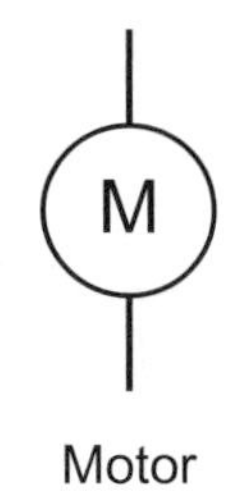

Motor

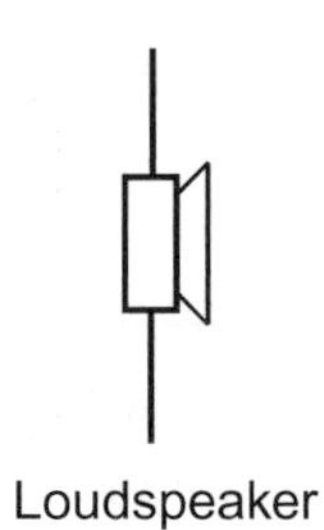

Loudspeaker

Bell

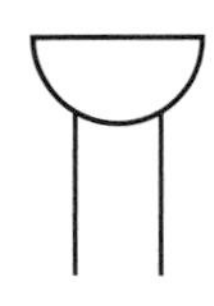

Buzzer

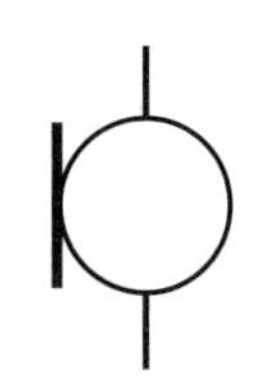

Microphone

Amplifiers

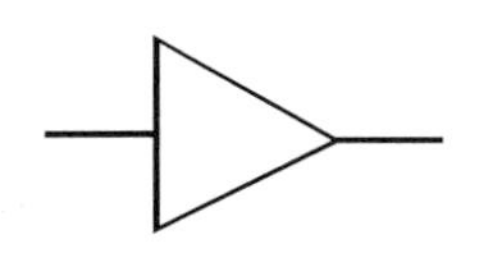

Amplifier

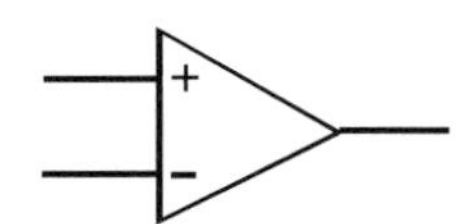

Operational amplifier

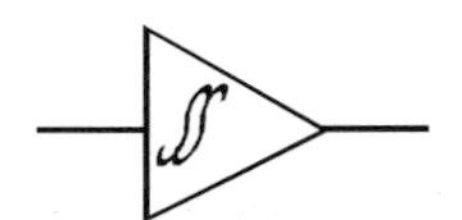

Schmitt trigger

Miscellaneous

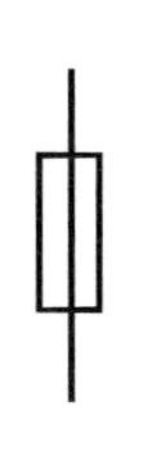

Fuse

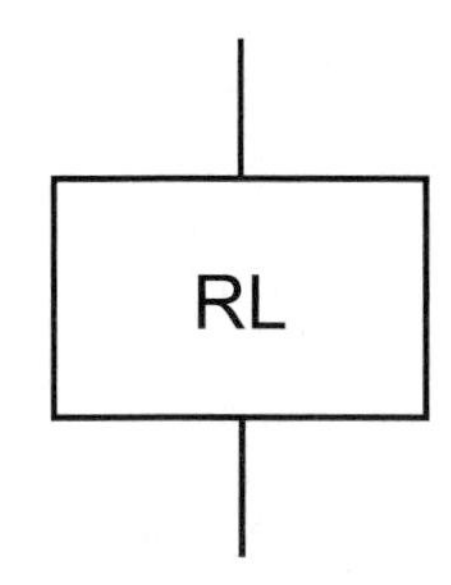

Relay coil

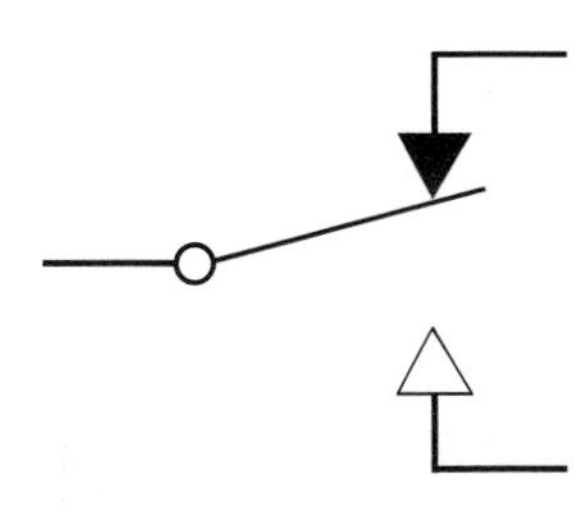

Relay contacts (SPDT for example)

B. SYMBOLS USED IN FLOWCHARTS

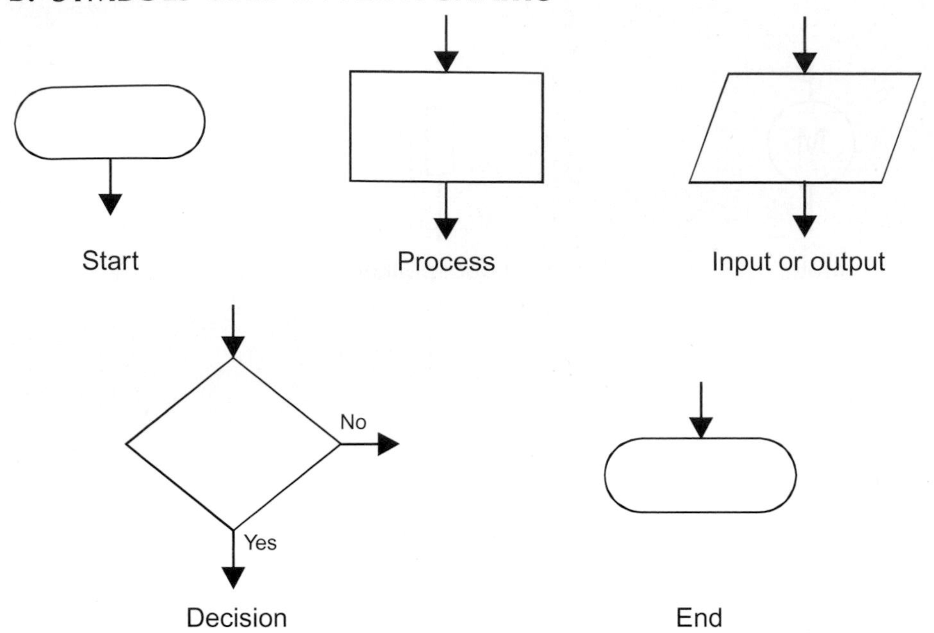

When using decision boxes, it is better for the 'No' arrow to point sideways, as shown above. The 'Yes' arrow continues the flow in the original direction.

C. ANSWERS TO QUESTIONS

These are the answers to numerical Self Test questions and to Self Test questions that are only a few words long.

Answers to 'Questions on ...' and multiple choice questions are available at our website:

http://www.elsevierdirect.com/companions/9781856176958

Page 4: They repel each other.

Page 4: They become charged like the plastic, so are repelled.

Page 4: Because they would repel each other.

Page 8: Silver, copper, aluminium, gold, carbon, rubber.

Page 11:

1 3
2 4.5 V
3 **(a)** Alkaline
(b) Silver oxide
(c) NiCad
(d) Lead-acid

Page 14:

1 396 W
2 43 mA
3 180 mW
4 500 mA
5 **(a)** 34 000 mA
(b) 1200 μA
(c) 0.0012 A
(d) 5.505 A
(e) 0.058 mA
6 **(a)** 4500 mV
(b) 11 000 V
(c) 0.675 V
(d) 0.521 mV
(e) 550 mV
(f) 0.000440 V
(g) 0.022 mV
(h) 3.300 kV
7 **(a)** 0.675 kW
(b) 33 000 kW
(c) 0.650 W
(d) 6 000 000 W
(e) 4.450 kW
(f) 2550 mW
(g) 79 000 W
(h) 33 000 kW

Page 19:

1 65.5 W
2 26.45 Ω

Page 22:

1 20 ms
2 256 Hz
3 sine wave, amplitude = 6 V, frequency = 20 Hz
4 1 ms

Page 26:

1 Good conductor
2 Flexible
3 Live, neutral, earth
4 Insulation
5 Cable can be bent

Page 28: To disconnect from live mains when fuse blows

Page 33:

1 0.544 W
2 25.912 Ω

Page 42:

1 50 Ω
2 885 Ω

Page 42:

1 2.5 Ω
2 7.5 mA
3 8.4 V
4 23 Ω

Page 44:

1 (a) 390 Ω
(b) 560 kΩ
(c) 1 MΩ
(d) 10 Ω
(e) 2.2 kΩ
2 (a) orange, orange, black
(b) red, black, yellow
(c) violet, green, brown
(d) yellow, orange, orange
(e) grey, red, brown

Page 49:

1 (a) 3520 Ω
(b) 60.6 Ω
(c) 1403 kΩ

2 110 Ω + 360 Ω, or
270 Ω + 100 Ω + 100 Ω

3 120 kΩ + 10 kΩ, or
110 kΩ + 20 kΩ, or
100 kΩ + 30 kΩ

Page 51:

(a) 20.34 Ω
(b) 1186 Ω
(c) 8.00 Ω
(d) 14.47 kΩ
(e) 50 Ω

Page 52:

1 (a) 39.3 Ω
(b) 928.5 Ω
2 (a) 1950 Ω
(b) 220 Ω

Page 52:

1 (a) 33 kΩ
(b) 4.7 MΩ
(c) 2.2 Ω
(d) 1.8 Ω ± 5%
(e) 27 kΩ ± 10%

2 (a) 47R
(b) 100 K
(c) 9K1
(d) 3K9J
(e) 750KK

Page 52:

The multiturn trimmer is 5 kΩ ±10%

Page 57:

1 0.25 F
2 10 C

Page 57:

(a) 1 nF
(b) 2200 nF
(c) 1 000 000 000 nF
(d) 0.047 nF
(e) 56 000 nF

Page 60:

1 556 μA, no current
2 Shorter

Page 65:

1 240 V
2 300 turns

Page 72:

(a) 400 Ω (or nearest, 390 Ω)
(b) 1.3 kΩ

Page 88:

1 0.7 V
2 Close to 0 V
3 120

Page 96:

1 250
2 1.25 S

Page 129:

1 4.2 V
2 Rises to 5.8 V

Page 135:

1 373
2 (a) Rl = 10 kΩ, R2 = 2.2 MΩ
(b) Rl = 10 kΩ, R2 = 120 kΩ
(c) Rl = 100 kΩ, R2 = 300 kΩ
(d) Rl = 1 kΩ, R2 = 1.2 MΩ

Page 137: 66 mV

Page 155:

(a)

TRUE gate

Input	Output
A	Z
0	0
1	1

(b)

NAND gate

Inputs		Outputs	
B	A	Z1	Z2
0	0	0	1
0	1	0	1
1	0	0	1
1	1	1	0

(c)

TRUE gate

Input	Output	
A	Z1	Z2
0	1	0
1	0	1

Page 156:

(a)

No equivalent

Inputs			Outputs	
C	B	A	Z1	Z2
0	0	0	1	0
0	0	1	0	1
0	1	0	0	1
0	1	1	0	1
1	0	0	1	0
1	0	1	0	0

(b)

TRUE gate

Input	Outputs		
A	Z1	Z2	Z3
0	1	1	0
1	0	0	1

Page 170:

(a) 1000 0100
(b) 0011 0011
(c) 0010 1001
(d) 0011 0110 0010

Page 211: 1

Acknowledgements

The author thanks these companies and organisations for permission to take photographs on their sites:

Bossong Engineering Pty. Ltd., Canning Vale, Western Australia.
Eastern Generation Ltd., Ironbridge, Shropshire.
Nuffield Radio Astronomy Laboratories, Jodrell Bank, Cheshire.

The author thanks Christine Bishop for permission to use her photograph, and also thanks Emma and Laura Bishop for acting as models.

The photograph of the low-cost hobby robot is from the author's book on practical robot building and programming, *The Robot Builder's Cookbook* (Newnes, 2006).

Index